A Practical Approach to Modern Imaging Equipment

Second Edition

A Practical Approach to Modern Imaging Equipment

Thomas T. Thompson, M.D. *Professor and Director of Radiology, University of Miami Hospital and Clinics, Miami, Florida*

4 Contributing Authors

Forewords by
Paul M. Stivender
Manager, Product Planning, Medical Systems Division, General Electric Company, Milwaukee, Wisconsin

Albert A. Dunn, M.D.
Chairman, Commission on Diagnostic Equipment, American College of Radiology

Little, Brown and Company
Boston/Toronto

To Catherine A. Poole, M.D.,
 Professor and Chairman of the Department of Radiology of the
 University of Miami School of Medicine
 for her unswerving support

Contributing Authors

Laurence P. Clarke, Ph.D.

Associate Professor of Radiology, College of Medicine, University of South Florida, Tampa, Florida

Jack S. Krohmer, Ph.D.

Professor of Radiology and Radiation Oncology and Director, Division of Radiological Physics, Wayne State University School of Medicine, Detroit, Michigan

Michael M. Raskin, M.D., M.S.

Clinical Assistant Professor of Radiology, University of Miami School of Medicine, Miami, Florida; Associate Director of Radiology and Director of Nuclear Medicine, St. Francis Hospital, Miami Beach, Florida

Martin Trefler, Ph.D.

Associate Professor of Radiology, University of Miami School of Medicine, Miami, Florida

Contents

Forewords to the First Edition
    By Paul M. Stivender   xiii
    By Albert A. Dunn   xiv
Preface   xvii

1. The Total Radiographic System   1
    Intensifying Screens   1
    Radiographic Film   7
    Film Processing   10
    Cassettes   13
    Grids   14
    Viewboxes   15

2. Modern X-ray Generators   17
    Terminology   18
    Kilowatt Ratings of Generators   18
    Three-Phase versus Single-Phase Generators   20
    X-ray Transformers   21
    Switching the X-ray Tube Load   30
    Interrogation Time   38
    Falling Load Generators   39
    Anatomically Programmed Generators   41
    Modular Generators   43
    Mobile Radiographic Equipment   43
    Generator Applications in the Radiology Department   48

3. Modern X-ray Tubes   53
    Construction of Rotating Anode X-ray Tubes   53
    Heating and Cooling Characteristics of X-ray Tubes   56
    Tube Damage Resulting from Improper Operation   62
    Design Characteristics of X-ray Tube Components   65
    Magnification Radiography   70
    How to Select an X-ray Tube   73
    Auxiliary Equipment   74
    Newer Developments   75

4. Image Intensifiers and Related Systems 81

    Imaging with Image Intensifiers 82
    The Eye as an Image System 82
    Design and Function of Image Intensifiers 83
    The Image Intensifier Tube and Housing Assembly 84
    Viewing the Radiographic Image 92
    Coupling Systems for Television Viewing 95
    Television Camera Tubes 98
    Basic Operating Principles of Television Monitors 104
    Special Circuitry for X-ray Television 109
    Automatic Brightness Control Systems 111
    Recording the Fluorographic Image 113

5. Automatic Exposure Controls (Phototimers) 127

    Types of Phototimers 127
    Limitations of Phototimers 131
    Performance of Automatic Exposure Controls 132
    Technical Considerations 135

6. Angiographic Injectors 139

    Types of Injectors 139
    Electrical Safety 140
    Components of Angiographic Injectors 142
    Disposable versus Nondisposable Syringes 146
    Commercially Available Injectors 146

7. Serial Film Changers 155

    General Considerations 155
    Types of Film Changers 156
    Radiographic Considerations 160
    Determination of Film Exposure Times 161
    Phototimer Applications 166
    Biplane Applications 167
    Future Applications 167

8. Computed Tomography   169
Martin Trefler

Data Acquisition in CT Scanners   169
Image Generation   173
Image Display Parameters   178
Image Quality in CT Scanners   181
Image Artifacts   184
Purchasing a CT Scanner   186
Inside the Multiformat Camera   187
Modern Developments   190

9. Special Considerations for Angiographic Suites   191
Speed Relationships in Angiography   191
Magnification Angiography   196
Digital Fluorography   202
Equipment Considerations   216

10. Instrumentation in Nuclear Medicine   221
Laurence P. Clarke

Scintillation Counters   222
The Gamma Camera   228
Single Photon Emission Computed Tomography (SPECT)   240
Data Processing Equipment   245

11. Nuclear Magnetic Resonance Imaging Instrumentation   253
Laurence P. Clarke

Basic Physical Concepts   254
NMR Imaging Systems   259
Instrumentation Choice and Bid Specifications   266

12. Diagnostic Ultrasound   269
Michael M. Raskin

Propagation of Sound   269
Focusing of Sound   271
Characteristics of Ultrasound   271
Interaction with Tissue   276
Clinical Applications   279
Selection of Equipment   282

13. Bid and Performance Specifications and Warranty and Service Contracts   285

Jack S. Krohmer

Guidelines for Purchasing Equipment   287

Performance Requirements and Acceptance of Equipment   288

Warranty Considerations   289

Service Contracts   290

Equipment Maintenance Records   292

Minimum Service Requirements   292

Appendix: Bid Specification Packet   294

Index   317

Forewords to the First Edition

Seventy-five years of clinical x-ray utilization have generated an immense amount of folklore. Dr. Thompson has cut through a lot of the diversions and has managed to concentrate on the fundamentals of the x-ray imaging system as they relate to the application of this system.

This book should be the twelfth or fifteenth edition of a work written 30 years ago. In the long history of medical x-ray technology, the equipment we use to produce the transparency we are all so familiar with has been treated very lightly, often fragmentarily, usually as an adjunct to books dedicated to planning the radiology department. On the other hand, the "leading edge" elements of x-ray technology—for example, focal-spot energy distribution or modulation transfer function analysis—have all been covered in great detail in the scientific and technical journals. In this book we find a very readable description of the complete "system-engineered" x-ray image-producing apparatus. Dr. Thompson first takes us through the image receptor we are most familiar with, the transparency we call the radiograph, and its associated cassette and screen subsystem. The theme of quality control is emphasized at this point. Then the text moves on to the generation and control of the electrical power needed to produce the image, followed by an excellent exposé of the mysterious and often misunderstood x-ray tube, which converts this power to useful photons. The development of these older inventions is described gradually in the text so that the reader can easily grasp the refinements that have taken place over the years in power development and reduction of focal-spot size along with reduction of exposure time, which have reduced both the number of x-rays needed and the number of patients requiring reexamination. We are even given a glimpse into the immediate future, with a look at what the microprocessor and its associated microelectronics can do for exposure control disciplines.

The relatively recent invention of the image intensifier tube is first covered from its dynamic visualization aspect, which is important to the diagnostic fluoroscopist; then we are introduced to the various trade-offs that are encountered when the recording of images is accomplished by using the intensifier as the primary image receptor. An objective overview of the various competing methods offered by different manufacturers is given, with an extremely well balanced and dispassionate comparison of products that constantly relates to the needs of the user/customer.

Angiographic film changers and pressure injector technology are treated in depth in order to give the user both an overview of the very limited choice of equipment available and some guidelines on how best to use this type of apparatus.

The recent explosion of apparatus designs following the introduction of Dr. Hounsfield's remarkable machine is described in a very readable chapter devoted to the computed tomography (CT) subindustry. Machine construction and image reconstruction are described, and the math of reconstruction is well explained and well diagrammed in the figures. A glossary of CT jargon emerges from the text.

The reader can best appreciate the real worth of this book by turning immediately to Chapter 13, which gives an overview of all the decision-intensive variables involved in the purchase selection of some common x-ray systems.

From a manufacturer's standpoint, we often look at what we have developed over the years with some degree of frustration. We see our engineers working to achieve more accuracy not just for the requirements of the medical field, but to stay ahead of the regulators. We see our scientists striving to obtain more contrast together with more resolution and lower dosage. We see our sales engineering groups striving to provide better utilization of equipment within the increasingly constrained hospital environments. But until now there has been very little written in the way of an objective report of the apparatus system as it relates to this continuous perfection of technology and its perceived utility in the diagnostic process.

All of us in the field of radiology, including hospital administrators, will appreciate this orderly dissection of the complex and diverse x-ray system apparatus. I believe that this volume will achieve the top of the "most-borrowed book" list in every radiology department within a few weeks after its arrival in the library.
Paul M. Stivender

For some time the American College of Radiology, specifically the Board of Chancellors and the Council, has been concerned about the performance, usage, and service of radiologic equipment. In 1973 the Board of Chancellors and the Council charged the ACR's Committee on Diagnostic Equipment to take action to educate radiologists and manufacturers in all matters related to equipment. Data about actual performance and service in the field are now in the process of being analyzed by the Committee.

The Committee realized that it was essential that radiologists be informed of the various systems and subsystems that go into the equipment they use. The increasing complexity and sophistication of radiologic equipment made the need for education even more imperative.

Dr. Thompson, a member of the ACR Committee on Diagnostic Equipment, has written a book that will prepare radiologists to make reasonable, informed decisions about what they need and what they should buy. This is the first such book written in an organized manner to explain to radiologists the basic elements and principles of the equipment they use.

The Committee feels that this book is a unique contribution to the education of radiologists. It has reviewed the book and enthusiastically endorses it.

A recognized authority in the field of diagnostic equipment, Dr. Thompson is to be applauded and commended for this important contribution to radiology.

The ACR Committee on Diagnostic Equipment recommends to radiologists most strongly that this book be included in their essential reading.

Albert A. Dunn

Preface

In the years that followed the publication of *A Practical Approach to Modern X-ray Equipment*, technology in the radiology profession has grown by leaps and bounds. When the first edition was started, computed tomography was in its infancy and since then the application of high-speed computers has allowed a variety of new applications in our profession—nuclear cardiology, magnetic resonance imaging, digital subtraction angiography and further advances in high-speed computed tomography. The advances in technology required the addition of new chapters on ultrasound equipment, magnetic resonance imaging and nuclear medicine instrumentation. Since the second edition covers more than diagnostic radiology imaging equipment, the title has been changed to *A Practical Approach to Modern Imaging Equipment*.

As it has been in the past, consumer groups, industry, and regulatory agencies continue to demand more accountability for the expenditure of health care dollars. More and more hospitals are being acquired to be operated on a "for-profit" basis and we can expect more competition in the marketplace in years to come. In time it will become more and more difficult for radiology to keep up with advancing technological equipment and still cope with the economic pressures of the health care market. Hence, the exhortations noted in the preface of the first edition still hold true—an informed purchaser will be at an advantage at the bargaining table.

I have been asked a number of times about the source of information used in compiling this publication. The informed reader will recognize that a considerable amount of the information contained is not referenced and is not retrievable from peer literature. I have been fortunate over the past fifteen years to have made a number of contacts with almost all major manufacturers and have been privy to information not contained in the literature. In many instances information has been provided to me from manufacturing internal research with the understanding that the source of the information will not be revealed. A good example of this type of information is the extended x-ray tube life in relation to percentage of heat loading. No x-ray department would have the funds to purchase radiographic tubes simply to see how long they would last as a function of heat loading. This information obviously came from a major tube manufacturer.

I should like to acknowledge the assistance of the contributors to this edition. Jack S. Krohmer is well known throughout our profession and I consider him to be one of the most knowledgeable individuals concerning radiology equipment in the United States. Larry Clarke is one of the more knowledgeable persons in the United States concerning nuclear medicine instrumentation and magnetic resonance imaging. Likewise, Marty Trefler, is well known as an expert in computed tomography equipment. Mike Raskin is an accomplished author in ultrasonagraphy. So, I have been fortunate in obtaining well-informed experts in these fields to contribute to the new edition. The combination of these brilliant minds provides for information not normally obtainable in a text of this type. For their valuable contributions, I give humble thanks.

Again, this publication would not have been possible without the input of the people at Little, Brown and Company, especially Barbara Ward and Bob Davis. Many thanks for the help of Mrs. Audrey Lessner in typing the manuscript and other duties related to this project. And, to my family for their understanding of the many long hours away from home.

T.T.T.

A Practical Approach to Modern Imaging Equipment

1 The Total Radiographic System

One of the problems in presenting information about modern radiographic equipment is that the reader tends to divorce the equipment from the total radiographic system. Radiographic equipment is only a link in a system of which the end purpose is to make available a record of anatomy, pathology, or physiology for interpretation by the physician. The total radiographic system comprises not only the radiographic equipment but also the x-ray tube, films, screens, cassettes, grids, processors, processing chemicals, and viewboxes. A system is only as strong as its weakest link; one may have available the best radiographic equipment in the world, but it is almost useless if, for example, all of the cassettes have poor screen contact. The purpose of this chapter is simply to make the reader aware of the total radiographic system and to mention some of the traps one may fall into if he is not aware of them. X-ray tubes are discussed in detail in Chapter 3 and thus will not be covered in this chapter.

Intensifying Screens

Intensifying screens play an important role in diminishing exposure of the patient to the radiation. In addition, the proper choice of screen speed directly relates to the choice of radiographic equipment; the faster the screen, the less output is needed from the x-ray tube. Most blackening of the film occurs from the interaction of light photons emitted from the intensifying screens; very little blackening occurs from direct exposure of the film to the x-ray beam itself.

Until recently almost all intensifying screens used in the United States were composed of calcium tungstate. Basically, three speeds were in common use: detail, medium-speed, and high-speed. The term "Par-speed" is commonly used in describing medium-speed screens; Par-speed is the registered trademark of duPont medium-speed screens. In Europe medium-speed screens are called "universal" screens. With the advent of rare-earth screens, it has become important for those in the radiology profession to know more about intensifying screens; hence, a basic discussion about intensifying screens is in order. A typical intensifying screen (Figure 1-1) is composed of a protective coat, a phosphor layer, a reflective coat, and a polyester or paper base. The protective coat is approximately 0.01 millimeter (mm) in thickness and serves the functions of preventing damage to the phosphor, preventing moisture from entering the phosphor layer, and providing a surface that can be

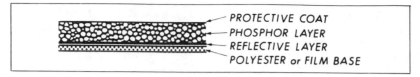

Fig. 1-1. X-ray intensifying screen. A typical screen is composed of a protective coating, a phosphor layer, a reflective coating, and a semirigid base material.

cleaned. Some of the newer phosphors such as cesium iodide, used in image intensifiers, are very hydroscopic and for that reason cannot be used in regular intensifying screens. The protective coat is especially important in film changers, in which the screens are apt to be scratched when the film is moved across them at very high speeds. A special antistatic coating is used in some cases to control static electricity buildup or discharge.

The radiation reduction effect and film blackening capability of an intensifying screen lie in the phosphor. The word *phosphor* comes from the Greek *phosphoros* meaning "light bringer" or "light bearer." In radiological terms, a phosphor converts radiation energy into light energy. If the light output from the screen persists for no more than a few microseconds ($< 10^{-8}$ sec), the screen is said to be *fluorescent*. If the light persists for more than 10^{-8} sec, the screen is said to be *phosphorescent*. A phosphor is composed of small heavy metal crystals dispersed in a solvent and binder. In the manufacturing of screens, every effort must be made to have all the crystals of the same size and to have them evenly dispersed in the layer. Even dispersion of the crystals is difficult to maintain during manufacturing, and there is a tendency for the crystals to clump, producing an artifact on the film known as *screen grain* or *screen structure mottle*.

As its name implies, the purpose of the reflective layer of the screen is to reflect the light produced in the phosphor emulsion toward the film. As is well known, more than 99% of the energy supplied to the x-ray tube is lost, and the reflective layer helps reflect as many of the light photons as possible toward the film, thus increasing the efficiency of the system.

Figure 1-2 is a simplified view of the radiographic image intensifying screen; it breaks down the various parameters that are collectively responsible for the efficiency or inefficiency of the system. When an x-ray photon interacts with the phosphor layer of the screen, a distinct number of probabilities of incidents can occur. There is a finite probability that an incident (incoming) x-ray photon will be absorbed by the intensifying screen; there is another finite probability that an absorbed x-ray photon will be converted into a light photon. In addition, there is a finite probability that the converted light photon will escape from the

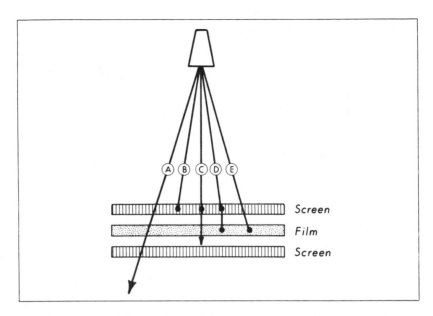

Fig. 1-2. Efficiency of the x-ray intensifying screen. An x-ray photon emanating from the x-ray tube has the probability of either passing through both screen and film with no interaction (*A*); being absorbed in the screen (*B*); producing a light photon in the screen, but one that does not contribute to the production of the latent image (*C*); producing a light photon that does contribute to the latent image (*D*); or producing a latent image in the x-ray film itself (*E*). The efficiency of the intensifying screen is therefore very low, ranging from 2 to 20% for calcium tungstate screens and about 50 to 60% for rare-earth screens.

screen in the right direction to hit the x-ray film, and finally, there is a finite probability that the incident light will expose a latent image in the film. A probability of 1 means that an event will occur 100% of the time. Since the efficiency of a screen is a product of these various probabilities, total screen efficiency is much less than 100%. The efficiency of calcium tungstate screens is very low: Generally the efficiency of detail screens is 2–5%, medium-speed screens 10%, and high-speed screens 20%. If one wants to utilize medium-speed or detail screens in a radiology department, he would have to acquire generators with a higher kilowatt capacity in order to produce a higher number of photons per second for equal film blackening as compared to a higher-speed screen system.

The efficiency of the radiographic imaging system is the product of the probabilities of adsorption, conversion, escape, and film exposure, and thus the efficiency of any system must be less than 100%. Since the art of technology does not allow the manufacturer to increase significantly the probability of exposure or of escape, efforts must be made toward increasing the probability of adsorption and conversion.

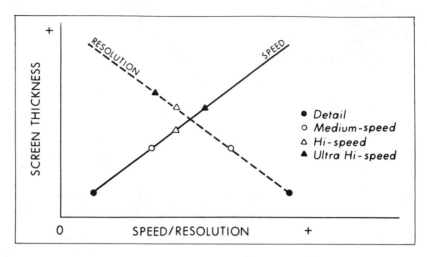

Fig. 1-3. Speed and resolution of intensifying screens. As the thickness of the screen phosphor is increased, the speed of the screen increases; however, the resolving power of the screen correspondingly decreases.

The speed (conversion and adsorption) and the resolving power of calcium tungstate intensifying screens are related to the thickness of the phosphor (Figure 1-3): The faster the screen, the thicker is the phosphor, and vice versa. On the other hand, the thicker the screen, the poorer is the resolution. As the phosphor layer increases in thickness, the image is spread throughout the layer, with the net result being that resolution is decreased. Calcium tungstate screens have been developed to the point that no significant improvement could be made in their speed without having a system with such poor resolution that it would be unusable.

Rare-Earth Screens

In 1970 Dr. Robert Buchanan of Lockheed Research, Palo Alto, California, startled the radiological profession with a paper presented before the Atomic Energy Commission in which he described a new phosphor and its activator, terbium, and illustrated the results that could be obtained with the then-available film emulsions [1]. Then, in 1972, the profession was startled again by Dr. Buchanan's announcement at the meeting of the Radiological Society of North America that a fivefold to sevenfold increase in speed over conventional calcium tungstate systems had been obtained (with equal resolution) using his phosphor and an experimental film produced by the 3M Company.[1] Since that an-

----

[1]3M Company, St. Paul, Minn.

nouncement, a number of "rare-earth" screens have been introduced commercially. The term *rare earth* is a misnomer, since not all of the new high-speed intensifying screens are made from rare-earth metals, but since it is a commonly used term, it is appropriate for use in this discussion.

The light spectrum emitted by rare-earth screens is considerably different in character from that of calcium tungstate screens. Calcium tungstate emits a broad spectrum of light of low intensity centered in the blue wavelengths of 420 nanometers (nm) or 4200 Angstrom units. The emission from the rare-earth screens consists of very sharp peaks of higher intensity, with the high intensity due to their higher absorption and conversion of x-ray photons into light. Almost all rare-earth screens presently on the market are composed of either gadolinium, a green-light emitter, or lanthanum oxybromide, a blue-light emitter. Radiographic film is manufactured to be most sensitive to the light emitted from the screen. In order to achieve maximum efficiency of rare-earth screens, green-light-sensitive film must be used with gadolinium screens and blue-light-sensitive film used with lanthanum oxybromide screens.

The major advantage of rare-earth screens lies in the speed of the system. Proper use of rare-earth screens will allow generators having less power to be purchased as additional or replacement equipment, with the result that the generator kilowatt race may be over. Since less radiation exposure is required for the rare-earth systems, the total population exposure to radiation can potentially be reduced. Another advantage is that rare-earth screens provide a higher resolving capability than do calcium tungstate screens for a given phosphor thickness, and at a relatively higher speed.

The rare-earth screens are not without problems. The screens respond in a nonlinear fashion to changes in kilovoltage, particularly at low kilovoltage ranges, as compared to the essentially flat response of calcium tungstate screens. The response time for most older phototimers is too slow to be used with the new screens. Automatic exposure control systems have to be redesigned to accommodate rare-earth screens. An evaluation of the rare-earth screens by equipment manufacturers indicates that both the ionization and photomultiplier types of automatic exposure controls will work, although some kilovoltage retracking may be necessary.

Since the rare-earth systems are so much faster than calcium tungstate systems, the *technique latitude* is proportionately small. This means that the technologist has little margin for error, and repeat rates can be expected to climb unless meticulous attention is paid to technique charts for conventional exposure control. Functional, well-designed automatic exposure control systems are the preferred solution.

The biggest problem of the rare-earth screens has to do with quantum mottle. *Quantum mottle* is defined as the statistical distribution of x-ray photons across the surface of the screen, with one area receiving more photons than an adjacent area. Quantum mottle can be controlled to a slight degree by decreasing the kilovoltage and increasing milliamperage, which allows a greater quantity of x-ray photons to leave the x-ray tube and gives a better statistical distribution of photons across the screen.

Some manufacturers advocate reducing the temperature of the automatic film processor to reduce mottle. This technique does reduce the visible mottle, but a better method is to choose a film-screen system with a mottle that is not objectionable to the person interpreting the film and then fully process the film.

For the most part rare-earth screens are fast replacing calcium tungstate screens. Conversion to rare-earth systems is expensive, but it is still far less expensive than having to purchase large generators. Some manufacturers lease screens to users of their films. For a given thickness of phosphor, rare-earth screens afford a faster speed and a higher resolution than do calcium tungstate screens.

Since the conversion and adsorption of photons is higher for rare-earth screens than for calcium tungstate screens, the use of higher kilovoltage techniques leads to a higher level of quantum mottle than is seen at lower kilovoltages; conversely, quantum mottle can be controlled to some extent by using relatively lower kilovoltages and higher milliamperage-seconds. Even though the use of lower kilovoltages allows a higher radiation exposure to the patient, the overall increase in the speed of the rare-earth screens still provides a total exposure to the patient that is relatively less.

Standardization of Intensifying Screens

Another point that should be considered in respect to intensifying screens has to do with standardization. When purchasing a medium-speed intensifying screen, one usually assumes that the speed of one manufacturer's medium-speed screen is similar to that of another manufacturer's. This is not necessarily true. Investigators have described a difference greater than 100% in the speed of screens that are labeled by manufacturers as being medium-speed screens [2]. They found that the exposure factors necessary to produce a net density of 1.0 required exposures varying from 0.55 to 1.6 millirad. This means that if different brands of screens are used in a radiology department, one cannot obtain consistent radiographs because of variability in screen speed. For consistent quality control, this variability is not acceptable. One must be certain that all categories of screens (e.g., detail, medium-speed, high-speed, and rare-earth) are standardized and that replacement screens match the speed of existing screens.

Care of Intensifying Screens

Needless to say, any radiology department should have a routine plan for cleaning and replacing radiographic screens. Screens that require more than ± 15% change in exposure factors to match a standard screen in the department should be replaced. Screen speed can easily be checked by making exposures with a step-wedge on each cassette in the department. Screen contact can be checked at the same time using a small-gauge wire mesh directly on top of the cassette facing the x-ray tube.

Radiographic Film

Medical radiographic film, together with its processing, is almost invariably the weakest link in the radiographic system. Little, if any, thought is given to the purchase specifications of radiographic film, although aside from the cost of labor, film cost is one of the major budget items in a radiology department. Quite often administrative and purchasing agents look only at the price of the film, neglecting for the most part the quality of the product. The evaluation and selection of radiographic film must be given as much thought as the purchase of radiographic equipment. The proper tools and knowledge should be available before any comparison of products is made. Both a film sensitometer and a film densitometer should be available, although the equipment does not have to be the expensive type used by the film manufacturers. Film characteristic curves should be plotted from sensitometric exposures and comparisons made from data rather than by a visual comparison between films. The reader should be cautioned about using radiographic step-wedge exposures to make sensitometric strips, since there may be considerable variation from milliampere (mA) station to milliampere station and from exposure to exposure, particularly with older equipment.

Radiographic film is composed of a base on which a film emulsion is coated on both sides. With a normal film base, light emitted from one screen will pass through the base creating an exposure on the opposite coat of film emulsion. This process is known as *crossover,* and its effect is to slightly degrade the image on the film. Crossover can be controlled by the addition of a dye (magenta for Kodak film) in the manufacture of the film base. This addition of dye to the film base slightly decreases the speed of the film but provides for sharper images.

The film emulsion coated on the base is composed of minute particles of silver bromide which, as seen under an electron microscope, are irregular particles. Screen speed, for the most part, is obtained by increasing the thickness of the phosphor coated on the film base. It is possible to increase the speed of the film by changing the shape of the silver bromide particles—from irregular particles to ones that are flattened. The effect of the flattened particles is to intercept more of the light from the intensifying screen, thus increasing film speed without sacrificing detail, as normally would be required when using a thicker phosphor in the intensifying screen. This new application of a well-known phenomenon is called *T-grain technology.*

Of all the major aspects of radiographic film, the important ones to be considered are base-plus-fog density, shelf life, the average gradient, film speed, and reciprocity failure.

Base-Plus-Fog Density

Base-plus-fog density is determined from the nonexposed portion of the radiographic film. This density represents the inherent blackness within the base of the film and the minimal chemical fog produced by processing the film. When base-plus-fog density is greater than 0.22, it becomes objectionable to the average radiologist and may interfere with resolution of fine detail on the radiograph. For example, in skull radiographs the average density is around 1.0, and if base-plus-fog density is 0.35, only 0.65 density units remain for the radiologist to use in interpreting the film.

As radiographic film ages, fog density increases. Although one may find an acceptable density level on fresh film, 3–6 months later this density may have increased to an unacceptable level, rendering the film unusable. Films that accumulate excessive base-plus-fog density over a short period of time are said to have a short shelf life and should be avoided. Since storage conditions may adversely affect shelf life, radiographic film should be stored at temperatures less than 27°C with 30–60% relative humidity, and it should be kept free from all extraneous radiation.

Average Gradient

The average gradient of the film is defined as the slope of the film characteristic curve, measured between net densities of 0.25 and 2.0. The average gradient denotes the contrast (or latitude) of the radiographic film. The contrast of a film determines its ability to record different shades of gray. High-contrast films are particularly useful for bone radiography but are undesirable for use in chest radiography by most radiologists. In contrast, wide-latitude films are useful for chest radiography but are not necessarily useful in depicting the trabecular pattern within bone structures. There are high-speed, high-contrast films that are not very useful with high-speed screens because of accentuation of mottle, which may interfere with visualization of radiographic detail.

Film Speed

Film speed is calculated at a net density of 1.0 on the film characteristic curve and relates to the amount of radiant energy required to produce that film density. It is extremely difficult to report film speed in useful terms, since it will vary according to methods of processing, chemicals, processing temperatures, and a host of other factors. However, the speed of a film relative to the speed of another film that has been exposed and processed under similar conditions is useful in allowing one to determine what exposure factors are needed to correct from one film to another. When film characteristic curves are formulated, data are most often plotted in terms of relative exposure. An exposure may be defined by one researcher as an exposure made at 70 kilovolts (kV), 10 milliamperes-second (mAs), 100-centimeter (cm) source-image distance (SID), on three-phase equipment, with the film processed under certain processing conditions; an exposure may be defined by another researcher in terms of entirely different exposure and processing factors. Thus, relative exposure varies from place to place according to the method of determining film speed, with the result that the user is unable to compare film speed from one brand of film to another unless he conducts his own tests. This deficiency may be overcome by using a sensitometer and a densitometer and plotting one's own characteristic curve. Unfortunately, duplication of processing methods in the field is not as accurate as it is at the factory, and one can determine the speed of one film relative to another only through experience. Film speed will vary from one mixing of film emulsion to another. Be aware of this variation and make sure it is insignificant; otherwise you may find that film from one emulsion varies so much from that of another that exposure consistency throughout the department is almost impossible to achieve.

Reciprocity Failure

With the advent of large-capacity x-ray generators that allow ultrashort exposures, film reciprocity failure becomes more important. The film reciprocity law, as it applies to radiography, states that there is a constant relationship between the intensity of radiation and the time during which the film is exposed. If the amount of radiation is increased, then exposure time can be reduced proportionately to allow matched film densities. For example, an exposure using 100 mA at 100 milliseconds (ms) should provide the same density as one made at 1000 mA at 10 ms, provided the x-ray timer is correct and all other factors are equal.

For all practical purposes, the film reciprocity law holds true for exposure times between 10 ms and 2 seconds (sec). For exposures that are longer or shorter, the constant relationship between milliamperage and time factors may not produce identical film densities. Film emulsions cannot be manufactured so that the response to all variations of time and milliamperage factors will produce equal densities for all radiographic examinations. Chest radiographs may be exposed at 6 or

8 ms on one end of the time segment; on the other end, hypocycloidal tomography may require a 6-sec exposure. If the radiographic film is manufactured to be more responsive on one end of the time segment, it necessarily cannot be as responsive on the other end. This means that a film that is suitable for ultrashort exposures is not necessarily suitable for prolonged exposures and will probably require an increase on the time and/or milliamperage factors to compensate for the lack of response. This means that more heat units are produced in the x-ray tubes for prolonged exposures, more radiation exposure is given to the patients, and possibly, more motion unsharpness is produced as a result of motion during the exposure.

Most radiographic film emulsion is formulated to have its maximum speed at an exposure time of 100 ms. This time frame encompasses most of general radiography, and reciprocity failure is negligible. However, one should be aware of potential problems of the kind just mentioned, particularly in unusual circumstances. The magnitude of radiation given to a patient for a chest radiograph is far less than that for an abdominal radiograph, and this factor must be taken into consideration.

Film Processing

Although the inadequacy of film processing is constantly seen as a major problem by the film manufacturers, the problem is often not recognized by most radiology departments. Since the advent of automatic film processors, the darkroom has been almost totally ignored, and this poses serious problems in film quality control. A hospital may spend millions of dollars for radiographic equipment, using all appropriate bid and performance specifications, and yet may pay little attention to the radiographic film processor itself. No one has been able to define the physical characteristics of a good-quality radiograph in terms useful to everyone, but both the radiology profession and industry agree that a major problem exists in a lack of consistent film processing. Often consultants have been hired to investigate "bad" radiographic equipment or "poor" film, and the problem has been found to be basically related to inadequate processing of the film. The secret of maintaining consistent sensitometric processing quality involves a constant awareness of the potential problems encountered with automatic film processors. Even an inexpensive processor will adequately process a radiograph if the basic principles explained in the following discussion are adhered to.

Radiographic film is made to manufacturing specifications and is intended to be processed in a specific type of chemistry. By altering or changing brands of processing chemicals, one may change the speed of the film, film contrast, or base-plus-fog density. This practice is sometimes satisfactory; in other cases, the results are quite unsatisfactory. The results depend on one's knowledge concerning the specifications and characteristics of the film and chemistry in question. There is no question that alteration of the film chemistry may allow one to create subtle changes in the processed film, but there are few people who possess the knowledge and ability to formulate these changes. The safest practice is to use the chemistry recommended by the film manufacturer.

Of equal importance in using optimal film processing systems is the understanding that a manufacturer's representative knows what film results are expected with his product in a given chemistry. Processing problems can often be easily detected by the manufacturer's representative. However, if an unknown factor is thrown into a problem (e.g., chemistry not suited for the film), the manufacturer may have difficulty in ascertaining whether the problem lies with the film or with the processing.

Many radiology departments forget that the major film manufacturers have the quality control standards, the technical capabilities, and the expertise to solve film and processing problems. All of the research and expertise of these companies is as close as the telephone. However, this expertise is geared toward a film processing system, and most companies are hesitant to become involved in a problem involving only part of the system.

The major problem encountered with most automatic processors is that of underdevelopment of the film. This problem is almost invariably the result of contaminated chemicals owing to lack of adequate cleaning, inadequate replenishing rates, and inaccurate temperature controls. One of the ploys commonly used by film manufacturers to get their products accepted by a radiology department is to offer film and chemicals on a trial basis. When their product is put into the radiology department, the processor is meticulously cleaned and adjusted. Since it is a well-known fact to those who are interested in film sensitometry that the major problem in adequate film processing is caused by a lack of cleanliness of the processor and inadequate replenishing rates, any film processed through a well-cleaned and adjusted processor will look better than a comparable film processed through a dirty, poorly adjusted processor. Figure 1-4 illustrates the difference between correctly and incorrectly processed film.

Each processor must be cleaned on a routine basis, depending on the amount of film processed. Cleaning must include transport racks, replenishing lines, tanks, and pumps. If one part of the processor is cleaned, but contaminated or oxidized chemicals are not removed from elsewhere in the system, the contaminated material can recontaminate the clean part of the system. System cleaners should be discouraged, since they may penetrate the porous portions of racks and rollers and provide a continuous source of contamination. Since most developing

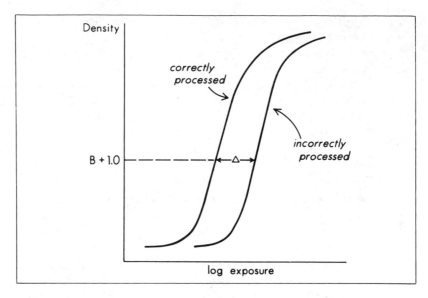

Fig. 1-4. Comparison of correctly and incorrectly processed radiographic film. Incorrectly processed film will generally show (1) a loss of speed, as evidenced by a shift of the film characteristic curve to the right of the log exposure scale, (2) an increase in the base-plus-fog level, (3) a decrease in top density, and (4) a change in contrast levels.

chemicals will oxidize within about three weeks after exposure to air, all old chemicals should be dumped and the processor cleaned at least every three weeks. After the cleaning has been completed, the chemical activity should be checked and balanced with the aid of sensito-metrically exposed test strips.

One should also be aware of inadequate replenishing rates in pro-cessors that are caused by variations in pressure transmission through diaphragm and magnetic drive centrifugal replenishing pumps. These types of pumps are greatly affected by head pressure transmitted from the replenishing tanks, and rates may vary as much as ± 50%, de-pending on the amount of head pressure. One may adjust the replen-ishing rate at 60 milliliters (ml) with the replenishing tank full and then recheck it with the tank half full and find that the replenishing rate has dropped to 30 ml. Overreplenishing is wasteful; underreplen-ishing may lead to underprocessing and require an increased film ex-posure to compensate for the lack of proper development of the film.

Another problem, in addition to the lack of adequate cleaning of the processor, is temperature drifts. The developer temperature may be high or low, depending on the accuracy of the thermostatic control and the pressure and temperature of the incoming water. Built-in ther-mometers should be ignored, and a low-cost bimetallic stem thermometer should be purchased and placed inside the processor. Processing film at higher than the recommended temperature increases the activity of the chemistry and overdevelops the film. The result is a darker, grayer film that has more mottle or grain than would be normally seen if the film had been processed properly. Processing with lower than optimal temperature results in a film with less density, which must be compensated for by increasing the radiation exposure to both the patient and the film.

In terms of unnecessary radiation exposure to the patient, inadequate film processing is still a major cause. The major cause of this inadequate processing is mostly neglect on the part of the user. Rarely is a processor adequately cleaned or maintained. One processor that can provide more adequate control of processing is the Kodak M-8. The M-8 was designed to solve the problems of inappropriate replenishing of chemicals and to gain rigid control of processing temperature. The processor is electronically controlled. Although expensive in relation to other processors on the market, the M-8 is desperately needed by the radiology profession.

Cassettes

Some may not consider cassettes to be related to the radiographic system. A number of years ago we experienced recurring problems with poor screen contact on some chest x-ray examinations performed with portable equipment. Since all the cassettes in the department were marked, the cassettes were pulled from service and rebuilt to improve screen contact. It was finally realized that the problem was that the cassette was bending under the pressure of the patient's weight. As the cassette bent, poor screen contact resulted. The original cassettes were replaced with more rigid ones, and the problem was eliminated.

Poor screen contact results when the intensifying screens are not in close contact to the radiographic film. As was pointed out in the section on intensifying screens, the screens provide the major portion of light to create the film image. When the light photons emitted from the screen strike an area of film that is not in close contact to the screen, they strike the film at a slightly greater angle and distance than they strike those areas that are in close apposition to the screen. This poor screen contact results in spreading of the light and thus a blurring of the radiographic image. Screen contact can easily be checked by placing a fine wire mesh on top of the cassette and making an exposure. The radiograph of the wire mesh can then be checked for areas of poor contact; if they are present, the back of the screen can be built up to eliminate the problem. One should be aware that if film-screen contact is too tight, pressure artifacts may be produced on the film and screen structure mottle accentuated. One should also note that a standard flat cassette will trap air when film is loaded into it, and this trapped air will produce a transitory poor screen contact. This problem can be eliminated by paying close attention when loading x-ray film cassettes or by acquiring cassettes with curved backs, such as the X-Omatic cassette[2] or the Cronex cassette.[3]

Another problem regarding cassettes is that there is considerable variation in the x-ray absorption in the front of the cassette. I stumbled onto this problem many years ago when trying to determine the cause of inconsistent exposures in a radiology department. Looking at four different cassettes from different manufacturers, we were able to demonstrate a 6-kV difference in absorption in the front of the cassette. If at all possible, all cassettes in the department should be obtained from the same manufacturer.

Grids

Grids are expensive and can easily be ruined if they are dropped or mishandled. They are composed of alternating strips of lead and interspaces made of plastic or aluminum. Grids are described according to their *grid ratio*, which is defined as the ratio of the height of the lead strips to the distance between them. The higher the ratio, the more radiation the grid will absorb and the more radiation is required to penetrate through the grid to the film.

Grids are also classified as being parallel or focused, and linear or crosshatched. In *parallel grids* the lead strips are parallel and perpendicular to each other and are useful only for long focal spot-to-film (source-image-receptor) distances or for very small pieces of film. Since the x-ray beam comes from essentially a point source, the x-ray photons are not parallel to each other at the plane of the film or at the image receptor unless the source-image distance is extremely long. In effect, only the photons at the center of the beam will be parallel to the grid; the photons that are off-center of the beam will strike the lead strips at an angle and thus will be wholly or partially absorbed by the grid.

To compensate for this lack of uniform absorption across a parallel grid, most grids are manufactured with the lead strips slanted toward the focal spot of the x-ray tube at a specified distance; the lead strips are "focused" toward the focal spot, hence the term *focused grid*. The distance between the focal spot and the plane of the grid, which must be specified, determines at what angle the lead strips are slanted. This specified distance is known as the *focal distance*, and the tolerances of the distances within which the grid may be used are known as the *focal range*.

---

[2]Eastman Kodak Co., Rochester, N.Y.
[3]E.I. duPont de Nemours & Co., Wilmington, Del.

The terms *linear* and *crosshatched* may be confusing. A *linear grid* is one in which the lead strips are straight, although the lead strips may be slanted in relation to each other. A parallel grid is in effect a linear grid. *Crosshatched grids* have two layers of lead strips with one set rotated 90 degrees to the other, as the term implies. Crosshatched grids may be either parallel or focused and provide better absorption of x-rays (clean-up) than linear grids.

In addition to the problem of obtaining grids without manufacturing artifacts, there are a number of problems associated with grids that directly affect radiographic equipment. Grid cutoff is a common problem and occurs when the radiograph is exposed at a distance outside of the focal range of the grid. This is a common problem with film changers in which the grid may have a focal range of 39–42 inches; grid cutoff occurs when the x-ray tube is moved or when focal spot-to-film distance is decreased to penetrate a large patient better. Another common problem occurs when the x-ray tube is moved off-center or tilted across rather than with the axis of the grid.

As a general rule, 12 : 1 and 16 : 1 linear grids should be used with kilovoltages greater than 100. Grids with 8 : 1 ratios are good for general radiography but require exact leveling and centering, which almost precludes their use for portable radiographic examinations. Crosshatched grids are useful only for special applications.

Viewboxes

Not much needs to be said about viewboxes, except that the light from the viewbox should be uniform across the viewing area and the hue of the light should be comfortable to the eye. The level of the light should be consistent with other viewboxes in the radiology department. Above all, a good viewbox should be used by the personnel responsible for quality control. It is not uncommon for those responsible for quality control to send back a radiograph for a repeat that appears perfectly normal on another viewbox; the problem rests in the variability of the light intensity of the viewboxes. The technologists should have viewboxes of equivalent quality in their review area as the radiologists have for interpreting the films.

Summary

This rather brief review of some of the problems that radiology department personnel should be aware of may help to solve major problems that were thought to originate in the radiographic equipment in the department but are actually centered elsewhere. Every one of the problems discussed in this chapter has, at one time or another, been thought to be caused by equipment failure. The reader must be aware that all problems in a radiology department do not result from equipment failure.

References

1. Buchanan, R. A., Finkelstein, S. I., and Wickersheim, K. A. X-ray exposure reduction using rare-earth oxysulfide intensifying screens. *Radiology* 105:185, 1972.
2. Stuermer, W. E. Rare-earth screens: A real progress in radiology? In *Proceedings of the Society of Photo-optical Instrumentation Engineers*, Vol. 70. pp. 209-211. Atlanta, Georgia, September, 1975.

Suggested Reading

Braun, M., and Wilson, B. C. Comparative evaluation of several rare-earth film-screen systems. *Radiology* 144:915, 1982.

De Smet, A. A., Ritter, E. M., Fritz, S. L., Martin, N. L., Chang, C. H. J., and Templeton, A. W. An evaluation of screen-film combinations for detail skeletal radiography. *Radiology* 143:259, 1982.

Ferguson, J. P., and Schadt, W. W. *Variability in the Automatic Processing of Medical X-ray Film*, BRH/DEP 70-13. Washington, D.C.: U.S. Department of Health, Education and Welfare, Public Health Service, Bureau of Radiological Health, June 1970.

Fuchs, A. W. *Principles of Radiographic Exposure and Processing* (2nd ed.). Springfield, Ill.: Thomas, 1958.

Rao, G. U. V., Fatouros, P. P., and James, A. E., Jr. Physical characteristics of modern radiographic screen-film systems. *Investigative Radiology* 13:460, 1978.

Robinson, T., Becker, J. A., and Olson, A. P. Clinical comparison of high-speed rare-earth screen and Par-speed screen for diagnostic efficacy and radiation dosage. *Radiology* 145:214, 1982.

Rossi, R. P., Hendee, W. R., and Ahrens, C. R. An evaluation of rare-earth screen/film combinations. *Radiology* 121:465, 1976.

Stevels, A. L. V. New phosphors for x-ray screens. *Medicamundi* 20(1):12, 1975.

Thompson, T. T. Selecting medical x-ray film, Part I. *Applied Radiology* 3:47, 1974.

Thompson, T. T. Selecting medical x-ray film, Part II. *Applied Radiology* 3:51, 1974.

U.S. Department of Health, Education and Welfare, Food and Drug Administration, Bureau of Radiological Health. Film-Screen Optics. In *Medical X-ray Photo-Optical Systems Evaluation*, DHEW Publication (FDA) 76-8020. Washington, D.C., October 1975.

U.S. Department of Health, Education and Welfare, Food and Drug Administration, Bureau of Radiological Health. Film-Screen Sensitometry. In *Medical X-ray Photo-Optical Systems Evaluation*, DHEW Publication (FDA) 76-8020. Washington, D.C., October 1975.

U.S. Department of Health, Education and Welfare, Food and Drug Administration, Bureau of Radiological Health. *First Image Receptor Conference: Film/Screen Combinations*, HEW Publication (FDA) 77-8003. Washington, D.C., October 1976.

2   Modern X-ray Generators

Over the past two decades there has been a gradual replacement of a large number of single-phase x-ray equipment with more sophisticated three-phase equipment. Three-phase equipment is generally more costly and complex than single-phase equipment, and it is important for the reader to know the basic principles of how different x-ray generators work. Certain types of generators are more suitable for certain applications than others, and much consternation can be prevented by purchasing the appropriate equipment for a specific application. In addition, the radiographer must understand the x-ray equipment to operate it properly. Many service calls for repairs are directly related to improper operation of the equipment; hundreds of dollars can easily be spent on repair bills simply because someone failed to push the right button or switch at the correct time. While it is the responsibility of the purchaser to acquire the appropriate equipment, it is the responsibility of the operator to use and maintain the equipment correctly.

Throughout this book the metric, or decimal, system is used as much as possible in describing a particular component of x-ray equipment. In relation to generators, the following conversion equations are helpful as a reference:

1 hertz (Hz) = 1 cycle/second

In a 60-cycle power line system as used in the United States:

1 cycle = 1/60 second = 16.66 milliseconds

1/2 cycle = 1/120 second = 8.33 milliseconds

Many other countries use 50-cycle systems.

To convert fractions of a second (sec) into milliseconds (ms), multiply the fraction by 1000. For example,

1/30 sec = ? ms

$$1/30 \text{ sec} \times 1000 \text{ ms/sec} = \frac{1 \times 1000}{30} = 33.3 \text{ ms}$$

To convert inches into the metric system:

1 inch = 25.4 millimeters (mm)
      or
      2.54 centimeters (cm)
1 mm = ~ 1/25 or 0.04 inch

Terminology

Before proceeding with a discussion of x-ray generators, it is useful to define some of the jargon peculiar to the subject. One of the terms used in describing all modern generators is *interrogation time*. Interrogation time, or phase-in time, as it is sometimes called, is defined as the amount of time required for the generator to react to a given stimulus or impulse. This time is calculated from the moment the exposure control is activated until the x-ray tube makes (or starts) an exposure; that is, interrogation time is the amount of time it takes the generator to "tell" the x-ray tube to make an exposure. In single-phase generators there is a minimum interrogation time of 8.3 ms. In some three-phase generators, interrogation time may be as low as 1 or 2 ms, and in constant potential generators with secondary exposure control this time may approach 0 ms.

X-rays are composed of a spectrum of energies. The minimum energy is determined by the filtration of the collimator, x-ray tube port, and insulating oil in the housing, together with the kilovoltage applied to the tube. The maximum energy is determined by the voltage differential between anode and cathode as provided by the x-ray generator. This maximum energy is commonly referred to as *peak kilovoltage*. The proper international designation of peak kilovoltage is now simply *kilovoltage*, which is abbreviated as kV, not KVP or PKV. The term *kiloelectron volt* (keV) is used to describe the quality of an x-ray beam spectrum and refers to the actual energy of x-ray photons coming from the x-ray tube. Since x-ray pulses are composed of a spectrum of energies, and since there are statistically more photons of less than maximum energy as limited by kilovoltage, average keV is always less than peak kilovoltage.

The term *kilowatt* (kW) is used in describing both x-ray generators and x-ray tubes. A *watt* is a measure of energy with 1 ampere (amp) under 1 volt (v) "pressure"; the term *kilowatt* (1000 watts) is used to describe the power output of an x-ray generator and how much power may be put into an x-ray tube.

Kilowatt Ratings of Generators

For three-phase generators, kilowatt ratings[1] are calculated according to the following formula:

$$\frac{kV \times mA}{1000} = kW$$

[1] In physics, $kW = kV + mA \times \cos \theta$, where θ is the phase angle between kV and mA. Since $\cos \theta$ is 1 only for purely resistive loads, actual kW is usually somewhat less than the formula indicates. The formula does suffice for this publication.

For single-phase equipment the formula is

$$\frac{kV \times mA \times 0.7}{1000} = kW$$

The 30% lower output from single-phase generators results from averaging the ripple factor from the rectified transformer output.

A number of factors can influence kilowatt ratings, including, for example, the size of the wire supplying power to the generator, the size of the feeder transformer, and the generator capacity and design. Moreover, when any electrical device is operated, the load placed on the electrical system causes a drop in the available power supply; the more energy required from the power line, the greater is the power drop for a given resistance of the power supply. The same thing happens with an x-ray generator: The more kilovoltage and milliamperage required from the generator, the greater is the drop in power supplying the generator. In modern generators the voltage drop compensation is provided by automatic electronic circuits. The evaluation and testing of radiographic equipment are performed under normal operating conditions, which are commonly referred to as testing *under load*. For convenience, kilowatt ratings of x-ray generators are determined at the 100-kV level. Kilowatt ratings of x-ray tubes are discussed in Chapter 3.

Kilowatt ratings of x-ray generators are important. Most users of radiographic equipment, and particularly those persons who purchase the equipment, are familiar with the terms *150-kV generator* or *125-kV generator*. The terms are meaningless by themselves and cannot be used to compare the capability of one generator to the other. These generators may have the capability of producing 500 milliamperes (mA), 1000 mA, or even 2000 or 3000 mA. When evaluating or comparing generators, one must think and speak in terms of the kilowatt rating of the generator, just as one should similarly consider the kilowatt rating of x-ray tubes.

When a generator is listed by the manufacturer as 150 kV *and* 1000 mA, it does not necessarily mean 1000 mA *at* 150 kV. One must get into the habit of asking, What kilovoltage at what milliamperage? For example, if one assumes that a 150-kV and 1000-mA three-phase generator is a 150-kW generator, he would be wrong. Generator kilowatt ratings are determined under load, which means that the operator cannot simultaneously obtain the individually listed maximum kilovoltage and milliamperage capabilities of the generator. For example, a typical 150-kV *and* 1000-mA generator tested under load would probably be rated 100 kV *at* 800 mA, or 80 kW—just about half of what one would expect had he assumed the 150-kV *and* 1000-mA output.

Table 2-1. Application of Two Typical Generators

Generator A Ratings	Generator B Ratings
150 kV at 800 mA	150 kV at 200 mA
125 kV at 1000 mA	120 kV at 1000 mA
100 kV at 1250 mA	105 kV at 1300 mA

One cannot depend entirely on kilowatt ratings to compare generators; the application of the generator must also be taken into consideration. For example, Table 2-1 compares two typical generators. At 150 kV the output from generator *A* is 120 kW compared to 30 kW for generator *B*, but at 100 kV, generator *B* has a higher kW rating. This difference in output becomes important in high kilovoltage techniques, particularly high kilovoltage chest radiography.

Three-Phase versus Single-Phase Generators

The term *generator* encompasses the operating console of the x-ray equipment along with its corresponding transformer (or transformers) and rectifiers. The type of transformer becomes important in respect to the output of the equipment. In full-wave rectified single-phase equipment, the transformer output begins at zero, builds to a positive peak, and then descends to zero during each half-cycle. The rectified output is therefore not constant but fluctuates from zero to a maximum. This fluctuation in output is called *ripple*. Ripple is the percentage difference from the maximum or peak voltage to the minimum kilovoltage. For example, at 100 kV and 13.5% ripple, the voltage fluctuates between 86.5 and 100 kV. For single-phase equipment the ripple is close to 100%, depending on the load applied.

In three-phase equipment, there are six pulses fed to the transformer during each cycle. The output from the transformer begins at zero potential, builds to a peak, and begins to drop toward zero just as the second pulse is building to a maximum. The second pulse begins to drop just as the third pulse is building to the maximum potential, and so on. The net result is that the generator output never drops much below the maximum potential. The ripple from three-phase equipment is much less than that from single-phase equipment; the amount depends on the type of transformer and rectification. Figure 2-1 shows a comparison of the ripple factors for single-phase and three-phase equipment. The important thing to remember is that the lower the ripple factor, the greater is the output of the x-ray equipment.

As was noted in the discussion of kilowatt ratings, when generators are operated under load, the ripple is accentuated in three-phase equipment. This accentuation of the ripple under load is known as the

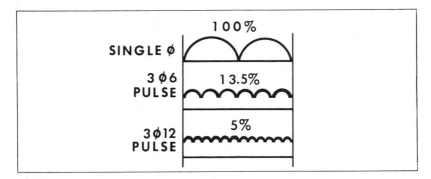

Fig. 2-1. Comparison of ripple factor x-ray generators. The greater output of three-phase 12-pulse equipment is obvious.

load ripple factor and is always greater than theoretical ripple. Some newer generators have adjustment facilities to normalize line resistance of all three phases; thus there is less accentuation of the ripple under load.

X-ray Transformers

Ideally, the output of an x-ray generator should approximate that of a pure direct current (DC) source, and the maximum output should be close to 150,000 volts (v). Some generators for special application may be limited to a lower output (for instance, a 100-kV maximum, as for angiography). One would like to take a common alternating current (AC) voltage, such as 115 v or 230 v normally found in any facility, and transform it into a higher, constant DC potential. In a battery-operated x-ray unit, DC voltages are obtainable from a set of high-energy batteries. Because of the relatively low current required from these units, batteries are adequate, but they are an inadequate power source to provide the high current and voltage levels required from modern, high-capacity x-ray generators. Therefore, high-energy batteries are not economically practical for routine use. (Incidentally, a battery x-ray unit converts DC to AC, transforms to high kilovoltage, and again converts to DC.)

A better source is the AC power that is available from standard power transmission sources provided by local power companies. Many years ago the power companies realized that it is more economical to transmit power in three phases as opposed to just a single phase. In most areas of the United States the three-phase power brought into the radiology department is typically in the range of 208–230 v or 480 v (480-v three-phase power should be the standard of future new departments). It is necessary to transform this voltage up to the level required to produce diagnostic x-ray energies and to change it from AC to DC. In order to do this, it is necessary to use a high-voltage transformer.

Three-Phase Voltage Transformers

Figure 2-2A is a diagram of a single-phase transformer. To refresh the reader's memory, the number of turns wrapped around the iron core on the primary side of the transformer is related to the number of turns in the secondary side and hence to the voltage step-up in the secondary. For example, if one had 100 v and 10 turns in the primary and 100 turns in the secondary, he would have an increase of 10 times 100 or a 1000 times increase in the voltage in the secondary. One should also remember that as the voltage is stepped up in the secondary, the amperage is stepped down proportionately, with the same power being maintained except for transformer losses.

Since radiology departments are switching to three-phase power, let us look at a three-phase transformer. A three-phase transformer consists of three sets of primary and secondary windings. Figure 2-2B is a drawing of a typical three-phase transformer. Three iron cores are joined together with three sets of primary coils and three sets of secondary coils, and again the same ratio applies; namely, there is a relationship between the number of turns in the primary and the number of turns in the secondary to produce the transformation or increased voltage output from the transformer needed for producing the desired energy.

Three-phase transformers are typically designed with the three sections of copper windings electrically connected in a *delta* or a *wye* configuration. The delta configuration is shown in Figure 2-3A. The ends of the copper windings are connected electrically so that, in essence,

Fig. 2-2. Single-phase and three-phase transformers. (A) Single-phase transformers have one set of primary windings and one set of secondary windings. (B) Three-phase generators have three sets of primary and secondary windings on the same metal core.

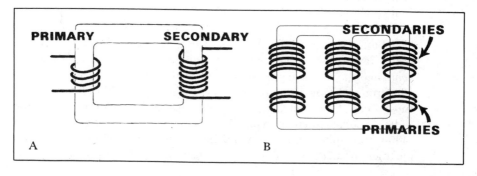

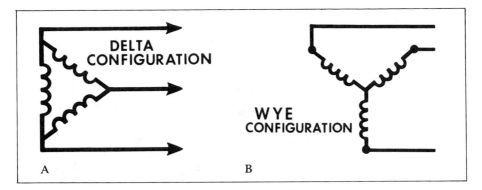

Fig. 2-3. Comparison of delta and wye transformers. In the delta transformer (A) the electrical windings are connected end to end, whereas in a wye transformer (B) common ends are connected together. The use of a delta/wye combination in the transformer secondary allows a 30-degree phase shift in the output, thus permitting a 12-pulse x-ray generating system.

one finds a delta (Δ) configuration (the Greek capital D) to the transformer. If the electrical connections are traced, it is noted that the bottom of one is tied to the top of the other, and so on; the result is the delta configuration.

The wye configuration, sometimes called the star configuration, is seen in Figure 2-3B. The tops or bottoms of each transformer coil are connected electrically so that the configuration is actually that of a Y. Both the wye and delta transformers increase voltage principally in the same manner, but phase shifting takes place between the two. Later on we will look at this characteristic difference between the two configurations and how this difference can be used to advantage.

As mentioned previously, the ideal voltage for an x-ray tube should be pure DC. However, the electrical power brought to the radiology room is three-phase AC, with only the three positive-going pulses being of any use to the x-ray tube; the three negative-going waves or pulses are of no use whatsoever. In an attempt to make the transformer output as close to DC as possible, rectification must be accomplished to convert the negative-going pulses to positive pulses, producing a full-wave rectified output.

High-Voltage Rectification

High-voltage rectifiers can either be the vacuum tube type (also called valves), or they may have solid-state composition, which is more common today. Solid-state rectifiers can be made of selenium or silicon. Selenium was the first solid-state material used for high-voltage rectifiers. The most advanced x-ray generators currently use solid-state silicon high-voltage rectifiers.

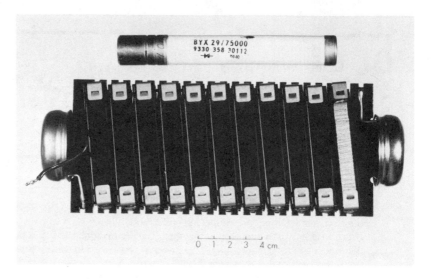

Fig. 2-4. A replacement-type selenium rectifier. Note the small, dime-size aluminum discs on the right, which have been exposed for better visualization. These discs are coated with selenium, which allows current to be passed to the x-ray tube in one direction, while resisting current flow in the opposite direction. The small cylinder at the top of the illustration is a solid-state selenium rectifier.

Selenium rectifiers, like the replacement type shown in Figure 2-4, are composed of thin deposits of selenium on dime-size aluminum discs. Each disc is a cell or *diode* that exhibits a low resistance to the passage of electrical current in one direction but resists a difference of potential of about 60 v in the opposite direction. By stringing enough of these cells together in a series, it is possible, when the voltage polarity is correct, to pass current to the x-ray tube and at the same time resist the high voltage presented in the opposite polarity by the x-ray transformer. A selenium rectifier has the disadvantage of having a greater voltage drop or power loss in the forward direction than a vacuum tube rectifier. Heat produced by the power drop lowers the life expectancy of the rectifier. The selenium rectifier can withstand maximum temperatures of about 266° centigrade (C). Proper design can allow successful use of this rectifier in x-ray generators.

In 1965 the General Electric Company introduced silicon high-voltage rectifiers, which are perhaps the most efficient and most reliable rectifiers available today for x-ray generators. Most American and European manufacturers are presently using silicon rectifiers in their high-energy generators. These rectifiers have a very low forward voltage drop and will resist an inverse or backward voltage of about 1000 v per cell, which is 10 to 20 times higher than the selenium rectifiers. Silicon rectifiers can tolerate considerably higher temperatures of up to

392° C, making them very durable and useful in x-ray rectification. In spite of their higher initial cost, silicon rectifiers are replacing selenium or vacuum tube rectifiers. The silicon rectifier is composed of a number of small diodes connected together, as shown in Figure 2-5.

In order to make a three-phase, six-pulse alternating current and to rectify the negative-going wave to a positive-going wave, it is necessary to incorporate at least six rectifiers in the high-voltage circuit. Figure 2-6 shows a delta primary transformer winding with a wye secondary transformer winding. Note the six solid-state rectifiers as demonstrated by the electronic symbols for such rectifiers. For simplicity, only one diode symbol per string of series-connected elements is shown.

Fig. 2-5. (A) A modern silicon rectifier. (B) The cylinder has been broken for better visualization. Note the rectangular diodes connected in series in the close-up view.

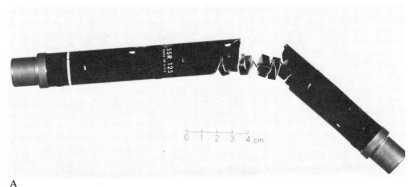

A

B

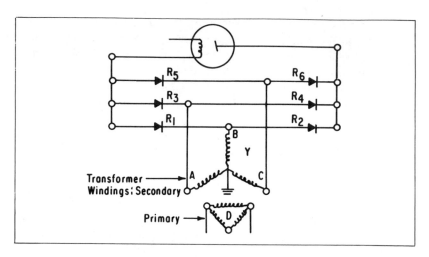

Fig. 2-6. Delta/wye x-ray transformer. The output from the wye transformer is rectified through a series of solid-state rectifiers denoted by the symbol ➤⊢. Current flow in the rectifier is in the direction of the arrow. X-ray output is six pulses per cycle, and the theoretical ripple is 13.5%.

Throughout this discussion of transformer-rectifier theory, the term *current flow* will be used. Current flow is the actual flow of electrons from negative to positive poles. Since one is often asked to explain current flow through this circuit, it might well be useful for us to do just that. The reader should remember at this point that current flow is in the opposite direction of the arrow in the rectifier symbol and that even though there are three electrical lines attached to the primary transformer windings, there is an electrical lag of 120 degrees between each leg or phase of the incoming power. At any given time the x-ray tube is receiving voltage from whichever of the secondary windings is above zero voltage potential. At a given moment in time, one winding is usually considered inactive. The two active transformer windings and their appropriate rectifiers form what would be essentially a four-rectifier bridge.

For example, let us assume that coils or transformer windings *A* and *B* (Figure 2-6) are the active windings and that winding *A* is negative in respect to winding *B*. Current flow would then proceed from transformer winding *A* through rectifier *R3* to the cathode of the x-ray tube and to the anode, then back through *R2* to coil *B*. During the next half-cycle, *B* would be negative in respect to coil *A*, and current would flow through rectifier *R1* to the cathode, to the anode, and then back through *R4* to winding *A*. Considering all points in time and the combination of the transformer windings and rectifiers, we end up with six pulses per cycle.

However, even the six-pulse output is still not as desirable as one would like, primarily because of the large ripple factor under load conditions. A method of improving the ripple factor of a six-pulse circuit is to add another set of secondary transformer windings, again hooked up in the same manner as the first secondary. Figure 2-7 shows a delta primary with two wye secondaries. To each wye secondary has been added a six-rectifier bridge, which gives a total of 12 rectifiers. The resulting DC voltage still has a 13.5% ripple theoretically, but the circuit is improved in that a better load ripple factor may be obtained, and it allows one to introduce under load conditions a direct reading DC-mA meter. This circuit also maintains fixed potential to ground, a very important design feature that is not possible with six rectifiers. This feature allows the manufacturer to build a transformer that provides a voltage from −75 to +75 kV to the x-ray tube, for a total of 150 kV, pole to pole. However, the problem remains with this six-pulses-per-cycle DC output that the voltage still has a theoretical 13.5% ripple under certain ideal load conditions.

In an attempt to achieve a pure DC output, manufacturers have gone a step further. Figure 2-8 does not look different from the previous diagram of six-pulse output (Figure 2-7) that gave a 13.5% load factor ripple, but a very significant difference is present in that the secondary is no longer a double wye connection; instead it is now a wye and a delta connection.

As mentioned previously, there is a characteristic difference between a wye and a delta transformer. Normally there is a 120-degree lag between the input and output of a wye transformer. With the same pri-

Fig. 2-7. Delta/wye/wye x-ray transformer. Twelve rectifiers are used in this combination. The output is six pulses, and the load ripple factor has been improved. Current flow through the transformer is illustrated.

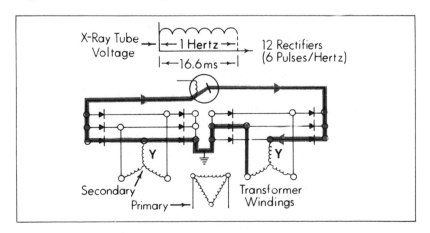

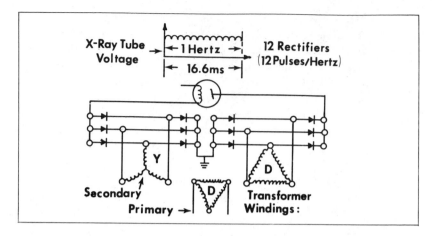

Fig. 2-8. Delta/wye/delta x-ray transformer. In this configuration there is a delta and a wye transformer secondary rather than the two wye transformers shown in Figure 2-7. This configuration allows an almost constant output and produces 12 pulses per cycle by utilizing a 30-degree electrical phase shift of the output between the two transformers; it also has the balance to ground and direct x-ray tube current measuring features seen in Figure 2-7.

mary voltage input into a delta transformer, there is an additional 30-degree electrical phase shift or difference between input and output. When delta and wye windings are connected together in the secondary, the output of the delta lags the wye by 30 degrees; that is, the phase timing of one secondary lags the other by 30 degrees. The result is that the output of one winding fills in the ripple of the other and results in a 12-pulse output rather than a six-pulse output, which means that every pulse from the primary is now doubled in the secondary by the addition of these wye/delta secondary connections. A double, six-rectifier bridge circuit is again needed to transform all negative-going pulses to positive-going pulses. We now have a 12-pulse output that theoretically has a 96.5% pure DC output, with an approximate ripple of only 5% under high load conditions.

Most modern 12-pulse generators are designed with one delta and one wye configuration in the secondary windings of the high-voltage transformer. In some cases the rectification may be different, or the means for turning the generators on and off may be different, but in every case there is principally one wye and one delta connection.

Medium-Frequency Converter Generators

In some parts of the United States there are continual problems with fluctuations in the power supply coming to the radiology department. This is particularly true in large cities such as New York and Miami.

The problem can be accentuated by inadequate power supply and distribution to the department from within the hospital or office. Although there is some variation from manufacturer to manufacturer, most generators are capable of handling $\pm$ 6 v variation of the incoming power. In my own department, we measured a voltage variation to the equipment of 0–280 v before a new power supply and voltage regulator were installed.

The above anecdote emphasizes the importance of proper control of the power supply to the radiographic equipment; equipment can handle only so much change before a variation in film density will result. Recently new type generators have been introduced, called *medium-frequency generators*. In a medium-frequency generator either single- or three-phase power input can be used, and the output is divorced from fluctuations in the power supply. In these generators the incoming power supply is fed into AC/DC/DC converters and then into a DC/AC inverter whose output is amplified and transformed so that the output is between that of a three-phase six-pulse generator and a three-phase 12-pulse generator. The output is a flat peak waveform similar to a constant potential generator. Medium-frequency converter generators are an answer to the varying voltage problem that does not necessitate the installation of a completely new power system or the addition of voltage regulators in each radiographic room. Figure 2-9

Fig. 2-9. Concept of a medium-frequency converter generator. Either a single- or three-phase power supply feeds an AC/DC converter, and the output is fed into a DC/DC converters. This high-frequency current is changed by a DC/AC inverter with the output being transformed into high voltage to supply power to the x-ray tube. This type of generator divorces the incoming power line and fluctuations from affecting the output of the x-ray tube. (Courtesy of Toshiba Medical Systems.)

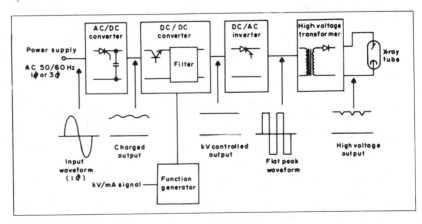

illustrates the principle of these new generators. The units are available from Siemens[2] and Toshiba.[3]

Switching the X-ray Tube Load

In the previous section it was shown how AC voltage is converted to an essentially pure DC voltage with only a very slight ripple in three-phase 12-pulse generators. The next task is to examine how the high voltage is applied to the x-ray tube; that is, how one turns the x-ray tube load on and off.

Today there are two basic methods of switching three-phase x-ray generators. Each has its own variations, but the major difference is in the method of switching or contacting the high voltage to the x-ray tube. One type of generator switches the voltage on or off to the x-ray tube by opening or closing a circuit on the primary or low-voltage side of the high-voltage transformer. This method is called *primary switching*. The other type controls the x-ray tube voltage with a circuit on the secondary or high-voltage side of the high-voltage transformer and is termed *secondary switching*.

Primary Switching

Primary switching is the least complicated method of contacting, primarily because it is much simpler electrically to turn 208–230 v or 480 v off or on than it is to switch 150,000 v. There are three approaches to this type of switching: electromechanical contactors, thyratrons, and solid-state silicon-controlled rectifiers.

Electromechanical contactors are seldom used today because they are relatively inconsistent in operation, have high maintenance requirements, are noisy, and at best are restricted to eight exposures per second. They are used primarily in low-cost generators. An electromechanical contactor works just like the doorbell of a house; that is, when a current is applied to an electromagnetic coil, it causes the plunger to move. The contacts are usually spring-loaded and strike the distal portion of the contacting mechanism, producing a circuit across the contacts. One can see that if the springs around these contactors are weak, the contacts will bounce; moreover, if there is not a sufficient voltage on the coil surrounding the plunger, the whole mechanism may have a poor electrical contact and produce noise. This noise is commonly heard in some of the old generators, the so-called contactor chatter.

The second method of primary contacting is with *thyratrons*. Thyratrons are actually large, gas-filled tubes used as electronic switches. A single grid in the tube controls the start of current flow after which

[2]Siemens Medical Systems, Erlangen, West Germany.
[3]Toshiba Medical Systems, Tokyo, Japan.

it is continuous. Thyratrons are being phased out in favor of solid-state devices because of their large size, cost, heat production, and maintenance requirements. They can be used in single- or three-phase systems. Two thyratrons would be required for a single-phase system and six for three-phase equipment. It would take a rather massive piece of equipment to contain the heat transfer from these tubes; in addition, the cost of changing a thyratron when one goes out is prohibitive in comparison to the expense of replacing a silicon-controlled rectifier.

The third method of primary switching used in most all new radiographic equipment is with solid-state semiconductors called *silicon-controlled rectifiers* (SCRs) or thyristors. A SCR (thyristor) is the solid-state equivalent of a thyratron. SCR contacting is generally the preferred method today and is gradually replacing all mechanical and hot filament devices for this purpose. SCRs are produced in mass quantity because they are used in many other fields, such as radar and variable-speed motor controls. As Figure 2-10 shows, an SCR contactor is small and looks very much like a solid-state, high-voltage diode. It is mounted in the generator or power module chassis. The electronic symbol of an SCR contactor is the same as that of a silicon rectifier except that there is a line drawn into the diode symbol, which represents the gate of the SCR.

At this point a quick review of how a thyristor, or SCR, works will be helpful (Figure 2-11). As a comparison, a typical triode contains

Fig. 2-10. A solid-state silicon-controlled rectifier (SCR) is shown at the bottom of the illustration, with a high-voltage solid-state diode seen above. Note the small wire at the center of the SCR, which serves as the electronic gate for the semiconductor.

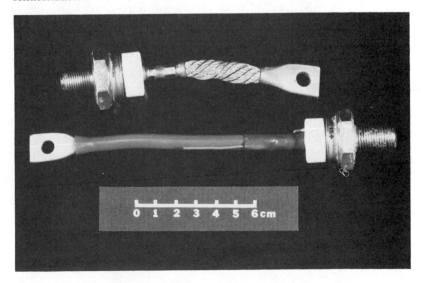

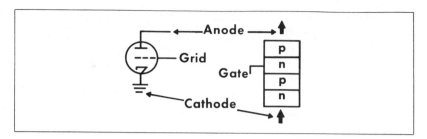

Fig. 2-11. Comparison between a vacuum tube triode (*left*) and an SCR (*right*). The vacuum tube contains three elements as noted. The SCR is composed of a number of semiconductors. In order for the SCR to conduct a current, an appropriate signal has to be applied to the gate of the semiconductor. n = negative side of the semiconductor; p = positive side of the semiconductor.

three elements: anode, cathode, and grid. Current flow is from cathode to anode. As long as the voltage on the grid is positive to the cathode but less positive than the anode, current will flow. One can control the current flow through the triode by controlling the voltage on the grid. On the other hand, a solid-state thyristor or SCR is composed of a series of semiconductor plates, and current flow is again from negative to positive. A semiconductor has the property of conducting current in only one direction and will not conduct until a signal is applied through a "gate" to a particular interface. It will then conduct as long as a positive signal is provided through the gate to that interface. To open the SCR switch, one simply has to remove the gate signal; when the anode voltage returns to 0, the switch will automatically open and stop conducting.

The first method of solid-state primary switching using a thyristor or SCR contactor was in a circuit called an inverse parallel circuit (Figure 2-12). The term *inverse parallel* means that there are two SCR gates that are back-to-back and connected in parallel, producing an inverse parallel circuit for each phase of the primary side of the transformer. The gates of the SCRs are normally controlled by a solid-state timer and an electronic memory bank in synchronization with the 60-Hz waveform, so that the high-voltage transformer will not saturate and distort the x-ray output. Normally at the beginning of each x-ray exposure, the gate of one or the other SCR, depending on the electronic memory, will be energized about 1 ms prior to the point where the anode becomes positive. This millisecond is needed to allow the gate to "get ready," since current flow starts at the 0 crossover of the AC waveform. The gate signal allows the SCR to begin conduction, which will continue until the input voltage goes through 0. The SCR will then conduct as long as its anode voltage is positive. For example, if the exposure switch is open, say at 4.1 ms relative to the 0 phase start, the

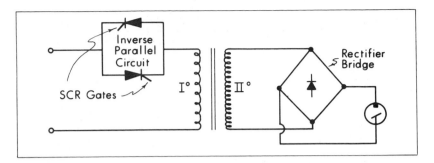

Fig. 2-12. Primary SCR exposure control for single-phase generator using an inverse parallel circuit.

SCR will not stop conducting until its current returns to 0. This cycle will result in an 8.3-ms exposure, since it takes 8.3 ms for the phase to travel from 0 to 0, even though the generator "told" the tube to turn off at 4.1 ms. Therefore, the time of an exposure can be controlled only in a multiple of 8.3 ms for a 60-Hz power supply. The fact that the exposure is controlled by impulses from a 60-Hz power supply leads to the terminology of *synchronous-type exposure;* that is, the exposure is initiated in synchrony with the 60-Hz power supply and cannot be terminated until the waveform goes through 0, unless forced electronic extinction is used.

The SCR inverse parallel circuit for three-phase generators has been superseded by newer techniques that make it possible to divorce the termination point of the exposure from the incoming line frequency, allowing a wider choice of exposure time increments in the short time range. This technique is called *SCR pulsed commutation.* (The word *commutate* means to produce an unidirectional current.) Used with *forced extinction*, this technique provides accurate x-ray exposures in the millisecond range, which are necessary for use with photospot cameras, rapid film changers, and all automatic exposure controls. This pulsed commutation extends the relatively simple primary SCR contacting system and makes it comparable to a more complicated and expensive secondary switch system in terms of exposure accuracy.

Figure 2-13 illustrates the principles of pulsed commutation. As one can see in the diagram, what really has been done is that a full-wave bridge using silicon rectifiers has been added to the ground side of three primary windings of the transformer. The rectifier circuits direct all six currents of the half-phases through a single SCR exposure-control circuit, which in turn is controlled by an exposure timer.

The secret of pulsed commutation, as far as interrogation time is concerned, lies in the electronic timer and electronic exposure control. Completely asynchronous initiation of the exposure, sometimes called

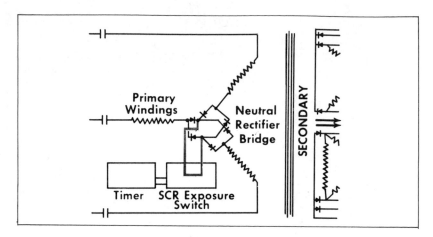

Fig. 2-13. Principles of SCR pulsed commutation. Note the six-pulse rectifier circuit that has been added to the star connection of the transformer, forming a neutral rectifier bridge. By the use of precontacting (not shown), the electromagnetic flux of the transformer can be controlled, and one no longer has to wait for transformer conditions to be "right" before the next exposure can be made. This circuit allows one SCR to control simultaneous turn-on and turn-off of all three phases.

millisecond interrogation, is not achieved using pulsed commutation alone; pulsed commutation is dependent on the electronic timer and exposure control. Normally, the electromagnetic waveform that produces hysteresis in the transformer must be kept moving in the same direction to prevent distortion of the output waveform, or possible damage to the transformer and tube may result. The electronic timer can "remember" in what condition the transformer was at the end of the last exposure, but it still has to wait for the proper conditions to exist in the transformer before the next exposure can be made. By adding *high-speed electronic precontacting* (that is, getting the transformer ready for switching) independent of phase conditions, one can reduce interrogation time down to 1 or 2 ms. Millisecond interrogation is therefore accomplished, and there is no longer a dependence on the incoming voltage waveform to start the exposure safely.

Millisecond termination or *forced extinction*, another form of exposure control, also requires a conservative design to prevent saturation and possible destruction of the transformer, and to produce a consistent x-ray output. The term *forced extinction* means that one can terminate the exposure at any time without waiting for the waveform to pass through 0 voltage potential. Short exposure accuracy using automatic exposure devices is usually specified as the minimum exposure time possible with a percentage (usually 10–15%) density change as measured on the film. Forced extinction techniques commonly allow

2-ms reproducibility of film density with automatic exposure devices. Forced extinction is almost mandatory with the use of rare-earth screens and in conjunction with automatic exposure controls. With forced extinction, a low DC voltage is applied to a capacitor network connected to the SCR circuit and, on command from the generator control, the capacitor will momentarily discharge to reverse the polarity at the SCR, forcing the SCR to stop conducting. Thus with forced extinction alone, one has to wait for the transformer to be "right" before the exposure can begin, but the exposure can be terminated at will. If one combines the electronic timing accuracy of forced extinction with high-speed electronic precontacting and the pulsed commutation technique as previously described, one has the capability of producing x-ray impulses of 1 ms on up with a negligible interrogation time and practically 0 termination time lag.

To summarize briefly, primary switching is a mechanism of turning on or off the current to the x-ray tube by opening or closing the primary transformer windings. *Synchronous switching* means that there is a relationship to the 60-Hz timing waveform, and these timing waves have to go through a 0 voltage potential before a timing impulse can be stopped or started. Nonsynchronous switching means that there is no relationship between the time an x-ray exposure can be made and the incoming timing impulse, which is important for rapid-sequence radiography; the exposure can be stopped or started anywhere in the waveform. Nonsynchronous switching can be accomplished by SCR pulsed commutation plus high-speed electronic precontacting; the exposure time can be reduced to the range of 1–2 ms, and longer maximum available exposure times can be achieved for high-speed film changers and high-speed (12 exposures/sec) spot-film cameras, since the exposure can begin the instant the film is in place.

Secondary Switching

The other basic type of x-ray generator uses secondary switching. This system controls the high voltage to the x-ray tube by switching in the secondary of the high-voltage transformer or at the x-ray tube rather than on the primary side of the transformer.

One subtype uses triode or tetrode switch tubes. These systems have a primary power supply rectified through a 12-pulse high-voltage transformer system. A capacitor filter network is added in addition to electronics for control of the secondary switch tubes. The voltage from the secondary of the high-voltage transformers is fed through an anode and a cathode switch tube to the x-ray tube. This type of generator, using this basic circuit, is commonly known as a *constant potential generator.*

A schema of a triode constant potential generator is shown in Figure 2-14. In the secondary windings are wye and delta secondary coils,

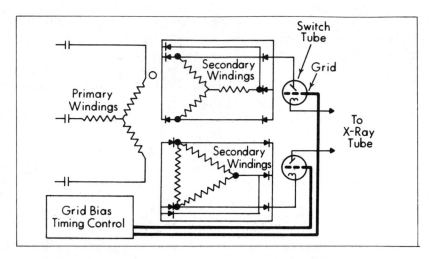

Fig. 2-14. Principles of a constant potential generator with triode secondary switching. Note that the output of the secondary windings is fed into triode control tubes. Through a comparator circuit, the anode-cathode voltage is exactly balanced. An exposure control circuit connected to the triode allows switching down to less than 1 ms, even for a 150- or 200-kW generator. Occasionally four-element "tetrode" tubes are used in place of the triodes.

again with six-rectifier circuits on each side. Note that the voltage from the wye windings is being fed into a switch tube that will go to one side of the x-ray tube; the delta secondary winding voltage is fed into a voltage switch tube that will go to the other pole of the x-ray tube. Note also the addition of the *grid basis timing control*, which feeds into the grid of the triode. The function of the grid bias control circuit is extremely important in the operation of constant potential generators. First, the grid bias control circuit must be able to compare and maintain the voltage output from both triodes to make absolutely certain that there is a constant existing voltage potential between the two tubes. When the voltage is exactly balanced, the output from the x-ray tube is almost pure DC without a significant ripple. Second, the grid bias control circuit is used to start and stop the x-ray exposure. This type of switch design will allow very rapid exposure sequencing, with the exposures as short as 0.5 ms and with essentially no interrogation time.

This circuit design does have some disadvantages, however. The length of the high-voltage cables leading from the switch tank to the x-ray tube must be kept short, preferably no longer than 30 feet. Cables longer than 30 feet will store energy; this energy extends the exposure after termination of exposure, detracting from the accuracy of exposure time at the very short end of the timer range. Because of this capacitance discharge there may be several milliseconds of anode heating and low-energy radiation produced by the x-ray tube, even after the high voltage has been switched off. Also, the hot filament tubes controlling the high voltage may have to be replaced at frequent intervals, and these tubes are almost as expensive as the radiographic tube. However, automatic failure indicators have made trouble-shooting and servicing much easier. The ultrashort timing, pure DC output, and automatic kV regulations of these generators make them useful in cinefluorography and angiography.

Another type of secondary switching employs a *grid-controlled x-ray tube.* This system has been in use for many years. The prime application for this type of switching has been in cinefluorography, in which the milliamperes are at a maximum of approximately 400–600 mA peak. The grid control system is similar to the tetrode or triode switching tube in which a grid bias voltage is used as a switch, but in this case the grid is actually the cathode cup of the radiographic tube itself. With this form of secondary contacting, the heating and postexposure radiation caused by high-voltage cable capacitance are eliminated.

Figure 2-15 is a schematic concept of a grid bias x-ray tube. Normally, as the filament is heated, an Edison effect, or electron cloud, is produced by the filament. If a bias voltage of 2–3 kV is applied to the focusing cup, this electron cloud can be controlled. If a high voltage is applied to the x-ray tube during an exposure, radiation is prevented because the bias voltage on the cathode cup prevents electron flow from the cathode filament. On a command from the exposure timer, the film changer or, in the case of cinefluorography, the shutter of the cinecamera, the bias voltage to the focusing cup is cut off, allowing current to flow through the x-ray tube and produce radiation. (A grid

Fig. 2-15. Principles of secondary switching via a grid-controlled x-ray tube. By applying a bias voltage to the cathode cup, one can control the x-ray exposure by an electronic timer, allowing as many as 500 exposures per second.

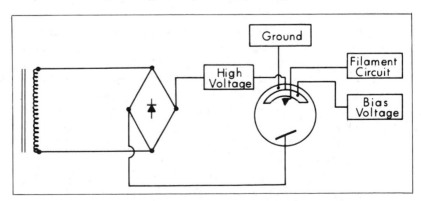

bias–controlled x-ray tube is not to be confused with a grid-focus x-ray tube, in which the actual electron beam emanating from the filament is focused down to produce a smaller effective focal spot.)

Contacting by use of a grid-controlled x-ray tube is extremely fast; filming rates as high as 500 exposures per second are possible. Because this system is of a secondary contacting type, it is completely divorced from the line frequency for interrogation or termination of the exposure. However, the high cost and complexity of this type of generator when there is a need for high milliampere loads make it likely that it will continue to be used primarily for cinefluorography rather than for high-energy angiography. A combination of primary switching for angiography and secondary grid control for cine techniques is very acceptable.

To summarize, secondary switching is the mechanism whereby an exposure is stopped or started by opening or closing the electrical circuits on the secondary side of the transformer. This process is accomplished by the addition of high-voltage switch tubes in the secondary or a grid bias voltage to the x-ray tube itself.

Interrogation Time

All the previous discussions have laid the groundwork for the reader to realize the importance of understanding modern x-ray generators and their applications in the radiology department. To reinforce the importance of interrogation time, the conditions under which interrogation time occurs should be reviewed. For example, a film changer will normally signal the x-ray generator randomly when a film is in place and the intensifying screens are compressed. When the generator receives this signal from the changer, there is a period of hesitation in conventional generators before the actual exposure starts, to permit the primary voltage and transformer to reach a proper condition.

With a conventional generator, "proper condition" means that two things must happen: (1) The line voltage must be passing through zero. (For a three-phase generator, line-to-line voltage between at least two of the three incoming power lines must be at zero.) (2) The voltage must be moving in the same direction as it was when the latest exposure was terminated. The time required to get the conditions right in the transformer for making the next exposure is called the *interrogation time*, which was mentioned at the outset. Even though the actual duration of the interrogation time can be any length between zero and a whole wave impulse (namely, 16.7 ms), one has to allow for the maximum of 16.7 ms in all generator applications using conventional synchronous exposure control circuitry. In asynchronous generators this waiting time is almost totally eliminated.

Falling Load Generators

With a conventional modern x-ray generator, the load to the x-ray tube is kept constant during an exposure. To preselect a quantity of x-rays (that is, milliampere-second), one has to choose a milliamperage and a time setting on the generator console. One then has to manipulate kilovoltage, milliamperage, and time to obtain acceptable exposure factors and still keep within the exposure latitude and loading characteristics of the x-ray tube. The more kV-mA-time selections available on the generator, the more flexible the x-ray equipment becomes. At the same time, the operator must consider the factors of motion unsharpness and blooming of the focal spot in respect to the quality of the radiographic image.

The modern x-ray generator can be designed with one of three types of exposure load control logic:

1. Selectable constant load (kV, mA, and time) for all techniques
2. Falling load (kV and mAs) for nontomographic techniques
3. Generators that allow the operator to select either of the above two modes of operation.

Both the generators employing *free selection* of kV, mA, and time and the *selectable constant load* type have the disadvantage that the user must determine an appropriate tube load and exposure time for each examination. On some modern generators, if all possible kV, mA, and time settings are considered, the operator may select one of a possible 19,280 combinations! It would therefore seem practical to have some simplified method for selecting technique factors.

In 1963 Philips[4] introduced their *falling load generator.* In theory, the operator can always achieve the shortest possible exposure time, since the x-ray tube focal-spot track is always operated at its maximum kilowattage. Because the milliamperage is rapidly reduced a short time after exposure begins, the x-ray tube target is protected against melting. When used with automatic exposure devices, the falling load generator logic functions as a tube safety device, thus eliminating the need for manual setting of the safety timer. Elimination of timer setting requirements and a manual load setting (mA) then made it possible to program the automatic exposure device for anatomical areas with a minimum of complexity.

Experience with falling load generators over the years has shown that starting every exposure at the x-ray tube limit has its drawbacks. In addition to focal-spot blooming at high mA, the instantaneous heating

[4]Philips Medical Systems, Eindhoven, the Netherlands.

of the focal-spot track to maximum with every exposure results in poor tube life. For this reason manufacturers of falling load generators may derate the initial load applied to the tube. In some of the latest designs the operator has no way to control the initial loads. These loads are set by the serviceman—one load for the large focus and another for the small focus. It is this same derating, together with the lowered kW at longer exposure times, that has caused the exposure time indication on falling load generators to be suppressed or deemphasized.

In the preceding discussion the expression *one load per focus* was mentioned. The load referred to is really the initial kW load applied to each tube focus. In practice, the starting mA component of the initial load is automatically lowered as the operator selects a higher kV; this is the so-called isowatt principle. In theory the kW for the initial load should be the same at, say, 150 kV as at 80 or 100 kV. In practice the "roll-down" of mA with kV is usually overdone to some extent. It is quite common to see a generator that is rated 100 kW at 80 or 100 kV but with an actual rating of only 50 kW at 150 kV.

The generator equipped with falling load exposure logic dynamically reduces the x-ray tube filament current after the start of the exposure. The resultant decrease in mA would cause the kV to rise at the same time unless some form of kV compensation were applied dynamically in concert with the falling tube mA. This compensation is accomplished several ways (Figure 2-16). The saw-toothed step method of compensation has given way to electronic methods of control in recent years. The most sophisticated method of kV control was pioneered by Siemens in their Pandoros generator.[5] This method results in stepless kV compensation by the use of high-voltage triodes in the secondary. With this secondary regulation technique, however, a generator rated for 150-kV exposures must be insulated for 200 kV in the transformer and triode switching tanks.

The advent of the microprocessor has opened the door to automation in setting all technique factors, including load and safety times in constant load generators. If the load is applied properly, there should be no need to derate the generator or to needlessly prolong the exposure time. With these devices the constant loads can be programmed for the part to be radiographed (e.g., maximum kW for chests, more conservative loads when practical, as in radiographing a thoracic spine). Statistically better tube life should result, since only those techniques requiring the maximum rated load are programmed for this kW. Microprocessors in constant load generators will also "look up" the proper safety times needed to protect the tube when automatic expo-

---

[5]Siemens Medical Systems, Erlangen, West Germany.

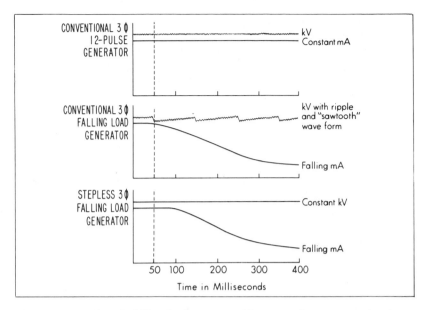

Fig. 2-16. Principles of a falling load generator (for comparison a conventional generator is also shown). With a conventional falling load generator, the milliamperage is kept constant at some high value for the first part of the exposure. During the exposure milliamperage drops dynamically, causing a voltage rise in the primary transformer. To compensate for this voltage dropping, resistance is added to the primary transformer at preplanned time intervals. The resulting kilovoltage waveform not only has the typical ripple but also has a saw-toothed appearance. In stepless or continuous falling load generators, the kilovoltage is kept constant electronically while the tube current is held at its initial value for 10–100 ms before being allowed to decay. The triode or tetrode switching system is also used to provide a constant kilovoltage.

sure devices are used and then effect their settings without operator intervention.

The falling load generator has its greatest application in conjunction with automatic exposure devices, where simplification of operator control is desirable. For angiographic applications these generators should be provided with a constant load selection feature at the kW rating specified, using the isowatt principle.

Anatomically Programmed Generators

As noted previously, one of the basic problems with conventional x-ray generators is the extremely large number of possible combinations of technique factors that can be chosen at any given time. The falling load generators eliminate some of these possibilities by restricting the operator to choosing kV and mAs for conventional techniques and kV only for automatic exposure controls, rather than kV-mA-time for all exposures. Some 30 years ago Picker Medical Systems[6] introduced a method whereby the operator chose only one technique factor, and the generator was preprogrammed to allow a preset kV-mA-time for that technique factor. This first anatomical programming never really caught on primarily because no good automatic exposure device was available then. In recent years European manufacturers have offered anatomical programming as a feature of their generators used in conjunction with automatic exposure controls. This method is used widely in Europe as well as in many institutions in the United States.

Like its predecessor, the new anatomically programmed module is an accessory for the generator. Equipment may be purchased that has anywhere from 8 to 40 or more anatomical programs. The purchaser decides within a given parameter what technique factors are to be dedicated to a particular technique button. For example, the technique factors for a radiograph of an elbow might be 55 kV, 100 mA, 100 ms, small focal spot. When the technique button for "elbow" is engaged, the generator is automatically set for those factors. As with falling load generators, the anatomically programmed generator is best used in conjunction with automatic exposure devices.

Programming of complete special procedure examinations is already available with some complex systems and will most likely be utilized more in the future. An ideal application of anatomical programming is in conjunction with gastrointestinal spot-film radiography. With conventional equipment, either the radiologist or the technologist has to change the generator control when going from spot filming of the gallbladder to spot filming of the gastrointestinal tract. With anatomical programming provided at the spot-film device, the radiologist would simply have to push the "gallbladder" technique button; when filming of the gallbladder was completed, the "GI" button would be activated. Normally, lower kilovoltage (70 kV) is used when radiographing the gallbladder as compared to the high kilovoltage (120–150 kV) required to penetrate the barium used in gastrointestinal fluoroscopy.

Microprocessors are replacing the relays and discrete transistors in the logic circuits of anatomical programmers. The microprocessor will provide access to a large but inexpensive memory on a single silicon chip. From stored data on this chip, the optimal milliamperage, safety time, and even nonautomatic exposure control fixed times can be added to the list of factors that are automated. Programmed selection of the tube load (kV and mA) may eliminate the need for falling load circuitry (see the section on falling load generators).

[6]Picker Medical Systems, Cleveland, Ohio.

Modular Generators

Two of the largest problems faced by equipment manufacturers are (1) to provide equipment capabilities to suit the needs of a large number of customers with markedly varied opinions as to what capabilities the x-ray equipment should have, and (2) to provide reliable, easily repairable equipment. In the past when a customer asked for a specific capability, it cost not only him but also the other purchasers considerably more, since all customers had to pay in part for the development and research that takes place within a company to provide for accommodation of the more sophisticated user-customer.

In trying to overcome these two major problems, a number of manufacturers have introduced *modular component generators.* These generators have been designed in such a way that the control console can accept a given number of interchangeable modules, thus greatly increasing the versatility of the equipment. If a customer wants no fluoroscopic capability, then one no longer has to purchase that capability, as would be necessary with most standard generators. The customer has the option of selecting or not selecting factors such as automatic exposure control, anatomical programming, falling load, load selection, and tomography.

In addition, most of the newer generators are of solid-state construction. Solid-state construction not only has decreased the physical size of the equipment, but it also allows service personnel to remove and replace entire components quickly, with the result that maintenance or repair time is vastly reduced. The introduction of solid-state electronics also allows the use of *multipurpose multiroom/multiple console generators.* With the advent of modern generators, it is possible to use one generator for a number of purposes and to control a number of adjacent radiographic rooms. Since the probability of two exposures occurring at the same time is statistically almost infinitesimal, it appears logical to purchase a multipurpose, high-priced generator to control more than one radiographic room. However, the reader should be aware that, in general, "time-shared" generators are not recommended for use in both fluoroscopic and radiographic rooms, since one would not be able to make an exposure in the radiographic room while someone was fluoroscoping in another room. Availability of service is another important factor, because if the one generator is not working, then all the connected equipment is not operable. It should also be recognized that "satellite" consoles are usually restricted in terms of function as compared to dedicated generator consoles.

Mobile Radiographic Equipment

As with other radiographic equipment, changes are also occurring in portable radiographic equipment. Webster's defines *portable* as "capable of being carried," which is obviously a misnomer when applied to radiographic equipment. What we really are talking about is *mobile* radiographic equipment, that is, radiographic equipment that can be moved. There is limited portable radiographic equipment being used by the military and private concerns in the United States, but most of the movable equipment is mobile rather than portable.

Until a few years ago most radiology departments had single-phase mobile x-ray equipment capable of operating at 100 kV and 100 mA. Most of these generators had a rating of only 7–10 kW. Occasionally 125-kV and 300-mA generators were seen; they were rated at 15–20 kW. The larger mobile units were bulky, heavy, and difficult to maneuver through the hospital. Selection of kilovoltage, milliamperage, and time factors was restricted due to the very nature and purpose of the equipment. Short exposure times could not be obtained if a thick body part was to be x-rayed, and thus a large number of the radiographs were technically unsatisfactory.

One of the main problems with older mobile units is related to the power supply to the area in which the x-ray exposure is to be made (i.e., wards and operating rooms), particularly in older facilities. Older hospitals were not constructed to provide uniform or adequate power to operate radiographic equipment with consistent results throughout the hospital, and one was forced to use the available power in the different areas. Newer hospitals to some degree do provide improved power capability, but the cost of installing special power lines for x-ray equipment throughout the hospital is very costly.

As with the power supply to a main radiology department, line resistance is extremely important for power lines supplying mobile equipment, perhaps even more so than for the main radiology department. When an x-ray exposure is made with any radiographic equipment, the incoming voltage drops by the amount necessary to transmit the load through the power line. When the load demand of the generator is high (i.e., when using high kilovoltage and high milliamperage) and the line resistance is also high, a large drop in kilovoltage will follow, lowering the output of the generator. Line resistance must be taken into consideration when mobile radiographic equipment is used. The generally accepted line resistance is 0.5 ohms for a 230-v line.

Another factor to consider is that line resistance may vary between different incoming power supplies that use two different phases of the main three-phase power supply. The line resistance may vary from room to room, ward to ward, and operating room to operating room according to the distance of the room from the power source. The variability may be significant; the x-ray equipment may not be able to adjust automatically to the changes, and as a result reproducibility of radiographic exposures would be next to impossible. The use of extension power cords, which effectively increases line resistance and hence lowers effective kilovoltage output, should also be taken into account. In general it is safe to say that mobile single-phase generators with more than 15–20 mA will not provide consistent reproducible output at various locations throughout the hospital.

Many of the problems associated with mobile radiographic equipment can be overcome by divorcing the equipment from the power supply during the exposure. This can be accomplished by using either *capacitor (condenser) discharge generators* or *battery-powered equipment.*

Capacitor Discharge Generators

In a capacitor discharge generator, current from the main power line is used as an input into a high-voltage transformer (Figure 2-17). Unlike other generators, the rectified output from the transformer is used to charge a large capacitor, or bank of capacitors, rather than to apply the power directly to the x-ray tube. As an exposure is made, the capacitor is discharged through the x-ray tube, providing the power for the exposure. X-ray tubes supplied with these types of units are usually grid-controlled in order to start or stop each exposure and to prevent radiation leakage. These generators operate at a nearly constant mil-

Fig. 2-17. Principles of a capacitor discharge generator. When the charge circuit is activated, voltage from the high-tension transformer charges a large (1-microfarad) capacitor. When the exposure control is activated, the capacitor is discharged through the x-ray tube. The length of the exposure is determined by a grid-controlled x-ray tube. The exposure curves show that the kilovoltage decays at a rate of 1 kV per milliampere. Only 30% of the exposure charge is useful because of this decay. The capacitor discharge unit is best used for low milliamperage exposures at short times.

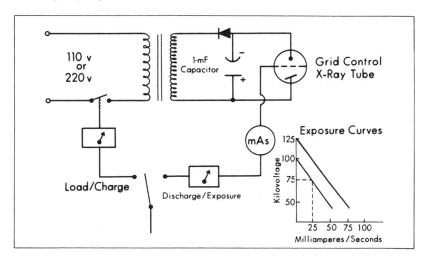

liamperage for a very short exposure time, and some of them are rated as high as 500 mA. (The focal-spot size of the x-ray tube must be large enough to handle this high instantaneous current.) Kilovoltage is variable, as are time/mAs selections.

There are many advantages to a capacitor discharge unit. The units are generally less bulky and are easily maneuvered through the hospital. Since they are charged from standard 110-v or 220-v lines, special power lines are not needed to charge the capacitors. If line resistance is high in one area as compared with another, the only effect this has on the generator is that it may take more or less time to charge the capacitor to its full capacity. The high milliamperage output with very short exposure times makes this unit suitable for radiographs of the chest and other small body parts. Capacitor discharge units for the most part are less expensive than other mobile radiographic units.

There are also disadvantages to capacitor discharge units. The major disadvantage has to do with limited mAs output. The maximum useful output from most commercially available equipment is in the range of 30–50 mAs, although more expensive and larger versions are becoming available. This limitation makes the unit inadequate for radiographs of heavy body parts, such as the spine and abdomen. With short exposure times (10 ms) the output is comparable to a three-phase 12-pulse system, but with longer exposure times the output drops off rapidly. This change in output is a result of the falling voltage and current during the exposure. As a rule of thumb, the kilovoltage drops 1 kV for each one mAs. For example, at an exposure of 100 kV, 30 mAs, the kilovoltage would drop to 70 kV during the exposure. There would also be a small drop in milliamperage due to the space charge effect of the x-ray tube. Depending on the main line resistance, the amount of remaining charge, and the size of the capacitor bank, recharge time may vary from a few seconds to a minute under usual conditions.

As long as short exposure times are used and the capacitors are kept fully charged, the capacitor discharge unit is an effective mobile radiographic unit. One can expect some contrast changes in the radiograph if an exposure is made when the unit is not fully charged or if long exposure times are used due to lowering of the kilovoltage output.

Battery-Powered Generators

The other method of divorcing mobile radiographic equipment from the main power line is with battery-powered units. In a battery-powered unit, power is provided by standard power lines to charge high-capacity, rechargeable batteries (Figure 2-18). These batteries may be similar to the lead-acid types used in automobiles or they may be of nickel-cadmium construction. Nickel-cadmium batteries, although they are more expensive than lead-acid batteries, have a distinct advantage in that they do not foster corrosion of nearby electronic parts

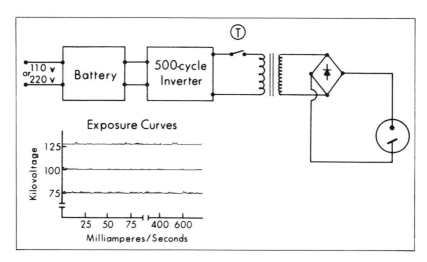

Fig. 2-18. Principles of a battery-powered x-ray generator. In this type of generator, a standard wall receptacle is used to charge or recharge a battery. When the exposure control is activated, current from the battery is fed through an inverter, creating a pulsed AC output that is transformed to high kilovoltage, rectified, and applied to the x-ray tube. The exposure curves show that the kilovoltage is held constant throughout the exposure. Since the batteries can hold a charge for a prolonged period of time, this type of unit is ideal for use during emergency power failures.

as lead-acid types do. (This is one reason that nickel-cadmium batteries are preferred for use in aircraft.)

The output from the batteries is fed into a DC inverter, which interrupts the DC voltage at a rate of, for instance, 500 times per second (other frequencies are also possible), producing an alternating current that rises to a maximum and falls to zero at a rate of 500 cycles per second. The 500-Hz alternating current is transformed and rectified to a potential voltage of 110 or 125 kV, 1000 pulses per second. This output combined with the inherent capacitance of the high-voltage system provides voltage ripple equivalent to a three-phase, constant load 12-pulse generator. In contrast to capacitor discharge units, the rectified output is applied directly to the x-ray tube as with conventional generators, but from a constant power source—the battery and a constant line resistance from the battery to the high-voltage transformer. The constant milliamperage applied to the x-ray tube allows the use of smaller focal spots in the x-ray tube, as compared to the capacitor discharge units.

The main disadvantage of the battery-powered unit has to do with the physical weight of the unit. The batteries require periodic charging, which can be done from a conventional wall receptacle anywhere in the hospital. In addition, a close watch must be kept on the batteries to make certain they are in proper operating condition.

One additional advantage of the battery-powered unit is that it allows a hospital to have radiographic capabilities if the main power lines are interrupted and emergency power fails. If the unit is kept charged as recommended by the manufacturer, a large number of exposures can be made with these units without recharging. (In an emergency, radiographic film can even be hand-processed in the tanks of an automatic processor!)

Generator Applications in the Radiology Department

In the preceding discussion, the basic concepts of x-ray generators were discussed. One can rightfully ask how these generators should be used in a radiology department. The selection of an x-ray generator depends on a number of factors, the first of which has to be the availability and quality of the maintenance of the equipment. It is senseless to spend $30,000, $40,000, or more than $400,000 for biplane cinefluorographic equipment and an x-ray generator, for example, if one knows full well that the closest service is hundreds of miles away. On the other hand, it does not make sense to purchase a generator from a local firm if one knows that the quality of the maintenance is less than acceptable. One can easily ascertain the quality of service provided by a local vendor simply by speaking with one's colleagues in the area who have equipment maintained by that firm. Questions such as How long is generally required to respond to a service call? Is the machine repaired in a reasonable length of time? How much does the vendor charge for his service? and Does the service engineer fix the problem on the first call or does he have to return several times to fix the same problem? are very important considerations. If a colleague has the same type of equipment, it would be useful to know what the service history is for that particular piece of equipment. Service should be the most important consideration in deciding what brand of radiographic equipment is to be purchased.

The second consideration involves the number and type of examinations to be performed with the equipment. If the work to be accomplished is primarily radiographs of extremities, then one normally does not have to worry about motion unsharpness and some of the other factors that come into play. However, one should not fall into the trap of looking at group data alone to ascertain how much work the department has to accomplish. Data should be developed as to how much work a particular radiographic room should be expected to perform. For example, raw data may indicate that 20,000 chest radiographs are to be done, which would indicate the need for an automated chest radiographic unit. However, a look at each patient as he arrives in the department may indicate that the referring physician is requesting a chest examination and a radiograph of the wrist, or a chest radiograph and a flat abdominal radiograph, and so on. It might be that the individualized data may show only a small number of patients arriving in the radiology department for chest x-rays only. One should then consider grouping equipment so that an automated chest radiographic unit is located in the same room with an automated radiographic table, and so forth. It is not always possible to maintain the features of quality control and convenience if the patient has to be moved from one radiographic room to another, and the efficiency of automated radiographic systems can be lost.

When one considers all the capabilities of x-ray generators, these capabilities can be classified into two major groups: the kilowatt rating of the generator and the timing capabilities of the generator. In comparing single-phase generators with six-pulse and 12-pulse three-phase generators, it should be obvious that the peak kilovoltage from each generator would be the same. If one selects 100 kV, the peak kilovoltage from any of the generators would be 100 kV peak. However, the *average* kilovoltage from a single-phase two-impulse generator is 70.7 kV; from a three-phase six-pulse generator, 95.6 kV; and from a three-phase 12-pulse generator, 98.9 kV. Kilowatt ratings at different kilovoltages should be taken into consideration, as noted previously.

The other factor has to do with timing capabilities. A single-phase generator can produce two pulses per cycle. These pulses are 8.3 ms (1/120 sec) apart, and thus the shortest exposure time one can obtain is 8.3 ms, but useful output equals only 3 ms per pulse (Figure 2-19).

Fig. 2-19. Useful output of a single-phase x-ray generator. In single-phase equipment, two pulses are available in 16.6 ms (1 cycle). The time required for the pulse to rise to a useful maximum is usually taken to be at 70% of pulse height. In the case of a single pulse, only 3.0 ms of the original 8.3-ms pulse is useful for producing an image. (The x-ray tube in a nonlinear resistance current is not directly proportional to applied voltage. Due to this fact, some transformers exhibit an effective and nearly constant radiation on-time waveform that approaches 4.0 ms rather than the 3.0 ms that is shown above.)

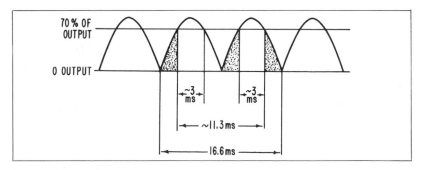

A three-phase six-pulse generator can theoretically allow a pulse of 2.7 ms, and a 12-pulse three-phase generator with special circuitry can allow pulses as low as 0.1 ms. Short timing impulses are important primarily in angiographic procedures, cine applications, and high-speed photospot camera applications.

An interrogation time approaching 1 ms is important in film changer applications, and near zero interrogation time is required in cine-fluorography. Primary pulse commutation switching can be used in most film changer applications, whereas secondary switching is required for cinefluorography. Forced extinction is necessary when high-speed rare-earth intensifying screens or automatic exposure devices are used.

One must assume that x-ray equipment will perform as specified by the manufacturer, although the experience of some purchasers would leave this open to question. Needless to say, before accepting an x-ray generator from a vendor, the purchaser should assure himself that kilovoltage, milliamperage, and time settings are accurate and that all output parameters of the generator meet the manufacturer's specifications. These factors are explained in more detail in Chapter 12.

As a general rule, single-phase generators are less expensive than three-phase generators, 125-kV generators are less expensive than 150-kV generators, and generators with lower kilowatt ratings are less expensive than those with higher ratings.

Table 2-2. Specific Uses for Types of Generators

Procedure	Generator
Routine radiography	150 kV, 500 mA, single-phase, solid-state, forced extinction primary synchronous switching, 50–100 kW three-phase
Skull examination	150 kV, 300 mA, single-phase, solid-state, synchronous switching, 50–80 kW three-phase
Radiography/fluoroscopy[a]	150 kV, 600 mA, three-phase, 6 or 12 pulse, 75–85 kW, solid-state contacting, forced extinction, primary synchronous switching
Angiography	
With cine	125 kV, 1000 mA, 12 pulse, secondary switching, 100–150 kW
Without cine	125 kV, 1000 mA, 12 pulse, asynchronous primary or secondary switching, 100–150 kW

[a]If film changers are to be used with these systems, 1000-mA capabilities are mandatory.

There are other differences between single- and three-phase equipment. For a given kilovoltage there is an increased radiation exposure
to the patient for single-phase units as compared with three-phase
equipment. This is true because a single-phase exposure has a spectrum of energies that are lower than a three-phase exposure, with the
average energies being 70% of those for three-phase. For a given kilovoltage exposure there is a difference in film contrast, with single-phase
exposures having more contrast than three-phase exposures. Again, the
same reason holds true. For example, with a 70-kV exposure, the effective kilovoltage for a three-phase six-pulse exposure would be determined by multiplying the percentage of effective kV by peak kV, or 90%
$\times$ 70 = 63; for single-phase, the effective kilovoltage would be 70% $\times$
70 = 49 kV. This difference in effective kilovoltage can produce a considerable difference in film contrast.

Table 2-2 provides a recommendation of the types of generators that
are suitable for specific examinations in the radiology department.
These recommendations should be used only as a guide, since work
loads and patient flow patterns also dictate the type of equipment
needed for a specific application.

3   Modern X-ray Tubes

According to the Bureau of Radiological Health, there are more than 100,000 rooms of x-ray equipment in this country. If one estimates that each room of equipment costs $129,500, then the value of the x-ray equipment in use today exceeds $10 billion! To think of it in another way, a hospital, a clinic, a physician, or a radiologist may spend $100,000 just to produce and control the energy from an x-ray tube. It is therefore surprising to find that in many cases considerable thought is given to the selection of radiographic equipment, yet little, if any, thought is given to the type of x-ray tube that will be used with a particular piece of equipment. The proper selection of a radiographic tube will allow the user to utilize fully the radiographic equipment. Proper use will extend the life of an x-ray tube to an acceptable economical level.

Construction of Rotating Anode X-ray Tubes

In the early 1900s an x-ray tube consisted of a simple diode, or *stationary anode tube*, in which electrons produced by the cathode were propelled to the anode by an electromotive force. The collision of the electrons with the anode produced heat and x-rays. Since the anode, or target, as it is usually called, could withstand only so many watts per unit area, x-ray tube ratings were accordingly quite limited. In the 1920s the idea was conceived of moving the target in front of the electron beam, with the result that a larger surface area could be presented and more watts absorbed, affording higher tube ratings for a given focal-spot size. Techniques of manufacturing x-ray tubes have improved since the 1920s, but the basic concept of a *rotating anode x-ray tube* has not changed since that time.

A rotating anode x-ray tube consists of an anode assembly, a rotor bearing system, a cathode assembly, a glass envelope encircling the anode and cathode assemblies, a stator, and a housing (Figure 3-1).

The *anode assembly* consists of a target and a rotor body. The *target* is a disc measuring either 75, 100, or 125 mm in diameter. The body of the target is generally made of molybdenum, although some companies have used carbon (graphite) as the basic substance. The periphery of the disc is beveled to a certain degree of angulation called the *target angle*, according to the specifications of the tube being manufactured. This beveled edge forms what is known as the *focal track*, which is coated with an alloy of tungsten and a small percentage of rhenium. The function of the rhenium is to inhibit microcrystalline disruption and surface roughening caused by the electronic bombardment from

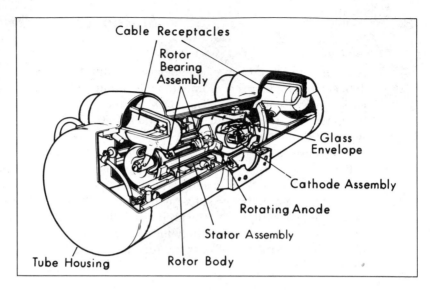

Fig. 3-1. The major components of a rotating anode x-ray tube.

the cathode and the resultant, nearly instantaneous, increase of temperature at the focal track. The purpose of the tungsten is to provide an efficient source of x-rays.

The target is mounted on a shaft that broadens out into what is known as a *rotor body.* The rotor body is in turn mounted on a bearing system on the anode shank in such a fashion that the rotor body and the target are free to spin on the bearings in the anode shank. The rotor rides on silver-plated bearings. These bearings are silver-plated to ensure an essentially frictionless surface, since there is no known type of liquid lubrication that will remain stable at high temperatures if used in a vacuum. All known lubricants break down and produce gas at high temperatures, and this gas will in turn interfere with the vacuum operation of the x-ray tube and will essentially destroy the x-ray tube. The rotor body is coated with a dark substance (usually carbon) that aids in the dissipation of heat from the anode assembly. The completed rotor body becomes the rotor of an induction motor.

The *cathode assembly* is composed of the cathode head and the filaments. The cathode head is made of nickel, with machined slots for exact placement of the filaments. The contour and exact location of the slots, the size and placement of the filaments in the cathode head, and the spacing between the anode and cathode will eventually determine and control the accuracy of the bombarded area on the target track and thus the focal-spot size. A spiral filament wound from tungsten wire becomes the source of electrons in the tube when the wire is heated to incandescence by a filament transformer. This process of heating a piece of wire to incandescence so that electrons are produced is known as *thermionic emission*. The cloud of electrons surrounding the filament produced by this thermionic emission is called an *electron cloud*, or *Edison effect*. The gauge of the wire, the pitch of the filament, and the number of turns in the filament are selected according to focal-spot size for a specific x-ray tube. During manufacturing, strains in the filament windings are removed by heating the filaments in a hydrogen atmosphere, which prevents oxidation and burning out of the filament during heating. Almost all rotating anode tube cathodes are provided with two filaments: a small one for detail recordings and a larger one for higher milliamperage for shorter exposure times.

The anode and cathode assemblies are encircled by an evacuated glass envelope. The glass envelope is made of Pyrex, which has been selected for its combination of optimal mechanical strength and minimal radiation filtration. Three grades of glass are used in joining the metal parts of the x-ray tube to the Pyrex glass because of the difference in expansion coefficients of metal and Pyrex glass.

Surrounding the anode assembly on the outside of the glass envelope is a *stator*, which serves the function of the armature coils of an induction motor. The function of the stator is to turn the rotor and target assembly in a rotating anode tube. The stator consists of two pairs of windings, or coils. When 60-hertz (Hz) power is applied to the stator, a standard speed rotation of approximately 3000 revolutions per minute (rpm) is achieved; when 180-Hz power is applied, high-speed rotation of approximately 10,000 rpm is obtained.

The evacuated glass envelope containing the cathode and anode assemblies is collectively called the *tube insert*. This tube insert is installed within a *housing*, also called a *shield* or a *casing*. The housing provides electrical and radiation shielding. A typical housing may be made of castings of extruded aluminum and produced in either one or three sections that have been cleaned and lined with lead for radiation shielding. Following the installation of the tube insert into the tube housing, the housing is filled with insulating oil, which serves the purpose of providing an electrical insulation. The housing incorporates an expansion bellows to allow room for expansion of the oil as the casing absorbs heat. If necessary, a safety device can be provided that inhibits generator operation if the housing temperature reaches an unsafe level.

Cable receptacles are made from powdered plastic in a forming machine under heavy pressure and high temperature. The term *horn angle* refers to the angulation of the cable receptacle in relation to the x-ray port or window. For example, a 90-degree horn angle means that the cable receptacle is at right angles to the x-ray port.

Throughout the manufacturing of an x-ray tube, the performance of the tube is monitored and recorded on an identification tag. The tag is kept on file as a record of the tube's manufacturing history and is referred to when the tube is returned to the manufacturer for replacement or warranty services.

Heating and Cooling Characteristics of X-ray Tubes

Aside from manufacturing error or electrical malfunction, the one thing that can destroy an x-ray tube is excessive heat. The most common errors in the use of x-ray tubes are the incorrect calculation of heat units and a lack of understanding of the heating and cooling characteristics of x-ray tube inserts and housing. Both radiologists and technologists should recognize the limitations of x-ray tube heat loading. On many occasions, particularly in angiography, a radiologist will ask a technologist to use certain filming techniques without fully realizing the damage that can occur to the x-ray tube when its ratings are exceeded. One must develop the habit of checking the exposure ratings and cooling characteristics of an x-ray tube before any exposure or series of exposures are made. Single-exposure ratings are governed by the focal-spot size, target diameter, target speed, and target angle. In addition, when making multiple or serialized exposures, one must also take into consideration the total anode mass and the anode heat dissipation rate. A large amount of heat can be produced in the x-ray tube in a very short period of time. If the maximum thermal capacity of the tube is reached through a series of exposures, it is necessary to allow the tube to cool before attempting additional procedures.

Most of the heat generated in the target from electronic bombardment is dissipated by natural radiation from the target's surface and thence through the glass wall of the tube insert into the insulating oil surrounding the insert. Although the target can dissipate up to 75 kiloheat units (KHU) per minute, the housing itself can only dissipate 15,000 to 18,000 heat units (HU) per minute, and under heavy operating conditions it is necessary to increase the dissipation of heat from the housing by a number of different mechanisms, which will be discussed later in this chapter. The heat storage or thermal capacity of the x-ray tube and housing must not be exceeded, or damage to the x-ray tube will occur. After heat is generated in the tube, a sufficient amount of time is required for that tube to cool to a safety area before more heat is produced.

Single-Phase and Three-Phase Heat Units

The term *heat unit* (HU) is used to identify the amount of heat produced in an x-ray tube per unit of time. Single-phase heat units are calculated from kilovoltage, milliamperage, and time factors. For single-phase (1ϕ) equipment, a heat unit is equal to the product of the kilovoltage times the milliamperage times the exposure time (kV × mA × sec). On three-phase (3ϕ) equipment the same formula is applicable; however, a correction factor of ~ 1.35 is used to account for the nearly constant multiphasic output from three-phase equipment (kV × mA × sec × 1.35).

For example, assuming exposure factors of 100 kV, 100 mA, and 100 ms:

$$
\begin{aligned}
1\phi \text{ heat units} &= kV \times mA \times time \\
&= 100\ kV \times 100\ mA \times 0.1\ sec \\
&= 1000 \\
3\phi \text{ heat units} &= kV \times mA \times time \times 1.35 \\
&= 100\ kV \times 100\ mA \times 0.1\ sec \times 1.35 \\
&= 1350
\end{aligned}
$$

Heat unit is a term derived for convenience and is used throughout the United States to describe x-ray tube power loading. In physics a current of 1 ampere (amp) developed across a differential of 1 volt (v) is equal to 1 watt (v × A = W). Since we use kilovoltage and milliamperage in radiology, the *kilo* and *milli* units cancel out, and therefore kV × mA = watts and kV × mA × time/sec = watts/sec. As the United States moves toward the metric system, the term *heat unit* will be replaced by the more correct term *watt.* To convert from watts (or watts per second) to heat units (or HU per second), multiply the unit by 1.35. To convert three-phase heat units per second to watts per second, divide by 1.35 or multiply by 0.7.

Tube Rating Charts

An x-ray tube can easily be destroyed by producing too much heat or producing heat too quickly; every effort must be made to keep within the rating of the tube. The thermal capacity of an x-ray tube can be determined from the tube rating chart that is provided by the manufacturer with each tube delivered to a user. When looking at x-ray rating charts, one must make certain that the exact chart for each specific model or tube type is used. For example, a tube rating chart for a 100-mm diameter target cannot be used for an x-ray tube having a 75-mm target. The tube rating chart for one manufacturer's tube cannot be used as a rating chart for another manufacturer's tube. Moreover, the proper single-phase or three-phase rating chart must be consulted, according to the type of equipment the tube is installed on.

Figure 3-2 is an example of a tube rating chart for a single-phase, fluoroscopic x-ray tube. The numbers on the left indicate the heat units stored in the anode, the curve marked "cooling" indicates the cooling rate of the anode, and the curves labeled "heat units per second input" indicate the rate at which heat may be put into the anode. Heat units and heat units per second are calculated as noted before.

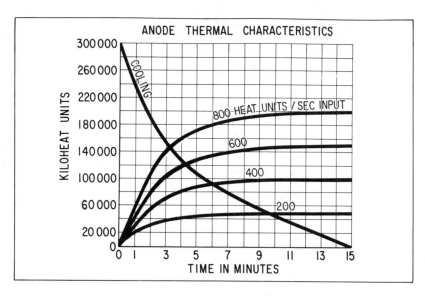

Fig. 3-2. Anode thermal characteristics in terms of heat units per second input into the target. This rating chart is used primarily for fluoroscopy. To determine which curve to use, multiply kilovoltage times milliamperage (for example, 120 kV × 5 mA = 600), then follow that curve along the time axis to determine how much heat will be produced in the target for a given period of time. In 4 minutes, using the 600 curve, approximately 120,000 HU would have been produced in the target.

As an example of how these rating charts are used, let us look at a typical gastrointestinal examination. For this examination let us assume that fluoroscopy is to be performed at 100 kV and 4 mA, that the examination will take approximately 5 minutes of fluoroscopic time, and that spot-film radiography will be made at 120 kV, 300 mA, 200 milliseconds (ms). After 5 minutes of fluoroscopy at 100 kV and 4 mA, approximately 90,000 HU will have been produced in the anode, according to the 400 HU per second curve in Figure 3-2. In addition, assume that during this period of time 15 spot-film exposures were made at a typical technique of 120 kV, 300 mA, 200 ms single-phase or 7200 HU per exposure. The heat generated from the spot-film radiography would contribute an additional 108,000 HU to the tube for a total of approximately 200,000 HU for the examination. Note that the maximum heat storage capacity on the chart is 300,000. If the examination were repeated immediately, the target would be heated an additional 200,000 HU; if this 200,000 is added to the original 200,000 we find that the anode heat storage capacity would be exceeded, resulting in tube damage.

Again in reference to the cooling curve in Figure 3-2, note that for this particular tube a period between 3 and 4 minutes is needed to allow the tube to cool from 200,000 HU down to a safe level of 100,000 HU. Repeated examinations can be made without harm to the x-ray tube as long as this waiting time is observed, the patient being examined is of average size, and single-phase equipment is used. If the technical factors were increased, the anode heat storage capacity of the tube would probably be exceeded, resulting in damage to the anode or even destruction of the tube. Therefore if more heat is generated per exposure or more exposures are made per examination, it is necessary to increase the waiting time between patients to prevent target warpage.

It should be noted that if a fluoroscopic x-ray tube is installed, calibrated, and used properly, little damage can occur to the tube under normal circumstances. However, if improper techniques are used, such as fluoroscoping a stomach or gallbladder at an oblique angle at low kilovoltages, the heat developed in the tube may be excessive and tube damage may result. The reader should also be aware that fluoroscopic kilovoltage controls only the kilovoltage used in fluoroscopy and not the kilovoltage used in spot-film radiography.

The chart in Figure 3-3 is for an x-ray tube with a focal-spot size of 1 mm and high-speed rotation of the anode. Kilovoltage is noted on the left side of the chart, and maximum exposure time is shown in seconds on the bottom. This chart is used to determine single-exposure ratings of an x-ray tube. To determine the maximum single exposure that could be made with this particular tube, one first examines what exposure technique could be used for the examination. For example, at 110 kV at 10 ms one could use a 900-mA station if it were available on the generator. If the maximum exposure time were 100 ms with 110 kV, the maximum milliamperage station that could be used is approximately 800. On the other hand, if the 1000-mA station were to be used at 50 ms, the maximum kilovoltage to be used is approximately 90.

Use of the tube rating chart is extremely important, particularly in angiographic applications, in order that the tube rating not be exceeded and the x-ray tube damaged. Figure 3-3 can also be used to determine the kilowatt rating of an x-ray tube. Kilowatt ratings are very important, particularly in comparing the tube of one manufacturer to that made by another. Again, one must make certain that the tube rating chart used is intended for the specific x-ray tube in question and that the focal-spot size, target diameter, target angle, and target speed, whether single-phase or three-phase, are all correct for that particular tube. To determine the kilowatt rating, locate in the three-phase chart the intersection of the 100-kV and 100-ms exposure lines, and at this point determine approximately what milliamperage could be applied with these factors (in Figure 3-3 this is approximately 850 mA). By moving the decimal point of the available milliamperage one place to the left, one can determine the kilowatt rating, which in this case

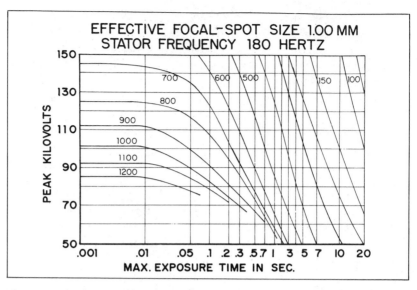

Fig. 3-3. Example of an instantaneous rating chart for 3φ equipment with 1.0-mm focal spot and high-speed rotation. This chart is used to determine if a single exposure can be made on the x-ray tube. The chart can be used to determine what milliamperage station can be used for a specific kilovoltage and time setting; what time setting may be used for a specific kilovoltage and milliamperage selection; and what kilovoltage may be used for a specific milliamperage and time setting. The easiest way to use the rating chart is to take the preselected technique factors and see if the combination of the factors lies to the left of the milliamperage curve. For example, with exposure factors of 100 kV, 600 mA, 1.0 sec, the combination of factors lies to the right of the 600-mA curve, and an exposure using these factors cannot be made without damaging the x-ray tube.

would be approximately 85 kW. This technique can be used to determine the kilowatt rating of any x-ray tube, but one must make certain that a three-phase, high-speed (180-Hz) rating chart is used.

Angiographic Rating Charts

In an angiographic series one must be able to initiate and terminate exposures as rapidly as two, four, or even up to twelve times per second in order to demonstrate the vascular system adequately. Heat is produced in the target very rapidly under these conditions, and heat generated at the surface of the target must be absorbed by the body of the target. If the thermal capacity of the target is exceeded or if heat is applied too rapidly at the surface of the target, uneven strains are created throughout the body, causing it to warp or crack. In preparing for angiographic studies one must first choose the factors that are to be used for a single exposure and then determine if that single exposure is within the single exposure rating of that particular x-ray tube. Again,

Table 3-1. Angiographic Rating Chart[a]

Exposures per sec	Total Number of Exposures				
	2	5	10	20	30
	Maximum Load in Peak kV × mA × sec				
1	37,000	24,000	16,000	10,000	7400
2	25,000	17,000	12,200	8000	6200
3	19,000	13,600	10,000	7000	5300
4	15,500	11,400	8600	6000	4800

[a]Effective focal-spot size is 1.0 mm; stator frequency is 180 Hz.

one must be certain that the correct tube rating chart is used (e.g., three-phase or single-phase equipment, high-speed or standard-speed rotation).

Tube manufacturers have developed rating charts for angiographic applications, from which a determination can be made as to whether or not a particular series of exposures is possible and what is permissible without causing damage to the x-ray tube. Table 3-1 is an example of an angiographic rating chart.

In reference to Table 3-1, assume that a theoretical angiographic series is to be made using 800-mA, 80-kV, and 100-ms exposure factors and three-phase, high-speed rotation. These exposure factors indicate that a single exposure would be allowable (referring to the single-exposure rating chart for that particular tube). The product of 800 mA × 80 kV × 100 ms × 1.35 is 8650 HU; this figure indicates the amount of heat that will be generated during each single exposure. If, for example, the radiologist has decided to expose film at a rate of two exposures per second, locate the exposure rate per second in the left-hand column in Table 3-1 and follow that line out until 8650 is located. In this case 8650 is located somewhere between 8000 and 12,200, and one can calculate by extrapolation that it would be possible to make approximately 18 exposures using these technique factors and at this filming rate. At four exposures per second using the same technique factors, one could make 10 exposures. Note that the number of possible exposures is not linear in relation to heat units input, just as the relationship between heat units per second input and cooling curves is not linear.

Single-Phase and Three-Phase Equipment

Reference has been made in previous paragraphs to single-phase and three-phase tube ratings. Tube ratings are affected by the power source that is used to operate the x-ray tube. In single-phase equipment, power is supplied in one phase, and the current through the x-ray tube fluctuates from 0 to a peak and drops again to 0 during each half-cycle. This fluctuation is called *ripple*, and as noted previously, this ripple is 100% in single-phase equipment. The mA meter on single-phase equipment indicates average rather than peak mA; this average mA is about 70% of peak mA.

In order to produce a specified value on the mA meter, the filament must be heated to peak mA. For example, in order to obtain 700 mA as read on the mA meter, the tube filament must be heated to produce ~ 945 mA, requiring ~ 5 amp of filament current. X-ray tube ratings are based on peak mA, and peak mA is in turn governed by filament temperature. Because the tube used in single-phase applications must be heated to peak mA, the x-ray filament must be operated and therefore rated at the higher value. A higher basic filament temperature leads to problems such as evaporation of tungsten from the filament.

In contrast, current in three-phase equipment is provided from multiphasic incoming power lines. Current supplied to the tube starts at 0 and rises to a peak; just as the first phase starts to drop toward 0, the second phase is reaching its peak. As the second phase begins to drop, the third phase is reaching its peak, and so on. The net result is that the current never drops to 0 and does not fall much below peak value. How much the average current drops depends on how many rectifiers there are in the three-phase circuit. For all practical purposes, average mA equals peak mA, and the mA meter on three-phase equipment reads peak mA. To achieve the same average mA from a three-phase generator as compared to a single-phase generator, the filament does not have to be heated to the same temperature level and the target load is more evenly distributed over the target focal track.

In single-phase equipment, kilovoltage waveforms resemble current waveforms and have a nearly 100% ripple at radiographic exposure currents. However, the kV meter reads peak and not average kV, which is the exact opposite of the mA meter; this means that although the kV meter reads 100 kV, the average kV output is 70. In three-phase equipment the multiphasic power input produces minimal ripple, and average kV is 95–98% of peak kV, depending on the type of generator (e.g., 6-pulse or 12-pulse three-phase, constant potential). The three-phase meter would then indicate 95–98% of peak kV. Thus the difference between three-phase and single-phase equipment lies in the more efficient output of the three-phase equipment, which results in an approximately 100% increase in output as compared to single-phase equipment. For example, a 100-kV, 50-mAs single-phase exposure would equal a 100-kV, 25-mAs three-phase 12-pulse output.

Tube Damage Resulting from Improper Operation

It would be wise at this point to divert our discussion and reconsider the previous points in terms of relevance. The question of relevance is being brought up by students and educational institutions throughout the country. One could rightfully ask what difference it really makes if he does not understand the heating and cooling characteristics of an x-ray tube. However, many people who operate x-ray equipment are not aware of the damage that can occur from the misuse and/or abuse of the x-ray tube. X-ray tubes cost anywhere from one thousand to several thousand dollars (and special tubes cost close to $20,000), so every effort should be made to preserve them as much as possible. The most common abuse of an x-ray tube is exceeding the single- or multiple-exposure ratings for a particular tube.

Anode Damage

Exceeding the instantaneous or maximum anode heat storage capacity will damage an x-ray tube in three major ways. First, excess heat produced by exceeding the maximum instantaneous ratings causes the focal track to be melted by the heat that is produced by the electronic bombardment from the cathode. Second, by applying massive loads (exposures above 500 mA) to a cold target, one can crack the target. This cracking occurs as a result of excessive heat being produced in a cold anode more rapidly than the cold metal can expand. To prevent cracking, the tube should be preheated using a standard warm-up procedure as specified by the manufacturer's instructions. The warm-up procedure should be repeated if the generator is idle for periods exceeding 60 minutes. Third, repeated exposures can overload the anode, causing it to warp because the heat simply cannot be radiated from the target body as rapidly as it is being generated. The rating chart for both single- and multiple-exposure ratings must be followed, or one will most likely end up with a damaged x-ray tube. Warm-up procedures may be eliminated in the future by the construction of segmental, slitted anode targets and by other anode metallurgical techniques.

Bearing Damage

Excessive heat can also cause damage to other parts of the x-ray tube. Rotating anode x-ray tubes rotate on bearings, which are normally constructed of silver-plated stainless steel. Volatile bearing lubricants cannot be used in a vacuum due to the possible production of gas, which would result in an unstable tube. Maximum expansion of the bearings coincides with the maximum tube ratings. When the gross thermal ratings of a tube are exceeded, some of the heat is transmitted to the tube rotor bearing assembly, causing an overexpansion and degradation of the bearings. The nonconcentricity of the bearings results in vibration of the target, binding of the anode, and a marked increase in noise. In a new x-ray tube, *coast time* (that is, the time a target will freely rotate on its bearings) from 3000 rpm until it stops is 40–60 sec. When coast time drops to 20 sec or less, significant damage may have occurred. It is dangerous to test coast time from 10,000 rpm because of harmonic vibration problems that occur in the 5000–7000 rpm range; for this reason, manufacturers rapidly brake the rotation of an x-ray tube from 10,000 to 3000 rpm.

Housing Damage

As the anode is heated, heat is transferred to the oil surrounding the tube insert. If the temperature of the oil is raised above its dielectric strength, it may carbonize or break down. This carbonization may result in an arc-over between the metal parts of the tube insert and the grounded metal parts of the housing, and the arc-over can puncture and destroy the x-ray tube.

Damage Caused by Excessive Filament Currents

Somewhat akin to the damage caused by excessive heat in the tube is the damage that can be caused by high filament currents. When a generator is turned on, the filament is heated to a low standby temperature but does not emit many electrons. A tube can stay at these low operating conditions for a long period of time without any significant damage occurring. However, when high-milliamperage exposures are made (for example, greater than 500 mA), tungsten evaporates very rapidly and is deposited on the glass envelope. This deposition of tungsten changes the electrical charge on the glass wall, which is normally negative, leading to a possible fluctuation of current during an exposure. As more tungsten is deposited, the tube becomes more and more unstable and eventually becomes totally inoperable. The most common cause of this excessive filament evaporation and deposition of tungsten on the glass envelope is the withholding of an exposure for long periods of time with the filament in the boosted condition, that is, when the operator holds down the "rotor start," or "prep," button without pressing the exposure button to release the exposure.

There are normally two buttons on an x-ray generator console that control the initiation of an exposure. These two buttons are operated sequentially. When the first button, called the *prep*, or *rotor start*, *button*, is depressed, two major things happen: The rotor is energized, and the filament is boosted with the proper amount of amperage required to produce the desired tube current. This tube current will be maintained until the second button, the *exposure button*, is activated.

In a normal situation, these buttons should be activated one right after another, since a built-in control will not allow an exposure to be taken until the anode is up to proper speed and the filament is heated to its proper level. This normally requires only approximately 1 sec. However, many technologists allow too much time to elapse between activating the rotor control button and the exposure control button, and most of this prolongation results from improper instruction and preparation of the patient.

There are situations that do require withholding an exposure after the rotor button is activated. Exposures have to be anticipated in crying children, and the delay may be too long to catch the exposure during the proper phase of respiration. The same holds true for elderly or uncooperative patients. In voiding cystourethrography, the urine stream must also be anticipated. Angiography is one of the most significant areas in which a prolonged rotor activation may be necessary, since the physician, monitors, injectors, film changers, and technologists must be coordinated and double-checked before the exposure is made. In most other instances, technologists can increase tube filament life and bearing life by preparing everything in advance and instructing patients properly. The longer the time interval between activation of the rotor start button and the exposure button, the longer will be the period during which the filament tungsten will evaporate and be deposited on the tube insert.

Tube life may be extended or at least the expected life may be obtained if the technician pays meticulous attention to tube rating charts, minimizes filament boost times, follows tube warm-up procedures if required, and minimizes high-speed rotation. These procedures require only a few minutes a day and can potentially save thousands of dollars in just one year's time.

Design Characteristics of X-ray Tube Components

The components of an x-ray tube are designed to accomplish specific purposes. Design modifications affect tube ratings, radiation field coverage, and the resolving capability of the x-ray tube. The most important factors are the design of the target and the design of the focal spot. The operating characteristics of an x-ray tube depend largely on these two factors, and we will discuss in this section how changes in design affect operating characteristics.

Target

The ratings of a rotating anode x-ray tube are a function of or are related to the target speed, target angle, target diameter, and focal-spot size.

Target Speed. The standard rotating speed of a target is approximately 3000–3300 rpm. This speed is obtained by applying 220 v of 60-Hz power to the stator of the x-ray tube. If 180-Hz power instead of 60-Hz power is applied to the stator, the tube will rotate at approximately 10,000 rpm (3 × 3300). The reason for high-speed rotation is to allow a greater area of the focal-spot track to be bombarded by the stream of electrons emanating from the cathode per unit time. By going from standard-speed to high-speed rotation, one can increase the effective rating of the x-ray tube by approximately 60–70%. However, when rotational speed is increased by a factor of three, wear on the bearings is increased to a factor nine times that of standard speed. Use of high-speed rotation in any radiographic application should be kept to a minimum. A braking circuit is added to x-ray tubes to slow the target after a high-speed exposure is made; today most manufacturers brake the speed to zero after an exposure has been completed. Ideally, rotation of the target during fluoroscopy should be minimized, and high-speed rotation is never used. A focal spot larger than 0.6 mm does not require rotation during fluoroscopy. Many generators sold today have an automatic control of high-speed rotation and programmed shutoff and/ or speed control during fluoroscopy.

Target Diameter. X-ray tubes can be purchased with a target diameter of 75, 100, or 125 mm. Increasing the target diameter provides a longer focal-track length available for bombardment by the electron beam. If the target diameter is increased from 75 to 100 mm, one could expect an increase in ratings of approximately 30%; if the diameter is increased from 100 to 125 mm, an additional 15% increase in ratings would occur.

Target Angle. The angle of the target not only governs the ratings of the x-ray tube but also limits the field of coverage by the tube. In a stationary anode x-ray tube with a target angle of 20 degrees, a focal-spot area with proportions of 1 : 3 is bombarded. In a standard rotating anode tube with a target angle of 15 degrees, an area with proportions of 1 : 4 is bombarded. This bombarded area, if projected toward the recording medium, is an area having a length-width ratio of 1 : 1. As the target angle is reduced toward zero, a greater area of the target track is bombarded and the instantaneous loading of the tube is increased. This increased loading is controlled by the filament length in the cathode. However, when the target angle number is decreased, there is also a reduction in the size of the field or cone of radiation coming from the x-ray tube. Thus, in order to obtain increased loading of the x-ray tube by reducing the target angle, one must give up field size coverage.

Figure 3-4 illustrates field coverage as a function of target angle. The field size is given on the left-hand side of the chart, and on the bottom of the chart is the source-image distance in inches. To determine field size coverage simply look at the target angle (for example, the 15-degree target angle) and determine, by following the line to the left, the field size this 15-degree target would cover at a specified source-image distance. With a 15-degree target angle and the commonly used 40-

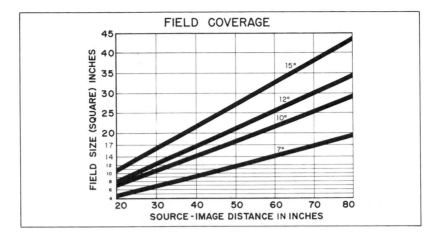

Fig. 3-4. Field coverage as a function of field size. To use this chart, first determine the target angle of the x-ray tube. Then for a specific source-image distance, locate in the left-hand column the field size coverage. For example, a 7-degree target angle will cover a 17-inch field at a source-image distance of 72 inches.

inch (1-meter) source-image distance, the field of coverage is about 22 inches. With a source-image distance of 40 inches, a target angle of 12 degrees barely covers a 17-inch field; a target angle of 10 degrees covers a 14-inch field. Field size coverage becomes important when one is selecting an x-ray tube for a specific application.

As far as the target angle number is concerned, another factor must be taken into consideration. As the target angle number decreases, that is, as the angle becomes smaller, there is more absorption of the x-rays within the anode itself. This absorption of x-rays within the anode produces an effect that is called the *anode heel effect* (Figure 3-5). The heel effect, which is caused by the anode providing more mass in which the x-ray beam can be absorbed, becomes more pronounced with smaller target angles. As Figure 3-5 shows, x-rays are produced as an electron strikes the target. Some of these x-rays are absorbed within the mass of the anode and some escape. The sharper the target angle, the more anode mass there is for the x-ray to be absorbed in; therefore, fewer x-rays escape in the heel or anode position of the x-ray beam.

Focal-Spot Size

There has been controversy in the past between prominent angiographers and x-ray tube manufacturers, centering around the measurement of the size of focal spots in x-ray tubes. Although this controversy may sound a little farfetched, the simple fact is that resolution is controlled to a large degree by the focal-spot size. This means that an x-ray tube with a focal spot measuring 0.6 mm is limited in its resolving

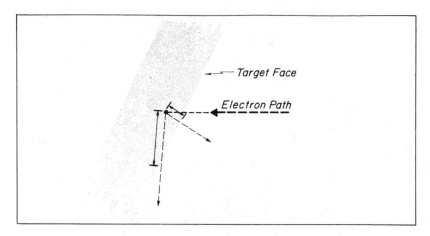

Fig. 3-5. Anode heel effect. As an electron interacts within the target, an x-ray produced by this interaction may escape perpendicularly to the target face or at any angle. The longer the x-ray must travel through the target, the more of its energy will be absorbed. Those x-rays escaping perpendicularly are less attenuated, and those escaping parallel to the target surface will be attenuated the most. As the target angle is steepened, more x-rays are absorbed in the target, thus increasing the heel effect.

capability. (See Chapter 9 for compromises in magnification angiography.)

Competition among x-ray tube manufacturers is centered on x-ray tube ratings. To be competitive, one manufacturer must offer an x-ray tube that is similar to another manufacturer's tube. X-ray tube ratings are governed by or are a function of focal-spot size, in addition to the target speed, target angle, target diameter, and target composition. The larger the focal-spot size, the higher is the tube rating assigned to that particular tube. Manufacturers' tolerances are set by the National Electrical Manufacturers' Association (NEMA) and are listed in Table 3-2. What happens in actuality is that a purchaser may request a focal-spot size of 0.6 mm from a supplier, and when the x-ray tube is received, the focal spot may measure anywhere between 0.6 and 0.9 mm and still be within the NEMA tolerances. This means that the resolving capability is reduced but power ratings are increased for this size focal spot (it is ethical for the supplier to do this). Focal-spot size is affected by a number of factors, including the number of windings, the pitch, and the diameter of the tungsten wire; the placement of the filament in the cathode head; the distance of the cathode from the anode; milliamperage; and, to a lesser degree, kilovoltage. If the manufacturer supplies a true 0.6-mm size for every 0.6-mm focal spot ordered, tolerances would be zero and the cost of an x-ray tube would be astronomical. The higher the tolerances, the less expensive the x-ray tube can be because there is less waste, just as with x-ray film or any other

Table 3-2. Focal-Spot Tolerance

Nominal Focal-Spot Size (millimeters)	Tolerances (percent)	
	Minus	Plus
<0.8	0	50
0.8–1.5	0	40
>1.5	0	30

Source: National Electrical Manufacturers' Association Standards Publication No. XR 5-1974.

commodity. In a normal manufacturing run there are obviously some x-ray tubes with true focal-spot sizes at one set of kilovoltage and milliamperage values, and most manufacturers will supply these true focal-spot sizes at reduced ratings and possibly at a higher price. However, there may be a delay in obtaining a tube with a true focal-spot size, since availability depends on when such a tube becomes available during production.

Another controversy involves how a focal spot should be measured. According to the most recent NEMA standards, focal-spot sizes of 0.3 mm or less are measured with the aid of a *star pattern* (also called a *resolution camera*). The typical star pattern is constructed in the form of a star or spoke pattern; it consists of a wafer of thin lead strips having a diameter of 45 mm and a thickness of 0.03 mm or greater. These lead strips converge into a vertex angle of 1.5 or 2.0 degrees, depending on the manufacturer. The star pattern can be used to measure both the resolution capability of an x-ray tube, in terms of line pairs per millimeter, and also the size of focal spots. The pattern is capable of measuring resolutions ranging from 0.8 line pairs per millimeter at the periphery to 15 line pairs per millimeter at the vertex. Focal-spot sizes and resolution can be calculated by inserting measurements into a standard, simplified algebraic formula supplied with the pattern. Some manufacturers supply nomograms with the pattern; in that case, one simply takes the measurements and reads the resolution or focal-spot size right from the nomogram.

If the focal-spot size is greater than 0.3 mm, a *pinhole camera* is used. A pinhole camera is a tubular structure measuring approximately 20 centimeters (cm) in length and 2.5 cm in diameter. It contains a thin gold plate through which a small hole has been drilled, usually with a laser beam. The diameter and depth of the pinhole are determined by criteria established for measuring certain sizes of focal spots. The end of the pinhole camera contains a slot for holding a piece of standard dental film.

For the reader who may be interested, the exposure factors determined by the NEMA are as follows: For both the pinhole camera and the star pattern, the kilovoltage selected is one-half of the maximum rated kilovoltage of the x-ray tube that is to be measured; for example, one should use 75 kV for a 150-kV–rated tube. If the maximum rated kilovoltage is less than 75 kV, the actual maximum rating of the tube should be used; for example, one should use 60 kV for a 60-kV–rated tube. The current in each case will be 50% of the maximum permissible milliamperage at an exposure time of 100 ms at the highest rated target speed (i.e., standard or high-speed rotation).

Magnification Radiography

In angiography the physician would like to see structures as small as 50 to 100 microns (μ), or 0.1 mm in diameter, since visualization of these structures where early pathological changes occur may allow an early diagnosis of disease processes to be made. In addition, magnification of certain body structures, specifically in radiographs of the hand, allows one to see early changes in some types of arthritis and particularly in the hyperparathyroid diseases. Good magnification of anatomical structures can be accomplished if one understands the basic premises and limiting factors of magnification. Since the size of a focal spot is related to the resolution of small structures, the reader should understand the basic concepts of magnification, particularly in relation to focal spots.

At the outset it would be well to differentiate between radiographic magnification and photographic, or optical, enlargement of an image. *Radiographic magnification* is a process in which the radiographic image is projected onto a recording medium located at a specified distance from the radiographic object. The degree of enlargement is related to the placement of the object between the x-ray tube focal spot and the recording medium. On the other hand, *photographic enlargement* is a simple magnification of a radiographic image, using a negative of the radiographic film as the medium of magnification.

Mottle

Certain problems are involved in photographic enlargement. Radiographic film contains inherent *noise*, or *mottle*, which is composed of three separate entities. One component of this mottle is *film grain*, which is caused by the inherent clumping of the film emulsion during the manufacturing process. The less clumping of the emulsion that occurs, the less grain there is within the film. Another component of mottle is known as *quantum mottle*; it is caused by the statistical distribution of photons emanating from the x-ray tube. As is well known, the energy from an x-ray tube is distributed in the form of photon "bundles," or *quanta*. Because of this distribution of energy, one area of the film may be struck with more photons than an adjacent area, and thus film blackening may be more prominent in one minute area than in an adjacent area. A minimal number of photons is required to produce a uniform blackening or density of radiographic film. As the number of photons resulting in exposure becomes smaller and smaller, quantum mottle becomes more marked, again due to energy distribution according to the quantum theory. The third component of mottle is *screen structure mottle*, which results from the grain size and clumping of the screen phosphor during the manufacturing process. Screen structure mottle is somewhat similar to the mottle caused by film grain but is much more prominent, particularly when rare-earth intensifying screens are used. Because of the increased efficiency of rare-earth phosphors, structural mottle combines with increased quantum mottle so as to put a practical upper limit on rare-earth screen improvement in terms of image quality, except for magnification techniques.

If the radiographic image is enlarged photographically, the inherent mottle is also enlarged proportionately. However, with radiographic enlargement (radiographic magnification), the radiographic projected image is enlarged but the inherent mottle is not.

Effects of the Focal Spot

In magnification radiography, a film is placed at some predetermined distance from the object so as to obtain a calculated image width or magnification factor. A focal spot as small as possible must be used, since the smaller the focal spot, the smaller is the object that can be resolved. Other factors have to be considered as well; one primary factor is that of *geometric unsharpness*. All across the focal spot, an infinite number of images are being projected to the image receptor. The extreme edges of the focal spot also project an image that contributes to a fuzziness or blurring around the edge of the projected image. This blurring is also called *penumbra*. The larger the focal spot, the larger is the penumbra; the larger the penumbra, the more unsharpness there is. The corollary also holds true in that the smaller the focal spot, the less unsharpness there is, and the larger the focal spot, the greater is the unsharpness.

The effect of the focal-spot size on geometric unsharpness can be minimized by increasing the source-image distance. The greater the source-image distance, the less effect the focal spot has on geometric unsharpness. At a source-image distance of 180 cm (72 inches), a typical 1.0- or 1.5-mm focal-spot size has little effect on the geometric unsharpness in chest radiography.

The size of the focal spot is of little concern if the body part to be radiographed is very close to the recording medium and reasonably long distances are used. However, different parts of the human body are at varying distances from the recording media, and an object-to-film distance of zero cannot be achieved, which means that a magnification factor is always present. Therefore, the further the radiographed body part is from the film, the more magnification there will be of a projected image. Thus, when doing any type of magnification work, one must take into consideration the focal-spot size, the source-image distance, the object-film distance, and the energy distribution of the focal spot.

Not only the size of the focal spot but also its energy distribution and configuration are related to the resolving capability of the x-ray tube. In a typical line focus x-ray tube, electrons bombard an area on the target track in a nonuniform manner. This nonuniform bombardment of the focal track occurs because more electrons come from the outer edges than from the center of the filament. The bombarded area has a configuration resembling a "double banana." When large focal spots are used, this energy distribution causes no particular concern, but as smaller and smaller focal spots are used, the double banana effect tends to form an object with a double image. This effect is particularly undesirable when one is attempting to do more than two times linear magnification.

Optimally, a focal spot should be round and should have a Gaussian energy distribution; this would mean that the electrons striking the focal-spot track would do so uniformly, and the double banana effect would then be eliminated. One method of providing almost homogeneous energy distribution is to apply an electrical negative biasing voltage of 75–100 v on the cathode focusing cup. The electrons can then be focused onto the target so that the focal spot is made smaller and rounder and there is a better energy distribution across the focal spot. This biasing is accomplished by adding an additional wire to the shockproof cable coming into the cathode end of the x-ray tube. The additional wire carries the negative bias voltage from a bias power supply unit connected between the generator and the x-ray tube. The term *biased x-ray tube* is used to describe an electrical biasing to the cathode that results in a smaller focal spot. However, when a smaller area on the focal track is bombarded, the rating of the x-ray tube is diminished. Biased focal spots are used primarily for magnification angiography. Since the focal spot is effectively smaller than with comparable line focus x-ray tubes, higher resolution can be obtained.

A final note on the subject of focal spots: One should be aware that a focal spot increases in size (blooms) as more milliamperage is applied to the filament. The enlargement is basically in width rather than in length. The focal-spot size also increases at low kilovoltages, primarily because the low voltage potential between anode and cathode is not sufficient to force the electrons to travel in a somewhat straight path from cathode to anode. Focal-spot enlargement varies from tube to tube but generally increases with increasing tube current or with decreasing kilovoltage.

How to Select an X-ray Tube

Not too many years ago it was quite easy to select an x-ray tube, simply because there were not very many available choices. However, x-ray generating equipment and procedures are currently more sophisticated than ever before, and more consideration has to be given to the selection of the x-ray tube to be used with a particular piece of equipment. In order to determine what would be the appropriate x-ray tube for a specific application, one must first determine the amount of work a particular piece of x-ray equipment might be expected to perform. As a general rule, one could expect to radiograph a maximum of 25 patients per radiology room per day; this extrapolates into approximately 50 examinations or 150 exposures per day. In a typical radiology department, approximately 50% of the exposures would be of the thorax, or chest. For loads lighter than this, a light-duty x-ray tube can be selected; for heavier loads, a heavy-duty x-ray tube may be needed. For heavy loads one might also want to consider a larger target, air circulators, or heat exchangers.

In selecting an x-ray tube, the following procedures should be observed:

1. Determine the power level of the x-ray generator to be used.
   a. For generators up to 500 mA, a tube with a 75-mm target can be used.
   b. For generators of 500 mA or higher, a target with a diameter of 100 or 125 mm is required. (The reader should be aware that 125-mm target tubes require more space and more weight support, are more expensive, and do not fit in standard radiographic/fluoroscopic tables. They should be reserved for special applications.)
2. Determine the maximum film size or field size to be used at the minimum source-image distance to be employed.
   a. For routine over-the-table radiography, select a 12-degree target angle. This target angle will cover a 17-inch field at a source-image distance of 40 inches.
   b. For fluoroscopy, with the tube installed under the table and with an average source-image distance of 25 inches, select a 10- to 12-degree target to cover a 25 × 25-cm spot-film device or a 9-inch image intensifier.
   c. For a 35 × 35-cm (14 × 14-inch) film changer using a 40-inch source-image distance, a 10-degree target can be used.
   d. For femoral arteriography performed with cassettes, which requires a wide field of coverage, e.g., 35 × 90 cm (14 × 36 inches) and a source-image distance of 163 cm (65 inches), a target angle of 15 degrees is required.

     e. For a 35 × 35-cm spot-film device with a source-image distance less than 30 inches (70 cm), a 15-degree target angle is required.

3. Determine the focal-spot size and kilowatt ratings as previously described. The focal-spot size should determine the highest loading for a specific application, but one should keep in mind the resolving capability of the tube.

4. For × 2 magnification on a 35 × 35-cm (14 × 14-inch) film changer with a standard 100-cm (40-inch) source-image distance, a minimum target diameter of 100 mm is required, and usually a 0.3-mm focal spot should be requested.

5. For × 3 magnification, consideration should be given to the use of a biased focal spot.

Auxiliary Equipment

Air Circulator or Blower

    An air circulator or blower should be purchased for any x-ray tube for which a medium or greater work load is anticipated. The purpose of the air circulator is to blow a stream of air over the housing and thus dissipate the heat from the housing to the surrounding air. A typical air-circulating fan will dissipate approximately 36,000 HU per minute to the surrounding air. However, even the air circulator is not sufficient to dissipate the heat from the housing as fast as it can be generated and dissipated from a heavy-duty target.

Heat Exchanger

    A heat exchanger is suggested for those radiographic applications, such as angiography, for which a very heavy load is anticipated. A heat exchanger works by pumping heated oil from the housing through a radiator in the heat exchanger. The oil is then cooled and pumped back into the housing. A heat exchanger can remove approximately 100,000 HU per minute from the housing. Heavy use of fluoroscopy and/or spot-film exposures also makes heat exchangers essential for achieving a good tube life.

Heat Monitoring Devices

    As an alternative to calculating the heat unit input to the target and its dissipation rate, as described previously, manufacturers are offering a monitoring device accessory that indicates the heat condition of the x-ray tube and how fast it is cooling. Two basic methods may be used to indicate the heat loading conditions within the x-ray tube.

    One method senses the target disc temperature by means of a fiber-optic light pipe or probe located in the tube housing in such a position that it "looks" at the back of the target. The probe transmits the color or redness of the target to an infrared sensor on the outside of the housing. The information is conveyed by a small cable and displayed on a meter that indicates the target heat buildup and dissipation during operation. An increasingly audible signal, or "beep," warns the operator when approximately 50–70% of the target capacity is reached. The monitoring device can be interlocked with the generator control to interrupt and prevent exposure if the target approaches the maximum thermal capacity; the device will then withhold additional exposures until the target has cooled sufficiently to resume safe operation.

The other method of estimating heat loading is with a simulator. These systems are designed so that when kilovoltage, milliamperage, and time factors are chosen on the generator, a simulated exposure indicates to the operator whether or not that exposure can be made without damaging the x-ray tube.

Newer Developments

The foregoing discussion relates to x-ray tubes as we know them and as they are commonly used. X-ray tubes are not the problem they once were, primarily because of the advent of rare-earth screens, which greatly decrease the kilowatt requirements for radiographic tubes. There are exceptions, of course; high-kilowatt-rating tubes are still necessary for high-speed cineangiography, digital radiography, and computed tomography. The bane of the radiographic tube is heat. High-speed rotation of x-ray tubes was developed to spread instantaneous heat over an effectively longer focal track and thus increase the kilowatt rating of the x-ray tube. Unfortunately, there is still a large amount of heat produced in the anode. This heat is transferred back into the bearings, with the set of bearings closest to the target receiving the most heat and, therefore, the most damage from the heat. One manufacturer warrants its x-ray tubes for 20,000 exposures at standard speed and only 10,000 at high speed. This, in itself, reflects the continual problem of heat acquisition within the x-ray tube.

To overcome the problem of heat acquisition, many manufacturers are constructing increasingly larger radiographic tubes, with anode storage capacities of 1.5 million HU not uncommon. These tubes are by necessity quite large and are more difficult to suspend on a tube carrier. In those situations requiring high heat loading, the purchase of these larger tubes can be quite cost effective. To determine the cost effectiveness of the tube, simply divide the purchase price by the anticipated number of exposures; this calculation will give you the cost per exposure and allow you to compare one product with another. Moreover, there may be considerable cost savings in comparing one company's product with another in terms of price, performance specifications, and cost effectiveness.

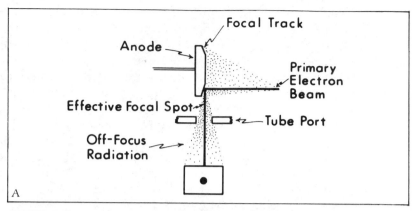

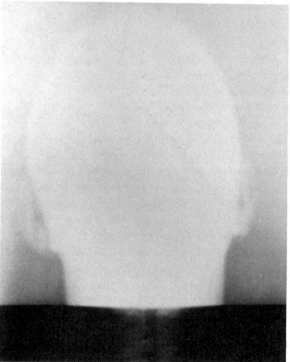

Fig. 3-6. Concept of off-focus radiation. (A) Although the electron beam coming from the focusing cup of the cathode is focused onto the anode, in fact some electrons strike elsewhere in the tube, producing radiation, some of which is emitted through the tube port with sufficient energy to produce an image on radiographic film. (B) An example of off-focus radiation. The image of the neck and skull is produced by off-focus radiation. Notice the collimator marks.

In the past few years we have seen the advent of a number of special-purpose radiographic tubes that have been designed to overcome special problems in radiology. Development of larger tubes as noted earlier is one example. Other special-purpose tubes include the stereoscopic angiographic tube, triple-focus tubes, metal-center-section tubes, and the magnetically supported high-speed tubes.

Metal-Center-Section Tubes

When conceptualizing the focal spot of a radiographic tube, one commonly thinks of the electron beam coming from the cathode being focused upon the anode producing x-rays that are "reflected" through the tube port as the effective focal spot. In reality, all of the electrons coming from the cathode are not focused into a narrow beam; there is what is known as a *Gaussian distribution*. In a Gaussian distribution of electrons, most of the electrons are clustered in the center of the electron stream, but others may be widely dispersed. The major portion of the electrons hits the focal track of the anode; others may interact on any metal surface within the tube or x-ray port, producing what is known as *stem radiation*, often called *off-focus radiation*. The exiting effective focal spot may consist of anywhere between 10 and 25% off-focus radiation. Additional off-focus radiation may be produced by high-energy x-ray photons as they exit the tube port and interact with the collimator or beam-limiting device. Figure 3-6 is an example of extremely bad off-focus radiation.

Off-focus radiation may be controlled to some degree by placing the first set of shutters of the collimator as closely as possible to the tube port or by adding a lead diaphragm at the tube port. Another method of controlling off-focus radiation is to insert an electrically grounded metal band into the x-ray tube at the time of manufacture; hence, the term *metal-center-section tubes*. Metal-center-section tubes effectively control most of the off-focus radiation coming through the x-ray tube port, and by the addition of a properly designed collimator, off-focus radiation can be greatly reduced. The control of off-focus radiation essentially allows one to produce sharper radiographs by the elimination of penumbra caused by the off-focus radiation.

Additionally, metal-center-section tubes have another advantage that might outweigh the control of off-focus radiation. One of the greatest causes of shortened x-ray tube life is the deposition of tungsten on the glass envelope of the x-ray tube. This tungsten deposition eventually destroys the electrical insulating characteristics of the glass envelope. Tungsten deposition is more commonly seen in high milliamperage applications such as angiography. By inserting a metal band in place of glass in the center section of the x-ray tube and by electrically connecting this metal section to ground, deposition of tungsten on the envelope can be greatly reduced, effectively increasing the life of the x-ray tube. These tubes should have a significant application in angiographic procedures.

Stereoscopic Angiographic Tubes

A typical radiograph is a presentation of three-dimensional anatomy in a two-dimensional plane. As such, a standard radiographic exposure does not show depth. Stereoscopic radiography is easily accomplished for almost any body part. To accomplish a stereoscopic radiograph, simply keep the patient in the same position and expose two films, moving the x-ray tube between exposures a total of 10% of *source–image receptor distance* (SID). For example, with a 40-inch SID, the first exposure is made 2 inches off-center in one direction and the other exposure is made 2 inches off-center in the opposite direction, for a total tube shift of 4 inches (10%). Stereoscopic examinations may also be exposed by changing tube angulation. In some cases, especially in angiography, it is important to be able to visualize depth in order to accurately locate pathologic conditions.

Methods have been devised for obtaining stereoscopic angiographic studies. Some have simply shifted the tube during the contrast-material injection; others have used two radiographic tubes positioned over the film changer, alternately exposing each tube and film. None of these methods has been successful. In the first type of technique, the time taken to shift the tube is too long, and the patient invariably breathes or the contrast material leaves the region of interest. In the second, with two radiographic tubes the exact angulation of the central beam is difficult to achieve in a way that will allow stereoscopic viewing. To obtain stereoscopic viewing there is a relationship between viewing distance, SID, and tube shift or angulation. As viewing distance increases, the necessary tube shift decreases, and as SID increases, more tube shift is required.

These problems are solvable by constructing a radiographic tube containing two focal spots placed sufficiently apart so that when the films are exposed and viewed, a stereoscopic image is seen. Although there is some variation between different types of tubes, generally a 4-cm separation of the focal spots is necessary in order to obtain stereoscopic images. The focal spots are fired alternately, with one focal spot prevented from firing by a grid similar to a grid-controlled x-ray tube. In general, exposures 100 ms apart are obtainable. Figure 3-7 illustrates the concept of a twin-focus radiographic tube.

Magnetically Supported Rotating Radiographic Tubes

Potential destruction of tube bearings by the excessive accumulation of heat and filament evaporation are two of the major problems encountered in radiographic tube design. As noted previously, as bearing destruction occurs, the coast time of the anode decreases, meaning that

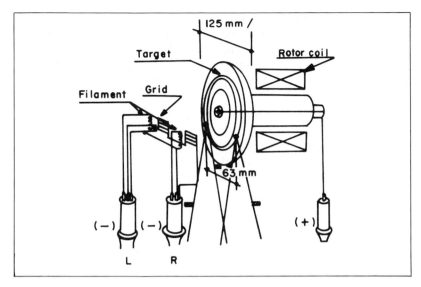

Fig. 3-7. Concept of stereoscopic radiographic tube. Two cathodes are used with each electron beam focused on different parts of the anode. The cathodes are fired alternately by a grid-controlled mechanism. (Courtesy of Toshiba Medical Systems.)

eventually the bearings will freeze and the tube will be destroyed. Again, because of limitations of design, rotation of the tube at high speed (180 cycles) is limited to about 10,000 rpm. Moreover, because of harmonics, the tube has to be rapidly decelerated to about 3000 rpm when going from high speed to standard speed in order to protect the bearings.

To overcome these problems, Toshiba Medical Systems[1] has developed the *magnetically supported rotating anode tube.* Instead of typical bearings, the anode is supported on a magnetic bearing without the anode actually touching the bearing. Eliminating the bearings essentially removes the limit as to how fast the anode can rotate, and many of the bearing problems are thereby eliminated. The magnetically supported rotating anode tube is currently undergoing field trials and may represent the biggest breakthrough in tube design since the development of the rotating anode. Removing the standard ball bearings allows a higher speed of friction-free rotation of the anode, thus extending the life of the x-ray tube and providing sharper images. These tubes should have applications in computed tomography and angiography.

[1]Toshiba Medical Systems, Tokyo, Japan.

Triple-Focus Radiographic Tubes

On occasion, there is a need for a greater tube loading or smaller focal spot than would normally be needed. Arthrography is a case in point. Triple-focus radiographic tubes, such as the one made by Dunlee,[2] consists of three focal elements in the cathode as compared to the normal two.

Suggested Reading

Doi, K., Loo, L-N., and Chan, H-P. X-ray tube focal spot sizes: Comprehensive studies of their measurement and effect of measured size in angiography. *Radiology* 144:383, 1982.

Milne, E. N. C. X-ray Tube Focal Spots. In *Medical X-ray Photo-Optical Systems Evaluation*. DHEW Publication (FDA) 76-8020. U.S. Department of Health, Education and Welfare, Public Health Service, Food and Drug Administration, October 1975.

Takahashi, M., Ozawa, Y., and Takemoto, H. Focal spot separation in stereoscopic radiography. *Radiology* 140:227, 1981.

---

[2]Dunlee Corp., Bellwood, Ill.

Image Intensifiers and Related Systems

Not too many years ago it was a familiar sight to see radiologists and even some technologists wearing red-lensed goggles and scurrying from one radiographic room to another. The red goggles identified those who were engaged in the morning's fluoroscopic activities. Fluoroscopy in those days was the domain of the radiologists, since few other physicians wanted to spend the time necessary to dark-adapt their eyes so as to be able to see the faint fluoroscopic image that was available then. The purpose of fluoroscopy is to visualize physiological motion, to discriminate between overlying anatomical parts, and, with the aid of contrast media, to visualize structures not normally seen without the aid of such media.

Prior to the advent of image intensifiers, conventional fluoroscopic screens consisted of a fluorescent material, usually copper-activated zinc cadmium sulfide, coated on a thin piece of cardboard. (The zinc and cadmium salts produce a spectral emission within the yellowish-green range to which the dark-adapted eye is more sensitive.) The fluorescent screen was backed by a fairly thick sheet of lead glass that attenuated the radiation not absorbed by the fluorescent screen, thus preventing the radiologist's eyes from being struck by raw radiation. The light level produced by the earlier fluoroscopic screens was equivalent to that of a moonlit landscape.

The obvious disadvantages of the early fluoroscopic screens were the low light level emitted from the screens and the need to dark-adapt one's eyes. In addition, one had to view the fluoroscopic image at a close distance in order to detect the faint image. Working at this close range made it difficult for the radiologist to teach the house staff and to point out anatomical detail to other observers. Some 20 or 30 minutes were required to dark-adapt the eyes in order to even see the image, and this time period was often negated by some careless person who would open the door of the fluoroscopic room, causing white light to strike the eyes and possibly ruining the red adaptation.

Patients were examined in almost total darkness, and it was difficult to see their reaction to palpation. Their anxieties and fears were augmented by the darkness of the examining room. Even with prolonged periods of adaptation, the eyes still could not see all the information that was displayed on the fluoroscopic screen due to the low light output. To increase the light level, the radiologist would eventually ask the technologist to increase the kilovoltage or milliamperage, or both, so that the light emitted from the screen would be increased. Excessive radiation exposure from fluoroscopic procedures was common in those days. Because of the poor perception, a considerable amount of time was required to perform a fluoroscopic procedure, which also added to the radiation exposure given to the patient. Even then, not all the information was perceived.

Imaging with Image Intensifiers

Imaging is the prime purpose of the radiology department. Ideally, viewing the image should come prior to recording. For this reason the *radiographic/fluoroscopic* (R/F) procedure was established, although many routine radiographs can be taken without prior viewing and/or positioning the area of interest with the fluoroscope. Viewing in the early stages of radiology was done with a simple fluorescent screen, but today viewing is almost exclusively carried out on a television screen via an image intensifier. The simple fluorescent screen of the past (less than $500) has been developed into an image viewing and recording system which may cost as little as $20,000 for a simplified mirror-optic viewing system or as much as $150,000 for a sophisticated television/photospot camera system. Today, about one-third of the cost of a radiographic/fluoroscopic room is related to the fluoroscopic viewing/recording system. Systems for imaging and imaging transmission will become more important (and expensive) in the future as radiology improves image quality and as improvements are made in computerizing the radiographic/fluoroscopic image.

It is most important that the radiologist and the radiographer alike understand the basic principles of the image intensifier system, although it is not necessary to understand all the complex technical detail.

The Eye as an Image System

As previously discussed, visualization of the early fluoroscopic screens required dark-adaptation. *Dark-adaptation* meant that the fluoroscopic viewer had to spend about 20–30 minutes wearing goggles to allow the rods of the retina to accommodate to the low light levels while maintaining cone vision. Low light levels are perceived primarily by the *rods*, which are located in the periphery of the retina where visual acuity is lowest. In the process of dark-adaptation, a photosensitive material (*rhodopsin*, or *visual purple*) found in the rods of the retina is allowed to regenerate. Exposure of the rhodopsin to light causes it to bleach to visual yellow, thus decreasing visual sensitivity. The longer the visual purple is allowed to regenerate, the more sensitive the rods will be to low light levels. The rods sense primarily 100% contrast, that is, they discriminate between white and black and very little between shades of gray. As the light level decreases, speed of perception also decreases. The longer one visualizes at low light levels, the more eye fatigue becomes a problem, which further decreases detail perceptibility.

On the other hand, the *cones* are color- and daylight-sensitive; they are more concentrated toward the center of the retina and less concentrated in the periphery, where the majority of the rods are located. The cones are most concentrated in the region of the *fovea*, which is the region of greatest visual acuity. Visual acuity allows one to distinguish between two different light stimuli in space and thus is related to distance and light intensity.

The purpose of the image intensifier is to increase the light level of the fluoroscopic image to the point where perception can occur with the cones rather than with the rods of the retina and thus to allow better visual discrimination. This means that the image intensifier also has to have the capability of enhancing contrast perceptibility. *Contrast* is defined in radiology as the difference between photographic densities, that is, differences between black and white. As a general rule, in order to perceive contrast there must be a difference in densities of approximately 10%. Contrast enhancement and visualization of the radiographic image are accomplished by the use of the image intensifier. Present-day image intensifiers provide a contrast ratio of more than 10 : 1 (gray scale).

In 1941 Dr. Edward Chamberlain predicted that for an image intensifier to be practical, luminance gain over fluoroscopic screens would have to be increased by a thousandfold in order for the light to be perceived primarily by the cones rather than the rods, thus eliminating the need for dark-adaptation [1]. This prediction was confirmed by Strum and Morgan in 1949 [2]. Modern-day image intensifiers increase the brightness level approximately 5000–10,000 times, and the need for dark-adaptation has been eliminated.

Design and Function of Image Intensifiers

There are three types of image intensification systems in general use today. Two of these systems are based on the concept of taking an image produced on a fluoroscopic screen and transmitting that light image through a series of lenses and mirrors to the input screen of a television pickup camera. The third system utilizes an x-ray image intensifier tube that converts the radiographic image to a light image via a series of electronic lenses and a complex electronic network. The image is intensified and presented for viewing either through a mirror-optic or a television system. The first two systems are seldom used. The x-ray image intensifier is commonly used and will be discussed in detail.

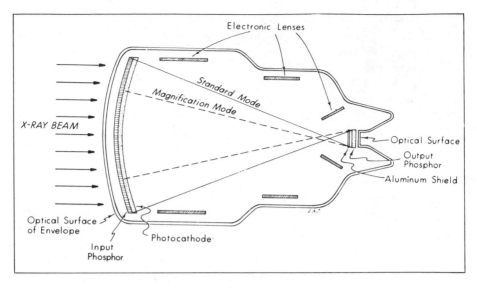

Fig. 4-1. Basic image intensifier. This drawing illustrates a dual-field image intensifier, whose mechanical components are similar to a single-field image intensifier. The device is composed primarily of an evacuated glass envelope that contains an input phosphor, a photocathode, focusing grids (electronic lenses), and an output phosphor. The output phosphor area of the tube is generally flat but is drawn in this manner for purposes of illustration. The magnification mode is activated by changing the voltage on the focusing grids.

The Image Intensifier Tube and Housing Assembly

Figure 4-1 depicts the basic components of an x-ray image intensifier tube. The tube functions as a triode or tetrode, the purpose of which is to convert an x-ray image into a visible light image that is bright enough to be seen without any special visual adaptation. The image intensifier tube is composed of an anode, a cathode, and one or two focusing electrodes enclosed in an evacuated bell-shaped glass envelope. The circular, spherical cathode/input screen assembly is located toward the mouth or large part of the glass envelope; the anode/output phosphor is located toward the smaller end. The third and fourth focusing electrodes are circular electrostatic lenses, which are spaced between the anode and cathode within the glass envelope. As with any vacuum tube, all surfaces of the tube must be round in order to withstand the pressure between the vacuum inside the tube and the ambient air. Surrounding the image tube is a "μ-metal" ring, which is designed to shield the image intensifier from stray magnetic fields (including the earth's magnetic field). Figure 4-2 illustrates a typical image intensifier tube.

In a typical situation the input screen or phosphor, which may measure 15–25 centimeters (cm) in diameter, is composed of a fluorescent

A

B

Fig. 4-2. A typical image intensifier tube. (*A*) The input area. (*B*) The output phosphor area.

material, which is vapor-deposited on an aluminum substrate. The process of *vapor deposition* allows the phosphor crystals to be deposited perpendicular to the surface of the aluminum substrate, thus making it possible to pack the phosphor crystals at a much higher density than with previous phosphors (powder packing with a binder was the old method of construction). The vapor deposition process also eliminates the use of a glass barrier between photocathode and phosphor, allowing the phosphor crystals to be packed much more compactly. This perpendicular deposition allows the crystals to act like fiberoptic bundles and essentially eliminates flare, that is, lateral and oblique transmission of light photons produced by the input screen.

In most modern x-ray image intensifying tubes, the input phosphor is composed of cesium iodide rather than the zinc cadmium sulfide used as the input phosphor until a few years ago. Cesium iodide phosphors allow a two- to threefold increase in quantum efficiency together with a two- to threefold increase in spatial resolution. The conversion of a greater number of x-ray photons into light photons also allows a significant reduction in the scintillation, or noise, that was seen in the cadmium phosphor tubes at a comparable exposure rate. Cesium iodide image intensifiers have a quantum detection efficiency of about 65%, compared to approximately 20% with cadmium phosphors. A further increase will create serious technical problems; for practical purposes, the efficiency is close to its maximum.

A *photocathode* composed of alkali and antimony is vapor-deposited directly onto the input phosphor, which provides ultimate image transfer between the two. This vapor deposition of the photocathode directly onto the input phosphor eliminates the need to have a separate interface between photocathode and input phosphor. This process alone increases the potential resolution of the cesium iodide image intensifier, as compared to the older zinc cadmium sulfide tubes. The function of the photocathode is to convert the light photons produced by the input phosphor into a proportional emission of electrons, and thus the photon image of the phosphor is converted into an electron image at the photocathode. The electrons can then be accelerated to effect an increase in the brightness level.

The electron image at the photocathode is accelerated and focused via the electrostatic lenses onto the smaller output anode phosphor. On the inside of the envelope, interspersed between the anode and cathode, are a series of focusing grids or electrostatic lenses whose function is to focus the electron image produced by the photocathode onto the output screen phosphor. Since the function of the electrodes is essentially to focus the electron image onto the output phosphor, they are called *electronic lenses*. The purpose of the anode is to provide a point of voltage differential for the cathode. The electrons emanating from the photocathode actually pass through the focusing fields and are then converged to a fine focus on the output screen. The output screen, usually measuring from 15 to 25 millimeters (mm), is composed of zinc cadmium sulfide. In most applications, the inside surface of the output phosphor is covered by a thin foil of aluminum to prevent the light backscattering from the output phosphor from interfering with the photocathode.

The image intensifier tube works in this fashion: X-ray photons of different intensities emerging from the patient pass through the dome of the intensifier tube and then through the aluminum substrate and interact proportionally with the input phosphor. Light photons produced by the phosphor then interact with the photocathode, and electrons are emitted in proportion to the number of incident light photons producing the electron image. The image is focused on the output phosphor, where a light image of higher intensity is produced; this image is inversed and reversed to the original image.

Image Tube Housing and Power Supply

A sturdy mechanical housing is needed to protect the large, delicate glass envelope. In addition, lead shielding is added to protect the operator and filming devices from stray radiation. The static voltages to the image tube (about 30 kV) can be supplied by an external power supply, but with new advances in electronics, many imaging systems have self-contained power supplies, thus eliminating the high-voltage cable. The housing also provides a means for mechanical support as well as interfacing for spot-film devices and the image distributors, which will be discussed later. The mechanical requirements for the housing are enormous, considering that it may have to carry up to 30–40 kilograms (kg) in the form of a spot-film camera, cinecamera, and/or television camera.

Because of the weight of the unit, most image intensifiers require ceiling suspension, especially if photospot and cine cameras are attached to the fluoroscopic tower. Some compact units without the attached cameras can be mounted directly to the fluoroscopic arm connected to the radiographic table. A ceiling suspension is a nuisance because it prohibits free movement of the radiographic tube ceiling suspension bridge. In recent years, R/F tables have been redesigned to accept the heavy weight of the image intensifier, including the new *large-field-of-view* (LFOV) units. In some cases the R/F table will accept the LFOV units but will require the addition of a ceiling suspension if photospot and/or cine cameras are added.

Characteristics of Image Intensifier Tubes

There are a number of terms and definitions that should be considered in a discussion of image intensifier tubes. An understanding of these basic terms and definitions (standards) is necessary in order to compare and evaluate the characteristics of an image intensifier by itself or in an entire system.

The term *gain*, or *intensification factor*, indicates how many times brighter the output of the image intensifier is in comparison to a reference or a standard CB-2 fluorescent screen. It is one of the most important terms used in describing image intensifier tubes. Using 1 as a reference, the modern image intensifier is capable of producing an increase of 10,000 to 20,000 times the brightness produced by a fluorescent screen with a brightness gain of 10,000 being optimum. One ingredient in achieving gain of brightness in intensifier tubes is the accelerating potential applied to the electrons emitted at the photocathode. As this anode-cathode voltage is increased, the electrons are imparted with a greater energy, which yields a greater brightness at the anode/output phosphor.

Since there are so many factors that influence gain, the term *conversion factor* (Gx) was coined. *Conversion factor* is defined as the ratio of the output luminance of an image intensifier to the input radiation dose rate. Since there are so many factors that affect the brightness output of the image intensifier, a conversion factor is needed so that one may compare one image intensifier to another. Factors such as the quality and quantity of the input radiation, the photocathode emission, the ratio of the area of the input screen to the area of the output screen, the anode-cathode voltage differential, and the luminance of the output fluorescent screen all affect gain. For modern image intensifiers, the conversion factor usually ranges from 45 to 100 and even to 200, depending on the specifications of the tube. As a rule of thumb Gx equals about 1% of the old term *gain*, or *intensification factor*.

Under international agreement, the conversion factor is expressed as the ratio of the luminance of the output screen to the input exposure rate and is indicated as candela per square meter (cd/m^2) and milliroentgens per second (mR/s) or $\dfrac{\text{cd/m}^2}{\text{mR/s}}$. This is the term used by manufacturers when quoting performance specifications for image intensifier tubes.

The gain factor is associated with what is termed the *minification factor*, or *electron demagnification factor*. It is common for a radiologist to use a minification glass when viewing chest radiographs. The purpose of this glass is to gather the light emanating from the chest radiograph and to focus it into a smaller area; thus, the minification improves the contrast gradient of the detail for better perception by the radiologist. An image intensifier accomplishes the same thing: It minifies the image from the input phosphor. The minification factor is the linear ratio of the diameter of the input phosphor to the diameter of the output phosphor. For example, a 22-cm (9-inch) input screen and a 22-mm (1-inch) output screen would have an input-output ratio of 10 : 1 and therefore a minification factor of 10 : 1.

Another factor that is directly proportional to the gain or Gx is the ratio of the input and output screen areas, which is termed the *area reduction factor*. This factor is important to know when one is switching modes or field size on dual- or triple-field designs. For example, let us look at the light emitted from a 9-inch image intensifier input phosphor that is focused on an output screen measuring 1 inch.

(Area of circle $= \pi r^2$)

$$\text{Area reduction factor} = \frac{\text{area of input phosphor}}{\text{area of output phosphor}} = \frac{(\pi r^2)}{(\pi r^2)_2}$$

$$= \frac{3.14(4.5)}{3.14(0.5)^2} = \frac{(3.14)\ 20.15}{(3.14)\ .25} = \frac{63.89}{.78} = 81$$

Therefore with a 9-inch image intensifier having a 1-inch output, one achieves a brightness gain of approximately 81 by simply reducing the size of the area.

Another important factor to consider is the *resolution*, or *resolving capability*, of an image intensifier tube. Resolution is usually expressed in terms of line pairs per unit length (e.g., millimeter, centimeter, or inch). A *line pair* is defined as an opaque line separated from another opaque line of equal width by a nonopaque substance. Resolution is measured with a line pair test pattern that has strips of lead and plastic of equal width. The smallest pair of lines of a line test pattern placed directly at the entrance of the image intensifier that can be visualized at the output phosphor determines the resolving capability of the system or the individual component (such as the intensifier tube). A number of factors influence the demonstrable line pair resolving capability of the image intensifier, including the quality of the radiation striking the input phosphor and the absorption or lead equivalency of the test pattern. Normally when image intensifier tubes are tested, soft, unfiltered, low-kilovoltage radiation and a test pattern with high irradiation absorption are used. Under such measuring conditions a high-quality image intensifier can be expected to resolve 3–5 line pairs per millimeter. Resolution would be lower under normal operating conditions. Resolution is measured by the use of optical or film camera techniques. Television viewing subsystems rarely reproduce all of the image tube resolution capabilities.

The resolution of an image intensifier system is stated relative to the area of the image (e.g., one-third center or two-thirds periphery). One should also always reference the complete measuring procedure under which a stated resolution can be viewed or recorded, including factors such as kilovoltage, dose rate, source-image distance, filtration, location of test object, focal spot, film and film size, and recording camera (television, cinecamera, or spot-film camera, including f-stop). As mentioned previously, measurements should preferably be taken according to established national or international standards.

Contributing to the resolution of the image intensifier is the contrast range. *Contrast range*, or *contrast ratio*, is defined as the ratio of the luminance of the center of the output screen when the input phosphor is fully irradiated to the luminance of the output phosphor when a central circular area of 10% of the input phosphor is shielded in the second of two measurements. A contrast range of 10 : 1 is normally found in the newer types of image intensifiers, and a ratio as high as 15 : 1 is found with fiberoptic output windows. The contrast range increases as the radiation striking the input phosphor is collimated to smaller areas of the phosphor. Contrast is affected not only by the physical properties of the image intensifier tube but also by penetration of the input phosphor by x-ray photons and their interaction at the output rather than the input phosphor and the backscattering of light photons from the output phosphor to the input phosphor. The effect of these events is to decrease contrast.

There are other factors involved in the evaluation of image intensifying systems, but they are less important now that cesium iodide is used as the input phosphor for image intensifiers. The first of these factors is lag. *Lag* is defined as persistence of luminance after the radiation has been terminated. Lag is common with any phosphor, but it is much less significant with cesium iodide. Any phosphor has a rise and decay time, meaning that it takes time to build up an image (brightness) and also for the image to disappear (decay). With the newer types of image tubes, this rise and decay time (lag) is less than 1 millisecond (ms); older tubes lag was on the order of 30–40 ms. Lag is much more important in respect to television pickup camera tubes, which will be discussed later.

Vignetting is a phenomenon seen in optical systems where the periphery of an image is darker than the central area. There are a number of causes of vignetting in image intensifier tubes, one being *photon flux*, the energy distribution of the beam across the input phospor. For the most part, vignetting is caused when the optical system couples the output of the image intensifier tube to the input of the television camera tube. In vignetting, off-focus light enters the lens at an angle. The effect of this off-focus light as seen by the television camera tube is less light on the periphery than in the center of the field; the television monitor presents this as less brightness (more darkness) on the periphery of the television screen as compared with the center.

Pincushion defect is a form of distortion of the image. The pincushion defect is another type of lens aberration in which the transmitted image is not faithfully reproduced. For example, if a test object consisting of a wire mesh is placed in the entrance field of the image intensifier, severe "pincushioning" would result in an image that is pulled in on the sides and extended at the corners, hence its name.

Although vignetting, pincushioning, and veiling glare pose no serious problems for the routine use of an image intensification system, these aberrations or artifacts can produce serious problems in digital fluoroscopy, since the computerized image must correct for these errors in order to present an accurate image to the examiner.

Dual- or Multiple-Field Image Intensifier Tubes

Minification or demagnification is accomplished in an image intensifier tube by focusing the electron image from the relatively large photocathode onto the much smaller surface of the output phosphor. When the static voltages on the electrostatic lenses (normally those located adjacent to the anode) are increased, only a portion of the input image will be projected onto the full diameter of the output phosphor. This variation in voltage results in a lowering of the minification ratio, thus creating a magnification of the image impressed onto the output phosphor. One of the most common dual-field image intensifiers is the 9-inch/6-inch (9/6), although formats such as the 9/5 and 10/4 are available. Triple-field formats such as 9/6/4½ are also available. Since only the most central portion of the image intensifier input phosphor is being used in the magnification modes, vignetting is decreased and resolution is increased slightly. For example, the performance data of one manufacturer's image intensifier tube show a resolution of 4 line pairs per millimeter with the 9-inch mode and 5 line pairs per millimeter with the 5-inch mode. Gain also drops by about 50% when switching from a 9-inch mode to a 6-inch mode and to 25% when switching from a 9-inch mode to a 4½-inch mode.

Multiple-mode image intensifiers are often misunderstood. Superficially, it appears that higher radiation doses are given to the patient when one switches to the smaller modes, because under normal circumstances the required radiation dose is dependent on the ratio of the area of the input to that of the output screen. Thus switching from a 9-inch mode to a 5-inch mode would mean increasing the dose rate, and switching from the 5-inch mode to the 9-inch mode would mean a dose rate reduction. However, regulations of the Department of Health and Human Services (HHS) require that the x-ray beam be collimated and the radiation coverage to the input screen be changed automatically when switching from one mode to another. With these equipment requirements, the integrated radiation dose to the patient should be close to equal from one mode to another.

The distinct advantage of a dual- or multiple-mode image intensifier over the single-mode tube is the increased quality of the image in terms of resolution and contrast for smaller modes. With the smaller mode of a dual-field image intensifier, the central portion of the image screen is used, and the peripheral fall-off in resolution and vignetting are essentially eliminated. Under ideal conditions, one might find that the 6-inch mode of a dual-mode image intensifier has equal or better performance characteristics than a single-mode 6-inch image intensifier. Recording camera subsystems using the 6-inch mode can record images of smaller areas, such as the duodenal bulb, closer to actual size.

The output phosphor of a dual-field image intensifier is often larger than that of a single-field tube, thus providing greater resolution capabilities. Six- or 7-inch image tubes are more compact and of lighter weight than the larger multiple-mode image tubes and may be adequate for most routine examinations. Large-field (i.e., 9- or 10-inch) single-mode image intensifier tubes are not considered to be suitable for routine use.

Conventional image intensifier tubes are constructed of a glass envelope with a convex input window. Practically, because of the high vacuum required and the thickness of the glass at the input window, conventional image intensifier tubes are restricted to a maximum of 9 or 10 inches. In the new LFOV systems, the size of the entrance window can go much higher: to 12, 14, or perhaps even 16 inches. This larger field of view is important for digital angiography, because a larger portion of anatomy can be seen without moving the patient or fluoroscope during the examination.

Instead of having the envelope constructed entirely of glass, the new LFOV tubes can be constructed of glass with an input window composed of metal or the entire tube can be composed of metal. Entrance windows can be composed of aluminum, titanium, or stainless steel. The use of a metal greatly reduces the weight of the intensifier tube, allowing it to be mounted on a fluoroscopic table without the use of a ceiling suspension. Most of these tubes are pentodes rather than tetrodes, which decreases pincushion distortion. Also, the metal input windows are concave rather than convex. Figure 4-3 illustrates the concept of an LFOV image intensifier. In comparison with standard image intensifier tubes, the new units have a greater quantum detection efficiency, higher contrast ratio, and better resolution. They are also considerably lighter than the old tubes.

Viewing the Radiographic Image

There are two methods whereby one can visualize the output of an image intensifier: One is with a *mirror-optic system* and the other is with a television camera or *television monitor system*. The advantage of the mirror-optic system is that it is comparatively simple and much less costly than a television viewing system. A disadvantage of the mirror-optic system is that generally only one or perhaps two people can

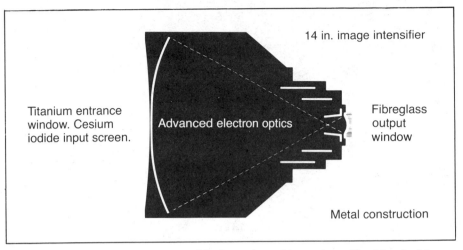

Fig. 4-3. Concept of a large-field-of-view (LFOV) image intensifier tube. In this tube the entrance window is composed of titanium with a cesium iodide input screen. The envelope is constructed of a lightweight, nonmagnetic metal. The fiberoptic output widow increases the signal-to-noise ratio. The electron optics provide for better focusing of the image onto the output window. (Courtesy of Philips Medical Systems.)

view the output of the image intensifier at one time, and then only if the viewer is willing to move with the image intensifier as it is moved. The obvious advantage of the television system is that a number of people may view the image produced by the intensifier; another advantage is that the contrast of the x-ray image can be improved by the television monitor. The disadvantage of the television viewing system is that it involves a much more complicated and costly piece of equipment than a mirror-optic system.

The Mirror-Optic Viewing System

As the preceding paragraphs have indicated, the output of the image intensifier tube has been considerably enhanced in terms of brightness, but the output image is so small that it must be enlarged again so that one may view it. Thus, the problem is to take a small output image approximately 20 mm in diameter and convert it to an image large enough to be seen with binocular viewing. This enlargement and transmission of the image can be accomplished through a series of lenses and mirrors. It should be noted that an image can be transmitted through a series of lenses and mirrors without a significant loss of brightness as long as there are no scattering filters; only a small amount of loss of brightness occurs due to absorption of the light.

Figure 4-4 illustrates the principles of the mirror-optic viewing system. The person who is viewing the fluoroscopic image must place

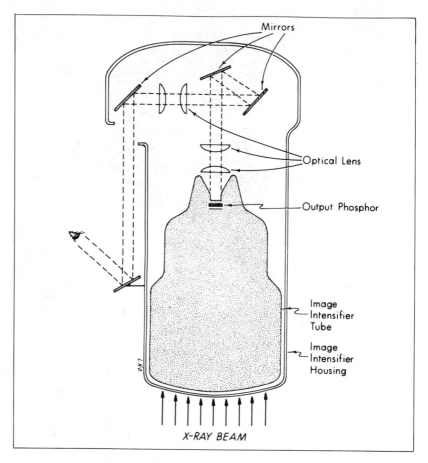

Fig. 4-4. A mirror-optic viewing system. The output from the image intensifier tube is focused through a series of lenses and mirrors onto a viewing mirror. There is no coupling system, as is necessary with television viewing systems.

himself in such a position that his eyes are in the light beam coming from the reflecting mirror. Some experience is necessary in order to keep one's eyes within the exit light beam, particularly as the fluoroscope is moved.

The quality of the reflected image is related to the quality of the lenses and mirrors used in the mirror-optic system. Moreover, the lenses must be kept in exact adjustment. Maladjustment of either the lenses or the mirrors may cause distortion of the exit level.

Television Fluorography/Fluoroscopy

Viewing the output of an image intensifier via a television chain introduces great flexibility into the use and viewing of the radiographic image. In addition to allowing the potential electronic recording of the radiographic image, it also affords a method by which the contrast of the radiographic image may be electronically enhanced. Use of the television chain releases the fluoroscopist from being fixed to a specific point or spot, as is required by the mirror-optic viewing system. Television viewing also introduces great flexibility into the use and viewing of the fluorographic image. Every year 1500–2000 new television viewing systems are installed, and new systems are constantly being developed. Unfortunately, the television image intensifier viewing system is much more complex and costly than the old fluoroscopic screen and the mirror-optic image intensifier system. In the following sections we will explore the major components of the television subsystem and point out the pertinent characteristics the reader should be aware of.

Coupling Systems for Television Viewing

As noted previously, the common size of the output phosphor of the image intensifier ranges from 15 to 25 mm in diameter. This output image must be coupled to a television pickup camera having a target face that is usually 15 mm in diameter but that may be as large as 25 mm or more in diameter. Interfacing the output phosphor of the image intensifier tube and the television pickup camera can be accomplished by one of two methods: an optical system or a fiberoptical system. Figure 4-5 depicts the two systems. In the future, other couplings may be introduced and the need for a separate coupling system may be eliminated.

Optical Coupling

At the outset it should be pointed out that when one speaks of an optical lens, it does not necessarily mean that there is only one element in the lens. A lens may be composed of a number of elements, each of which is designed to perform one or more specific tasks. These elements may be concave, convex, flat, or any other variation of a typical optical lens configuration. Optical coupling in medical x-ray systems is usually accomplished by using two sets of lenses. The first lens, called the *objective lens*, takes the light emitted by the output phosphor of the image intensifier tube and changes that light into a light beam with an infinity focus. If one considers the output phosphor of the image intensifier tube as being a point source and the light emerging from that source as being in the form of a cone, the purpose of the objective lens is to take this cone of light and convert it into a parallel beam rather than a cone of light. This *infinity beam* (or *parallel beam*, as it is sometimes called) is necessary so that a reflecting mirror may be inserted into the beam. This is done in order to split the beam and allow the

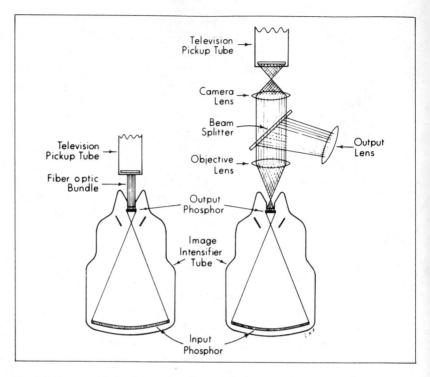

Fig. 4-5. Two coupling systems for television viewing: a fiberoptical coupling (left) and a tandem lens system (right). The fiberoptical coupling shortens the fluoroscopic tower, and the quality of the image is better than with the tandem lens system. The tandem lens system allows the introduction of a beam splitter, which is necessary for multiport viewing or recording.

introduction of other types of recording systems, which will be discussed later. This reflecting mirror is called a *beam splitter.*

   The objective lens is in tandem with a *camera lens.* The purpose of the camera lens is to refocus the light transmitted through the objective lens onto the input plate or phosphor of the television pickup camera tube. As with any other lens system, the exact focusing of the elements of the lens system itself is extremely crucial, with tolerances as low as fractions of a millimeter. It also stands to reason that if the lens system is handled roughly, the focusing can be thrown off, with resultant loss of resolution of the system.

   As with any other optical system, the amount of light transferred from the periphery of a lens is not exactly the same as the amount of light transmitted through the center or axis of the lens system. This decrease of peripheral light is known as vignetting, as previously mentioned. This loss of light transmitted through the lens system and focused on the input phosphor of the television camera can be as high as

75%, although it is usually about 10–30%. The result of vignetting on the television image is a loss of intensity on the outer edges. With a high-quality television system, vignetting is of no serious consequence because either extra elements are added to the lens to compensate for the loss of light, the television camera system may have antivignetting circuitry, or both.

As previously noted, one can expect the line pair resolution of an image intensifier tube to be somewhere between 3 and 5 line pairs per millimeter. A typical optical subsystem can resolve approximately 8 or 9 line pairs per millimeter.

Fiberoptic Coupling

The other method of coupling the output phosphor to a television camera is with fiberoptic bundles. Fiberoptics have the capability of transmitting light with a minimal loss of intensity and also with minimal flare. In this coupling system, a very thin fiberoptic disc (only a few millimeters thick) forms a vacuum seal and thus enables ultimate contact between the output phosphor of the image intensifier and the television camera tube. Adjoining the small fiberoptic disc of the image intensifier is the television camera tube, which also has an integral fiberoptic disc that serves as the image carrier from the output phosphor to the scanning target of the television camera tube. The fiberoptic components consist of a bundle of fine, individually shielded square or hexagonal glass rods, with up to several hundred rods per millimeter or several thousand per square millimeter. The special advantage of the fiberoptic system with regard to image quality lies in the fact that its transfer qualities are independent of location and contrast, as opposed to the tandem lens system. Again, in contrast to the tandem lens system, the fiberoptic coupling system gives uniform illumination of the image, and vignetting caused by loss of brightness toward the periphery of the image field does not occur. Fiberoptic coupling allows an exact match between the output of the image intensifier tube and the pickup tube of the television camera. The use of a fiberoptic bundle allows the overall system to be shorter and lighter, thus making the unit much more mobile. However, fiberoptic coupling does limit the versatility of the image intensifier subsystem to some extent. The possibility of using a beam splitter within the coupling system is eliminated, and thus the fiberoptic system is useful only in television applications. Spot-film or cinecameras, which record indirectly from the output phosphor, cannot be used with a fiberoptic coupling system unless an electronic recording device is used. This means that one must utilize the video signal to record off the image subsystem, which puts a practical limit on resolution. Also limiting the resolution of these subsystems is the size of the individual fibers in the coupling component.

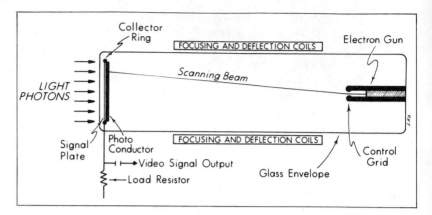

Fig. 4-6. Operational components of a vidicon camera tube. The basic components include an evacuated glass envelope that contains a signal plate, a photo-conductor, and an electron gun. As light photons strike the photoconductor, electrical resistance is changed in proportion to light intensity. Electrons from the scanning beam are collected by the signal plate in proportion to the resistance developed in the photoconductor. A portion of the composite video signal is developed from the signal plate and the load resistor circuit, which controls the aperture grid of the electron gun of the television monitor.

Television Camera Tubes

There are two main types of television camera pickup tubes used in x-ray applications today: the *image orthicon* and the *vidicon*. The more common tube is the vidicon. There are several types of vidicons; the most notable type is the Plumbicon, which is discussed later in this section. All the vidicon tubes are similar in operation, with the main difference being in the composition of the target material. Several experimental pickup tubes are being tested; some of them show great promise, especially the return beam vidicon. Although these new tubes are not in general use today, advances in camera tube technology can potentially make remarkable improvements in the television chain.

The Vidicon Pickup Tube

The *vidicon television camera pickup tube*, called vidicon for short, is a cylindrical tube measuring approximately 15 cm in length and 3 cm in diameter (a 1-inch vidicon). Larger tubes are also available. Figure 4-6 is a diagram of a typical vidicon, and Figure 4-7 is a photograph of an actual tube. The vidicon is composed of a target and electron gun enclosed in an evacuated glass envelope. The electron gun provides a source of electrons, which are focused into a fine, pinpoint stream and accelerated toward the target by a voltage differential. Deflection of the electron beam is accomplished by a series of magnetic coils that provide a push-pull mechanism, resulting in a horizontal-vertical deflec-

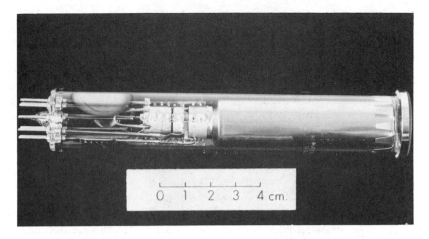

Fig. 4-7. A standard vidicon television pickup tube used in x-ray applications.

tion of the beam. The push-pull mechanism provided by the magnetic coils results in a scanning of the target in a linear fashion, starting in the upper left-hand corner and ending in the lower right-hand corner. This scanning produces a trace pattern called the *raster.* The pattern of the raster is somewhat akin to the movement of a typewriter (namely, moving across one line, indexing, moving back to the next line, and subsequent similar actions) (see Figure 4-10).

The face plate of the vidicon is composed of optical glass that is free from defects, or a fiberoptic disc as previously described. Attached to the inner surface of the face plate is the target of the vidicon. The target is composed of two layers, a *signal plate* and a *photoconductive layer.* The signal plate consists of a very thin sheet of metal, which is thin enough to allow light photons to penetrate but still thick enough to hold an electrical charge. The signal layer is connected to a collector ring and in turn to a load resistor and voltage source. The voltage on the signal plate is kept positive with respect to the electron beam. Applied to the signal plate is a photoconductive layer, which in most cases is composed of antimony trisulfide. By *photoconductive* it is meant that the electrical resistance of the material changes in proportion to the intensity of the incident light striking the material. This means that the higher the intensity of the light coming from the optical coupling system and striking the face of the target, the less resistance there will be of the photoconductive material; in other words, the brighter the image, the less resistance there will be, and the darker the image, the more resistance there will be.

The vidicon works in this manner: As an image is projected from the output phosphor of the image intensifier through the optical coupling system onto the target of the vidicon, the resistance in the photoconductive layer is changed in proportion to the intensity of the light. This change in resistance discharges the charge between the photocathode and the signal plate. The amount of the charge is dependent on the light intensity and the time this intensity is applied (e.g., 33 ms in a standard 525-line television system). As the electron beam scans this photoconductive layer, it strikes an individual image point every 33 ms for a duration of only microseconds, with the electrons recharging the target to near equilibrium. The amount of current required to accomplish equilibrium is measured at the collector ring and is then amplified and transmitted to the television monitor via a coaxial cable as a video signal. The video signal is thus in proportion to the recharge current of an individual image point on the photoconductive layer.

One of the characteristics of the vidicon of some concern is that of *lag* or *persistence*. In proper technical terms, this lag may also be called the *rest image* or *remaining image*. The initial scan only neutralizes the charge on the photoconductive layer by a certain percentage, approximately 60–70% in vidicons and 90–95% in Plumbicons. Thus an image is remaining after the first scan, while a new image is added during the next 33 ms. The second scan, therefore, consists of a mix of the new image and the previous rest image, and the third scan consists of three images (the last image is always the strongest, most contributing visual image). From this discussion one can see that the lag is dependent on the percentage of maximum recharge of the photoconductive layer by the electron beam.

On the fluoroscope, lag presents as an image that remains for a short period of time after the fluoroscope has been moved from one area to another. While this may present no significant problem for most fluoroscopic work, it does create a problem in cardiovascular fluoroscopy. The lag of the vidicon does have some advantages, however, in that it tends to smooth out scintillation noise in the television image presentation.

Another problem that sometimes surfaces has to do with adjustments of the vidicon. In order to increase the sensitivity of the vidicon, the target voltage and beam current can be increased. However, these adjustments, while increasing the strength of the video signal output from the vidicon, may have the opposite effect on resolution. When the beam current is increased, the diameter of the beam is also increased. Since there is a direct relationship between beam width and resolution, increasing the beam diameter will decrease resolution. In addition, a higher signal voltage is obtained when the target voltage is increased. However, increasing this voltage will also result in increasing the dark current. *Dark current* is the signal voltage that is generated even when the target is not illuminated. Increasing the dark current enhances the visibility of manufacturing defects in the photoconductive layer; these defects appear as blemishes on the television monitor image.

The *Plumbicon*[1] is a special type of vidicon that uses lead monoxide as the photodiode target material. Unlike that of the vidicon, the signal current generated by the Plumbicon is almost proportional to the level of illumination; hence, there is no significant reduction of contrast with the tube. Moreover, there is essentially no dark current generated. The most significant difference between the lead monoxide vidicon and the standard vidicon is that there is considerably less lag with the Plumbicon, leading to a marked decrease in motion blurring as compared to a standard vidicon. The advantage of the lag in smoothing out scintillation noise is not present with a Plumbicon so that the image developed by a Plumbicon tube is more noisy than that of the standard vidicon, but the Plumbicon presents an image of higher resolution. One should be cautioned that the lead monoxide vidicon may be slightly larger than some standard vidicons, and hence the two may not be interchangeable. However, vidicon and Plumbicon camera tubes are interchangeable in many systems, and the circuitry for both allows adjustment for maximum performance.

The Image Orthicon Tube

Until a few years ago, the *image orthicon television pickup tube* was popularly used in fluoroscopic television systems. However, complex circuitry and high maintenance made it impractical for use in x-ray applications. With improvements in the vidicon, the image orthicon has been largely eliminated for use in x-ray television systems and is presented here only for completeness, and because some variation of this concept may be used in future camera tube development. The image orthicon is more complex, more sensitive, and larger than the standard vidicon, and its image capabilities are far superior to those of a vidicon. There is essentially no lag, and the light sensitivity is almost equivalent to that of the human eye. However, because of its extreme sensitivity, the image produced by the image orthicon tends to be much noisier than than of a vidicon.

Although these tubes vary in size, a typical tube is approximately 45 cm long and 7 cm in diameter at the face plate. Figure 4-8 is a drawing of a typical image orthicon. The tube is composed of an optical face plate on which is deposited a photocathode. Each point in the photocathode emits electrons in proportion to the amount of light focused onto that point from the image intensifier. These image electrons emitted by the photocathode are collected onto a target. The target is composed of a thin piece of glass whose forward electrical resistance to electrons is negligible, but whose lateral resistance to electrical flow is quite high. On the anode side of the glass target is placed a very thin

[1]Philips Medical Systems, Eindhoven, the Netherlands.

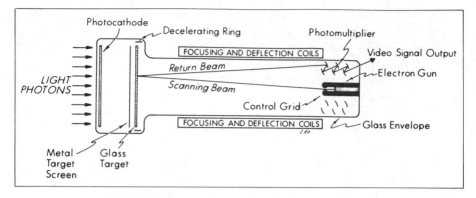

Fig. 4-8. Operational components of an image orthicon camera tube. The basic components include a photocathode, a metal/glass target, an electron gun, and a photomultiplier. As light photons strike the photocathode, electrons are emitted; these electrons strike the glass target, causing secondary electrons to be emitted, which are then collected by the metal target screen. A positive space charge is developed on the glass target corresponding to the image presented to the photocathode. This electronically stored image is scanned by the electron beam, which neutralizes the positive space charge; the excess electrons are then returned to the photomultiplier, where an electrical charge is developed and amplified to form a portion of the composite video signal.

wire mesh through which the image electrons pass before reaching the target. As the image electrons reach the target, secondary electrons are emitted and are collected by the wire mesh. The result is that a positive space charge is developed on the glass target, corresponding to the light image cast by the photocathode. The thinness of the glass target and the wire mesh cooperatively prevent any significant tendency for the space charge to dissipate laterally. The higher the light intensity on the photocathode per point, the greater is the space charge on the corresponding point of the glass target. This stored image is scanned by an electron beam produced by an electron gun and accelerated by a low voltage, but it is decelerated to almost zero velocity just before the beam strikes the target, thus essentially eliminating the production of secondary electrons. When the scanning beam hits a point on the target, the electrons neutralize the stored positive charge until equilibrium is obtained, at which time the excess electrons are deflected back toward the electron gun. When this return beam reaches the area of the cathode, the electrons strike a photomultiplier, which increases their number a thousandfold. The current developed by the photomultiplier is passed from the tube to an amplifier and is then converted to a video signal.

Other special television camera tubes are the *image isocon* and the *return beam vidicon*. These two units are similar to an image orthicon in operation. With the image isocon, use is made of the scattered beam rather than the reflected beam. Since the reflected beam is made up of both image signal and electronic noise, the use of a scattered signal from the photocathode eliminates signal noise. Isocon units are also known as low-light-level image tubes. A return beam vidicon is also similar to an orthicon, except that the target operates on the same principle as a typical vidicon. The return beam vidicon and the image isocon are extremely sensitive television camera tubes, as is the *extended red Plumbicon.*

Recently, a number of new vidicon television cameras have been introduced. These cameras are more sensitive and have higher resolution that the standard vidicon used in fluoroscopic television viewing systems. The main difference between the new cameras and the older vidicon is in the material used in the face plate. Silicon dioxide, the major material used in construction of transistors, is being used in some of these cameras and in others multiple layers of solid-state material is used.

Comparison of Television Pickup Tubes

Vidicon tubes are all similar except for the target material. With the exception of the Plumbicon, in which lead monoxide is used in the construction of the target, antimony trisulfide is the most commonly used photoconductive material. As far as size is concerned, the vidicon is smaller than an image orthicon. When focusing and deflecting coils, a housing, and other appropriate electronics are added, the vidicon camera is about one-fourth the size of a completed image orthicon camera. A Plumbicon or a standard vidicon system can be designed to operate completely automatically; this option was not available in earlier orthicon systems.

The image orthicon is much more sensitive than a vidicon, but this sensitivity has a disadvantage in that the signal-noise ratio is increased. This electronic noise appears as snow on the television monitor. For this reason, the orthicon cannot be operated at its full potential as an x-ray television pickup tube. Because of this need to operate the orthicon at less than its maximum sensitivity, the functional sensitivity of a vidicon and particularly of the Plumbicon is equal to that of the orthicon.

One of the major problems with any pickup tube has to do with persistence or lag, as previously discussed. In early vidicons this lag was quite noticeable, and any object motion or movement of the fluoroscope resulted in a smeared image for a short period of time. The image orthicon has almost no persistence, and the lead monoxide vidicon is a close second. A small amount of persistence improves the overall image because persistence tends to average or integrate out the signal-noise effect and presents a more pleasing picture on the television monitor. In most fluoroscopic examinations, physiological movements do not occur so quickly that a good vidicon camera cannot image them.

As a rule, the standard high-resolution vidicon can be used in almost any application. However, in examination of fast-moving objects such as the heart, the lead monoxide vidicon or solid-state vidicon may be the preferred choice.

The lead monoxide vidicon is capable of transmitting the radiographic image without any significant loss of contrast. However, the standard vidicon will loose 20 or 30% of the image contrast that is presented by the image intensifier.

The cost of the television camera must be taken into consideration. The lead monoxide vidicon is more expensive than a standard vidicon, but in a total camera system, this cost may be insignificant. As far as tube life is concerned, the lead monoxide vidicon has a life expectancy similar to that of a standard vidicon.

Basic Operating Principles of Television Monitors

Up to this point the radiographic image has been changed into different formats a number of times. The image intensifier has received an image on its fluoroscopic screen, changed that image into an electronic image, and converted it back to a light image at the output phosphor. The television camera pickup tube has received the light image and converted it into an electronic signal. The output of the television camera is in the form of a video signal, which will be combined with other signals and transmitted via a complex signal control and amplifier network to the television monitor screen with the aid of an electron beam similar to the one in the camera. The purpose of the television monitor, then, is to display the radiographic image in a visible form. The synchronized signals generated by the television camera network are coupled to the television monitor through a coaxial cable.

Television monitors are also known as *cathode ray tubes* (CRTs) and *kinescopes*. A monitor is similar in almost all respects to a home television set, except that there is no channel selector or audio capability. All television monitors have one basic signal system in common. In television, the image is converted into individual minute image points by scanning the image. The resulting linear (time-based) signal is transported from the television camera to the monitor or other recording areas, and the image dot is placed in relation to where it was picked up at the camera image. (In contrast, in optics all of the image points are transmitted parallel to each other and at one time. The image is transferred in toto.) In television, each individual point is transmitted consecutively, one after the other. A television image may consist of as

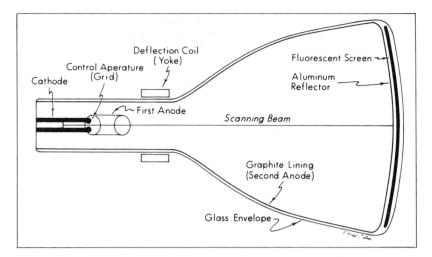

Fig. 4-9. A monochrome television picture tube or monitor. The composite video signal controls the synchronization with the 60-Hz timing impulse, the voltage on the control aperture grid in relation to the voltage signal developed by the camera tube target face, and, through the deflection coils, the location of the scanning beam in relation to the scanning beam of the television camera tube.

many as 1 million points per image and 30 million points per second, depending on the line rate and the bandwidth of the system.

The modern television monitor is composed of an evacuated, funnel-shaped glass envelope, an electron gun, deflection coils, and an output phosphor (Figure 4-9). The electron gun serves as a source of electrons. The heating coil in the electron gun heats a thermionic metal, the *electron-emitting cathode*, to the point where electrons are emitted. Just anterior to the electron-emitting cathode is the *control aperture*, or *grid*, of the electron gun. The function of the grid is to limit the number of electrons that are allowed to leave the electron gun and strike the output phosphor. This grid is controlled by a portion of the *composite video signal*. The more negative this signal is, the smaller is the number of electrons allowed to leave the electron gun.

Located around the throat of the television tube are *focusing coils*, whose function is to form the electrons leaving the electron gun into a pinpoint scanning beam. Also located outside of the tube are sets of *deflecting coils*. A synchronization pulse of the video signal synchronizes the electron beam deflection coils. The television camera, therefore, tells the television monitor scanning beam exactly where it is supposed to be. The magnetic coils then duplicate the exact scanning raster of the television pickup tube.

The inside of the television monitor is coated with graphite, which serves as a second anode with a voltage of approximately 15,000 volts (v). Inside the tube at the mouth is the output phosphor. The output phosphor is really a mixture of phosphors designed to give maximum light output with minimum oblique or lateral scattering. In some tubes this phosphor is faced with a thin layer of aluminum, which is impervious to the electron beam but which acts as a light reflector to enhance the output brightness level. The color of phosphors used in medical applications is subject to very tight specifications so as to present an image comparable to radiographs viewed on an illuminator.

There are two controls on the television monitor about which the operator should be concerned; these are the "brightness control" and "contrast control" knobs. For obvious reasons, no other control on the monitor should be touched by anyone other than a service representative. The television monitor has the capability of improving contrast, thus offsetting the contrast loss of the image intensifier and television pickup tubes. When the contrast control on a television monitor is increased, the video input signal is increased in amplitude, thus allowing a longer scale of gray on the monitor presentation. When the brightness control of the monitor is increased, the anode voltage is also increased, thus increasing the gain. Both controls have to operate in consort to gain the maximum benefit of contrast. Increasing the brightness control alone will decrease contrast, and vice versa. The proper way to adjust the brightness and contrast controls is to turn up the contrast until background noise is just visible and then to turn down the brightness level until the dark part of the picture becomes black.

The electron scanning beam bombards the phosphor of the television monitor from left to right and from top to bottom just as with the television pickup tube. This scanning pattern as it appears on the television monitor is called a *raster* (Figure 4-10). As the raster reaches the lower portion of the phosphor, the electron beam is deflected back to the top of the screen. During this period of time when the beam is moved from the bottom to the top of the screen, no signal is introduced and there is no illumination of the monitor. The process of stopping the video signal is called *blanking*, and the movement of the beam back to the top of the screen is called a *vertical retrace*. The term *horizontal retrace* refers to the movement of the electron beam from the end of a raster line back to the start of another line. During retrace, the beam is also moved vertically two lines in an interlaced system or one line in a consecutive scanning system.

Many readers will be familiar with the movies produced by the Hollywood studios several decades ago. These old movies were known as "flicks," because motion was jerky and the light flickered. In order to record and present smoothness of motion, a film has to be exposed and projected at a rate of 24 frames per second; to eliminate flicker the film must be projected at a rate of 48 frames per second. The x-ray television

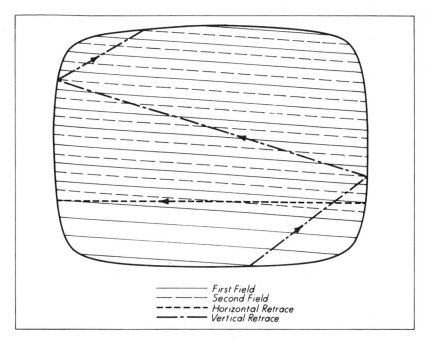

Fig. 4-10. This typical scanning raster depicts a 2 : 1 interlace, with half of the television frame or picture produced by the first field and the second half by the second field. Horizontal and vertical retrace are shown. The video signal is blanked during retrace movements.

system must also maintain these minimum standards for the same reasons.

In television, a single picture consisting of two half-pictures is produced in 33.3 ms synchronized by the 60-Hz pulse wave generated by the electric companies. This presentation of 60 half-pictures per second eliminates the problems of motion discontinuity and light flicker. The first half of the picture is composed of a raster scan starting in the upper left-hand corner of the screen and ending in the lower right-hand corner; the second half of the picture is produced by having the next raster scan placed between the lines of the raster of the previous scan. This means that the total picture is reproduced in 33.3 ms, which provides motion continuity, eliminates flicker, and provides good resolution.

Each half-picture is called a *television field*. The combination of two television fields produces a complete picture, which is called a *television frame*. The process of placing one raster line between two others is called *interlacing* (see Figure 4-10). The system of using two television fields to create one television frame is called a *2 : 1 interlace system*; it is by far the most commonly used interlace system. A *3 : 1 in-*

terlace system would be one in which three television fields were used to create one television frame.

Another term used in the television industry is *line rate*, which refers to the number of raster lines per television frame. X-ray television monitors are generally available with line rates of 525, 625, and 875 lines or more per frame. Future monitors may have higher line rates when a combination of normal scanning for reviewing and slow scanning for recording is utilized. For the purpose of discussion, let us examine a monitor with a line rate of 525 lines per frame in terms of resolution, since this monitor is the most commonly used today in medical x-ray imaging systems.

Vertical Resolution

In a 525-line system, 525 raster lines are available per television frame. About 7% of the raster lines are used for retracing, leaving approximately 490 lines for the frame. Since in a 2 : 1 interlace system, two television fields are used to compose one television frame, each television field has available 245 raster lines. These 490 raster lines per frame are all that is available, regardless of the size of the monitor screen. On a 15-inch diagonal monitor, these lines are distributed over a 9-inch vertical height; on a 10-inch diagonal monitor, over a 6-inch vertical field. Two general points should be stated:

1. The higher the line rate, the higher is the vertical resolution, assuming the required bandwidth is designed into the system.
2. The larger the monitor, the less detail is perceived, unless the monitor is viewed at an optimal distance.

It is important to know that the largest monitor does not always provide the best image. The combination of the large size, short viewing distance, and the low 525-line system allows the raster to become visible to the viewer, thus decreasing perception of detail. A smaller (14-inch) monitor presents at a normal viewing distance a raster-free image with higher contrast than, for instance, a 17-inch, 19-inch, or 20-inch monitor. Now one can see why a picture on a small home television set looks better than one on a larger television set, with all other things being equal, and the same thing holds true for x-ray television systems.

Horizontal Resolution

In the preceding paragraphs it was noted that vertical resolution is a function of the line rate. Horizontal resolution is a function of bandwidth. *Bandwidth* refers to the number of times an event can be cycled per unit time. In television, bandwidth refers to the number of times the scanning electron beam can be turned off and on per second so as to reproduce an image point in its true value from black to white. For a home television set, the bandwidth is 3.5 megahertz (MHz). One hertz is one cycle per second; thus the electron gun can be pulsed at a rate of 3.5 million times per second, or 0.3 ms per pulse. In a 525-line system with 30 frames per second, approximately 222 pulses or line pairs are available for each raster line. For example:

$$\frac{3,500,000 \text{ Hz}}{30 \text{ frames/sec} \times 525 \text{ lines/frame}} = 222 \text{ cycles per line}$$

Since commercial television is concerned with both form and detail, the 3.5-MHz bandwidth may be acceptable. With x-ray television the quality of the picture has to meet medical x-ray imaging standards, which are different from the commercial standards. Bandwidth is a most important factor in meeting this x-ray television performance standard. Bandwidth is related to horizontal detail, and most x-ray television systems use a bandwidth of at least 5 MHz, thus allowing 350 pulses per raster line.

Although theoretically the best resolution that could be obtained is with a television system having a very high line rate and bandwidth, in actual practice this is not true because parts of the television chain cannot be ideally matched to a high bandwidth. It appears that with the present cesium iodide image tubes and television pickup tubes, the best possible combination for high-detail fluoroscopy is an 875-line system with approximately a 15-MHz bandwidth. For routine use, the 525-line system with the 5-MHz bandwidth is adequate.

Special Circuitry for X-ray Television

The x-ray television systems of today have many unique features to provide high performance. Almost all cameras have automatic gain control in addition to the automatic radiation brightness control (which will be discussed in the next section). The gain control helps to maintain constant brightness and limits the radiation dose to a certain maximum level. If this maximum is reached and a call for more radiation is given, the television system adjusts automatically to a higher sensitivity. Other circuitry includes circular blanking of the camera and circular blanking of the video image, which improves overall performance and provides a clean circular image presentation at the monitor, with no mechanical masking required. The camera may have signal clipping circuits to prevent blooming and may completely cut off the video signal during standby.

Some cameras may have *automatic iris control* or *automatic fixed iris (f-stop) control*, which is important for parallel viewing while an image is being recorded on cinefluorographic equipment, photospot

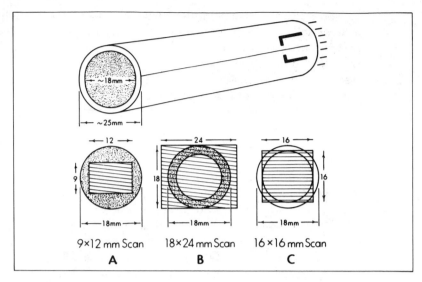

Fig. 4-11. Scanning rasters for television. A typical television pickup tube has a circular input phosphor measuring about 18 mm in diameter. Commercial television utilizes a 9 × 12–mm area of the target surface (*A*), which provides the standard 3 : 4 height-width ratio as seen on home television sets. In industrial television the target is overscanned, but the 3 : 4 ratio is still maintained (*B*). In square scanning of a circular image (*C*), the scanning raster is more closely fitted to the circular input phosphor of the television camera tube; the height-width ratio is 1 : 1, and overall resolution is improved.

camera, video tape recorder, and/or video disc recorder. The circuitry is too complex to describe in detail.

Another term used in television systems is *square* or *rectangular raster* (scanning), which may also include circular camera beam blanking. Standard home television has a rectangular raster scan of a 3 : 4 ratio, with the horizontal direction being 4 and the vertical direction being 3 (Figure 4-11). Most medical x-ray television systems with 525 lines apply the international standard, which also includes the 3 : 4 raster scan.

A significant improvement in resolution and overall performance of the camera pickup tube and the complete television system can be achieved by changing to square raster scanning with optional circular blanking (Figure 4-11C). Such an approach makes a television system a nonstandard system; it therefore cannot be interfaced with another standard system. In the future, nonstandard 875-line, 1200-line, and even 2000-line "slow scan" systems may use the square raster principle and other desirable features to improve television performance in conjunction with the image intensifiers, the reason being that the image output from the image intensifier is a circular image rather than a rectangular image as used in commercial television. These developments are aimed toward higher resolution in recording the video signal rather than viewing applications.

Automatic Brightness Control Systems

Up to this point we have assumed that the density of the anatomical part being irradiated is uniform and there is no variation in density (*radiopacity*) from one anatomical area to another. In reality this is not true, and some method must be available to compensate for this variation in density in order to maintain a constant brightness level in the image intensification system. There are two major ways to control the brightness level: by controlling the electronics of the image system, in particular the gain of the television system (*automatic gain control* [AGC]), and by controlling the radiation input into the image intensifier (*automatic dose-rate control* [ADC], which is sometimes called *automatic brightness control* [ABC]).

Electronic Brightness Control Systems

The electronic methods of brightness control work by varying the gain or sensitivity of the television system. The circuitry for such systems is fairly simple and thus is relatively inexpensive. The major disadvantage of the electronic gain control is that as the gain is increased, the noise in the system is correspondingly increased. If the brightness level is controlled by varying the sensitivity of the pickup tube or the gain of the television amplifiers, no consideration is given to the dose rate to the patient; thus, unnecessary radiation overexposure is potentially allowed. The ideal situation would be to provide the image intensifier with the exact amount of radiation required to produce an automatic acceptable brightness level at all times. Manual selection of various levels of video gain can be used effectively in concert with automatic radiation level controls to permit the reduction of image plane radiation for some very heavy patients, where "just getting an image" is important. Of course, the reduction of image plane dose causes a noisy image, but there are some large patients who cannot be seen well at any legal tabletop radiation level unless a noisy, high-gain video technique is used.

Brightness Control Achieved by Controlling Entrance Radiation

Mechanisms whereby the brightness level of the image intensifier is maintained within established limits are known as automatic brightness control systems or automatic dose-rate controls. There are three methods of maintaining brightness levels at the output phosphor, all of which vary the incident radiation to the patient. Control of constant entrance radiation and thus the brightness level in the image intensifier (related also to film blackening) can be effected by one of three major methods, namely, controlling the kilovoltage, the milliamperage, or the pulse width (time). All three systems make use of a *comparator circuit*. The function of a comparator circuit is to establish a reference against which entrance radiation or light levels are compared. Any variations from the constant reference point are compensated for by adjusting the entrance radiation to the point that the exit light level is equal to the reference.

The Kilovoltage Automatic System. For control of the kilovoltage, a variable voltage device is required, which varies the voltage on the primary or secondary side of the high-voltage system, thus maintaining a constant radiation level into the image intensifier. With this method the kilovoltage is varied in proportion to the reference as determined by the comparator circuit. The milliamperage may be constant, or it may taper off automatically to maintain the radiation dose within certain limits. The major advantage of the kilovoltage automatic system is that it provides a wide dynamic range of 20 : 1 or even more. Kilovoltage automatic brightness controls also maintain good radiographic contrast throughout a range of anatomical thicknesses, provided an adequate tube current and proper exposure times are maintained. If the milliamperage is maintained too low, the kilovoltage is driven too high, with a resultant loss of contrast. A primary kilovoltage automatic system is rather simple and inexpensive. On the other hand, secondary-controlled kilovoltage automatic systems are complex and very expensive. The kilovoltage control system is used occasionally in cinefluorography.

The Milliamperage Automatic System. In this brightness stabilization system, the kilovoltage is maintained constant and the milliamperage is varied to meet the needs of the comparator circuit. The main advantages of this system are its simplicity and relatively low cost. The main disadvantage is its sluggish reaction time, which is caused not by the control circuit but by the time required to heat or cool the filament of the x-ray tube. The image from this system tends to be noisier with small patients, due to the relatively small amount of milliamperage required at higher kilovoltage levels. The contrast level of the image cannot be improved, since the kilovoltage remains at a constant high level. With smaller patients, the operator should reduce the kilovoltage manually to maintain adequate contrast. The most sophisticated of these systems utilizes peaks of the video composite signal to control the fluoroscopic brightness. This method makes it practical to use very small fluoroscopic field sizes without upsetting the dose levels. The milliamperage automatic system is almost never used for control of cine exposures due to its slow response, but it is useful for video tape recording at high photon flux levels, providing it is coupled to a video gain control circuit.

The Pulse Width Automatic System. The third type of automatic brightness control system is related to controlling the time during which an exposure is made. This control is used primarily in filming and is ideal for pulsed fluoroscopic modes using video disc or electronic video storage systems. The pulse width automatic system operates with constant kilovoltage and milliamperage that are preselected prior to viewing or recording, and the pulse width is varied according to the needs of the comparator circuit. The main advantage of the system is that it allows the use of minimum exposure times for a fixed ideal kilovoltage and high constant milliamperage. Motion unsharpness is reduced during film recording, and thus this type of automatic system is desirable for cine and photospot applications. Cine pulse width control is generally used only with three-phase generators and secondary switching. The main disadvantage is that this is a costly, complex system requiring the use of a grid-controlled x-ray tube or constant potential generator with triode or tetrode switching. Since this system is more complex than the other two, it is more costly from a total systems point of view and requires more service maintenance, thus increasing the effective cost of the system.

In a comparison of the different systems, the kilovoltage automatic is probably the system of choice for television fluoroscopy and cine applications. It has dynamic range from about 60 to 125 kV. The pulse width automatic system should be restricted primarily to high-speed film recording and pulsed video recording.

Combined Systems. Today there are a number of fluoroscopic units that use a combination of brightness control systems. For example, one unit primarily controls milliamperage; however, when the maximum milliamperage is achieved, the kilovoltage is automatically increased to maintain proper brightness levels. This type is commonly called a *milliamperage automatic system with kilovoltage override.* The kilovoltage remains at a high level until the milliamperage has dropped below its maximum level; the kilovoltage then drops until it reaches a minimum preset level. For all practical purposes the milliamperage automatic type provided with a convenient kilovoltage control is probably the best all-around system for the conventional fluoroscopic room.

Recording the Fluorographic Image

The image intensifier has made it possible to visualize fluoroscopic procedures better, but a permanent record must be made to confirm the fluoroscopic findings. Permanent and semipermanent recordings may be made with spot-film devices, photospot cameras, cinecameras, and magnetic recordings such as video tape or video disc recorders.

Spot-Film Devices

The advantage of direct film radiography is that the recording system is limited only by the resolving power and load capabilities of the x-ray tube, the geometry, and the characteristics of the recording film and the intensifying screens. In addition, there is more versatility in film format size. With direct film recording systems, one can reasonably expect to resolve up to 5 line pairs per millimeter, not taking into account various kinds of object motion and geometric enlargement.

A spot-film device is interposed between the patient and the image intensifier. The spot-film device with the attached image intensifier, accessories, and weight support appurtenances is called collectively the *fluoroscopic tower.* Spot-film devices are available in which film cassettes can be loaded either from the front or rear of the device or automatically from magazines in other equipment. In any event, the cassette is held in the "ready," or "park," position and is shielded by lead until such time that the cassette is brought into the exposure position to record an image. One should note that after a cassette is brought from the ready position into the exposure position, a number of events must occur before an exposure can be made. These events include switching off fluoroscopy and integrating the radiography control system, i.e., boosting the tube filament to a correct preset milliamperage and usually changing to the large focal spot of the x-ray tube. Technique factors for fluoroscopy usually include 100 kV and 1–3 milliamperes (mA), whereas radiography usually requires the same kilovoltage but milliamperage settings of 300–400 mA or higher. All these events take a finite time (usually about a second), and during this time the x-ray tube rotor also has to come up to full speed.

Because of this time requirement, two major drawbacks of direct radiography come to light. First, the exposure is not made during the same time fluoroscopy is being performed, which means that the exposure is always made after the last event has been seen with the fluoroscope. Due to this time delay of about one second, what is recorded on film is an event that occurred after the last event seen on fluoroscopy. An indirect factor is that if the cassette is brought into position at a rather high speed, this force of movement will shake the fluoroscopic tower, resulting in vibrations that will affect image resolution or sharpness if a proper time delay is not provided. *Servopositioning* techniques are now being used to shape the acceleration and deceleration curves of the cassette; these techniques allow for very rapid transfer with time left over for vibration damping. Spot-film access times using these techniques are reduced to about three-quarters of a second.

Spot-film devices are now available wherein the operator does not have to return the cassette to the protected tunnel between exposures. Some of these units can expose four quadrants of cassette with only about 0.5 sec between exposures. Cervical esophagograms at approximately two per second are theoretically possible, provided that four exposures are enough. This same feature makes it possible to record a complete cervical and thoracic esophagogram on two half-film areas of a small cassette with a single barium swallow.

For years the standard cassette used in spot-film devices has been the 9½ × 9½ inches. Recently, multiformat spot-film devices have been introduced and shortly will become the standard for spot-film devices. In a multiformat spot-film device, almost any sized cassette may be used. Film format size is controlled by a microprocessor, as are the exposure controls and collimator shutters. Figure 4-12 is an example of the multiformat sizes available from Medicor.[2] For the most part, two exposures per second are obtainable from these multiformat spot-film devices.

The introduction of the image intensifier with the infinity beam optical coupling between intensifier and television camera allows the use of a multiport output system feeding not only the television system, but also a photospot or cinecamera, and in some cases, a mirror-optic system (Figure 4-13). As noted previously, only 10 or 20% of the light transmitted by the infinity optical system is used to provide the light input for the television system when the dose is increased for high photon flux recording. The remainder of the light can be used to expose radiographic film in a camera. This division of light is accomplished by a *beam splitter*, or *image distributor*, which is nothing more than a reflective mirror placed in the infinity beam with a portion of the light from the infinity beam reflected toward the television camera (10%) and the remainder (90%) transmitted toward the recording camera. In all cases the beam splitter is automatically driven into place when the "record" mode is activated. At the same time, the radiation input is increased to provide more radiation dose for film recording. An iris diaphragm may simultaneously be placed in front of the television camera to provide the proper light input for the camera. An automatic iris (f-stop selection) can be replaced by the simpler built-in automatic gain control. In all cases the photon flux used for recording must be increased over the flux used per frame of video. The human eye integrates the succession of noisy video images, but a single recorded frame of film at video photon flux levels would show quantum mottle due to the statistical fluctuation of photons. For this reason, photon fluctuation levels of 1–3 for video are increased to 20 or more microroentgens

[2]Medicor, USA Ltd., Columbus, Ohio.

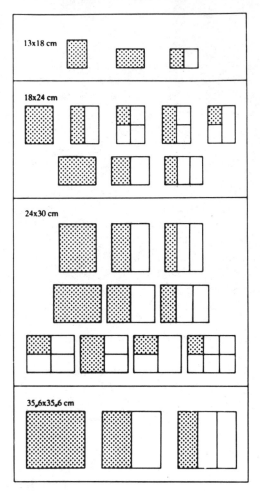

Fig. 4-12. An example of the various formats available from a multiformat spot-film device. (Courtesy of Medicor Medical Systems.)

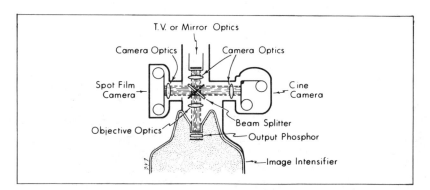

Fig. 4-13. A triple-port viewing/recording system showing the importance of the beam splitter. The beam splitter directs a portion of the light received from the image intensifier tube to the appropriately selected viewing/recording port.

(μR) per frame for cine techniques and a minimum of 100 μR per frame for photospot recording.

Photospot Cameras

The term *photospot camera*, or *spot-film camera*, is commonly used to describe cameras that record from the output phosphor of an image intensifier and use a roll-film or cut-film format of 70, 100, or 105 mm. The 100-mm size usually means cut film, and the 70- and 105-mm sizes are roll-film types. Regardless of the film format size, *all of these* cameras record from the output phosphor of the image intensifier. As such, the maximum resolving capability is limited to the resolving capability of the image intensifier and that of the optical system, which is today close to 3–5 line pairs per millimeter. Since the cameras record from the output phosphor, simultaneous viewing and recording take place. However, single-spot events cannot be observed clearly because the exposure times are generally too short (i.e., less than 50 ms, and often even shorter time frames).

Automatic exposure control for photospot recording is at its simplest when the generator is using the same automatic exposure device circuitry used with spot-film devices and for Bucky radiography. The modern photospot exposure controllers are useful down to 2 ms, with good density reproducibility. Other photospot exposure control methods include combinations of the methods used for cine film brightness control, usually with the extra cost and complexity of secondary contacting.

In comparing recording cameras with spot-film devices, the resolution capability of the film exposed with a spot-film device is theoretically higher than with a recording camera, but in practice the resolution of the two systems may be equal or even better with the camera. This is explained partly by the camera's capability of recording at much shorter exposure times, thus decreasing motion unsharpness. In addition, the cameras are cassetteless, thus providing far more convenience in operation.

Today, film frame rates of photospot cameras range up to 12 frames per second. Because one of the characteristics of these cameras is that only a small amount of film needs to be transported for each frame, camera filming rates can be increased to as much as 20–24 frames per second. However, such high frame rates have doubtful usefulness. In actual use, frame rates of 2–3 per second are all that is required for routine radiography, except for special applications such as cardioangiography.

When one compares the large and small format recording cameras, the larger format cameras appear more acceptable than the smaller ones. The larger films are also much easier to handle and to visualize on the standard viewing box than the 70- or 90-mm formats. Standardization and format size seem to indicate that either the 100-mm or 105-mm film size would be the superior choice. It should be noted again that the 100-mm size usually refers to cut-film cameras, whereas the 105-mm size refers to roll-film cameras using 105-mm perforated roll film. The construction of the roll-film cameras allows for patient identification between films, and therefore the entire square format area (except for perforations) can be utilized for images. The 100-mm cut-film types require a circular mask or front platen to hold the film clip flat during the exposure. The corners of this mask contain the patient identification. Thus the size of an anatomical part seen on 100-mm film is smaller than the size that is seen on 105-mm roll film, to a degree somewhat greater than is suggested by the film size alone.

Regardless of the film format size, it should be noted that the amount of radiation required to expose a film in a recording camera is lower than that required with direct radiography. In most instances, the amount of radiation is only 10–30% of that required for a spot-film exposure, depending, of course, on the calibration of the equipment and the mode of operation.

At the present time one of the major disadvantages of using a recording camera is the difficulty in processing the film. Most film processors are designed to handle large cut film, and it is inconvenient in most cases to have to bother with a film format size that is different from the rest of the film used in a radiology department. In large installations where a significant volume of recording camera film is exposed, it is advantageous to purchase a processor specifically for the processing of camera film. It is also advantageous to have an automatic unload feature. In most cases, camera roll film could be taped to a standard-size film and run through a standard processor, provided the chemistry and processing requirements of the standard- and small-film formats are identical. Another potential disadvantage of using a recording camera is that a 100- or 105-mm camera usually weighs from 7 to 25 kilograms (kg), which requires an appropriate counterbalance.

Comparison of Photospot Cameras and Spot-Film Devices

Some comments are in order concerning recording cameras and spot-film devices. As far as reliability is concerned, the recording cameras are just as reliable as spot-film devices. There were problems with earlier models of photospot cameras, but these cameras have now been perfected to the point that they are as reliable as spot-film devices. With the recording camera, film costs are about one-sixth those of the spot-film device. In addition, the use of a recording camera negates the need for periodic replacement and acquisition of the cassettes and screens used with spot-film devices. When all of these factors are summed up, it appears that a recording camera is probably more economical than a spot-film device. As noted earlier, spot-film devices are available with either front or rear loading of the cassette, or on occasion, a front and rear cassette load feature. One can normally expect more problems with a front-loading mechanism than with a rear-loading one (this is rather obvious, since more mechanical steps are required to transport the film cassette from the front part of the spot-film device to the "park," or "ready," position). However, the convenience of front cassette loading as compared to rear loading may more than offset the increased complexity and cost.

It should be pointed out that there are numerous occasions in which spot-film devices are added to a fluoroscopic tower in addition to a photospot camera. With the equal reliability of the two devices, it is questionable as to whether one should pay the extra $15,000 or $25,000 for an additional recording method just to have a backup in case of failure. A trend toward more spot-film camera imaging can be predicted.

Cinecameras

Like recording cameras, a cinecamera is normally installed on the fluoroscopic tower and records from the output phosphor through a series of lenses and mirrors. The camera shutter controls the x-ray exposure and again, as with recording cameras, radiation exposure is increased when one changes from the fluoroscopic to the "cine record" mode. In addition, total radiation exposure is directly proportional to the frame rate being used. A beam splitter is used just as with a recording camera, thus allowing simultaneous viewing and recording.

The shutter of most cinecameras is timed by the 60-Hz power supply. This means that a cinecamera is *phase-locked*, which is an important feature providing synchronization of television and cine systems. Most cinecamera shutters are open half the time, 180 degrees, and closed half the time, the other 180 degrees[3]. Since a cinecamera shutter is controlled by the 60-Hz power supply and synchronized to the 30-frame-per-second video, cine film rates are multiples or divisions of 30 (e.g., 7½, 15, 30, 60, 90, 120, 150, and so on). The cinecameras can use either 16-mm or 35-mm film.

In older x-ray equipment, the cine film was exposed from a continuous x-ray beam, with the result that the patient was continuously

---

[3]The cine exposure time must always be less than the camera shutter-open time. Shutter-open time for a 180-degree shutter in milliseconds equals $\dfrac{1000}{2 \times \text{frame speed}}$.

irradiated regardless of whether the cinecamera shutter was open or closed. This meant at least double the radiation to the patient, and a shutter-open time equal to the motion-recording time (i.e., 16.6 ms for 30 frames per second). This type of cinefluorography is generally not used with modern x-ray equipment. Modern cinecameras use a pulsed x-ray beam produced either by a constant potential generator or grid-controlled x-ray tube. These types of controls produce a square wave output from the x-ray tube. Brightness control and thus film density are provided as previously described.

Because the cinecamera records information supplied by the image intensifier, film resolution is again not as good as with the best direct radiography, with cine resolution being limited by the image intensifier. Radiation exposure per frame in fluorography is lower than with direct radiography. However, with a multiple series of exposures obtained by cine techniques, the total exposure to the patient far exceeds that obtained by single-exposure direct radiography. On the other hand, radiography does not provide the same information as that received by dynamic recording, and image quality improves due to the integration power of the human eye when the cine film is projected at high frame rates.

As with spot-film devices, a number of events must occur after a cinecamera is activated. In addition to the time required for the beam splitter to be driven into place, the radiation level boosted, the anode to come up to speed, and the focal-spot size changed (if required), there is also the time required for the cinecamera to come up to the predetermined speed. For most cinecameras, this time ranges from 0.35 to 1.0 sec, depending on the filming rate. Therefore, after the cinecamera is activated, there may be up to 1 sec of lost information.

One of the biggest considerations with cinecameras is whether one should use a 16-mm or a 35-mm format. The problem with both formats is that there is a relatively long span between the time the exposure is made and the time the film is ready for viewing. In most installations, although processing may be available immediately, there are generally no facilities for making work prints; thus the original print is projected and the film is scratched as it moves through the projector film gate.

For the most part, 16-mm film is less expensive and easier to handle than 35-mm film, and above all, it can be projected for large groups with a much less expensive film projector. There are 35-mm projectors available, but they are expensive, and the flicker in the projected image is difficult to control. On the other hand, the format of the 35-mm film allows more information to be recorded, since the area of a 35-mm frame is four times the area of a 16-mm frame. This increased filming area results in higher resolution. In some cases, 35-mm film is recorded but is reduced to 16-mm size on fine-grain copying film for better projection capabilities.

The reader should be aware of the term *overframing*, which refers to how much of the image is recorded on the film. Overframing means that since the image is magnified, only a certain percentage of the image is recorded on the film frame. The amount of overframing is controlled by the cinecamera lens and in some cases can be changed by the service representative by interchanging lenses.

Recording the Video Image

The video tape recorder is one of the most frequently purchased auxiliary pieces of radiographic equipment, but it is probably the least used. In addition to being expensive, earlier video tape recorders available for x-ray fluorography were temperamental, and it was common to have more "down" time than "up" time. Common problems included frequent plugging of the magnetic recording head by oxide removed from the video tape, breakage of the recording head, and most common of all, slippage of the drive belt.

The advantage of recording the video signal is that it is immediately available for reviewing a fluoroscopic procedure without having to repeat the procedure completely. Since fast events can be recorded, more attention can be paid to the event than to operating the fluoroscopic system at the same time. Simultaneous video and cine recording is possible, with the video tape being immediately available to determine whether the required information has been recorded on the higher-resolving cine film. The output from the television camera is ideally suited for electronic recording devices; however, since the electronic signal is going through another electronic interface, there is a slight loss of signal, and electronic noise can be introduced into the image. The need for increasing the photon flux when recording has been described previously. During video taping, some equipment boosts the entrance radiation over the normal fluoroscopic value, thus providing a better-quality video tape recording because a better image is presented to the camera by the output phosphor. On some systems it is also possible to "document" fluoroscopy at the lower fluoroscopic levels, if the user will accept the noisy image on playback.

Figure 4-14 is a diagram of a video tape recorder (also known as a *tape deck*). The video signal is written into and read from the video tape by a *recording head* located on the upper edge of a large rotating disc. Most x-ray television tape recorders use a *helical recording system*, in which the magnetic tape is wrapped in a spiral fashion around the recording drum so that the tape exits above the leading edge. The effect of this type of threading is to allow the recording head to run across the magnetic tape in a diagonal rather than a linear fashion, which

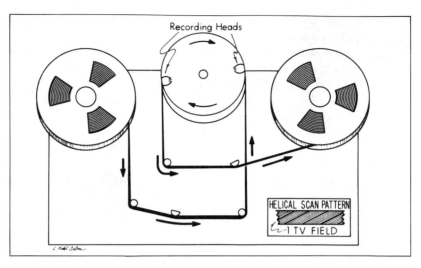

Fig. 4-14. Principle of helical video tape recording and/or playback. The video tape moves from the reel on the left side obliquely across the recording drum to the reel on the right side. The end result is that one television field is recorded with each pass of the recording head, as noted in the lower right-hand corner of the illustration.

results in a remarkable increase in the recording time per length of the tape. As the recording head crosses the magnetic tape, one television field is recorded. To maintain synchronization, the movement of the recording head and magnetic tape is timed so that one pass of the recording head across the magnetic tape occurs every 1/60 sec (or 16.6 ms). The magnetic tape is usually moved at a rate of 10 cm per second, and the recording head moves at 30 meters (m) per second across it. Needless to say, the relatively high speed between the recording head and the tape requires a very exact adjustment in order to maintain synchronization during replay. Some video tape players also have audio channels, which are of little use in x-ray fluoroscopy.

The resolution of the video tape player depends on the quality of the magnetic recording tape and the gap in the recording head. The higher the bandwidth of the video signal, the smaller the gap has to be in the record/play head, and the smaller the lead oxide crystals have to be in the magnetic tape. One should be cautioned to use extreme care in cleaning recording heads, since they can easily be damaged and the gap in the head widened, resulting in loss of the video signal on replay. The reader should also be reminded that the video tape recorder must be compatible with the bandwidth of the television pickup tube and television monitor used in the image intensifier system.

Video tape recorders are usually supplied with recording widths of ½ inch, ¾ inch, 1 inch, or 2 inches. The magnetic tape may either be reel-to-reel or cassette. The width of the tape is related to the fidelity resolution of the system, since more information can be recorded in the larger area.

Continuous Loop Video Tape Recorders. In addition to reel-to-reel and video cassette recorders, tape recorders are available that use a continuous loop of magnetic tape. These recorders are designed for recording and playing back short sequences of fluoroscopy and are not intended for permanent storage of the image. The advantage of the continuous loop tape system is that it is an enclosed unit, and one does not have to worry about constantly changing cassettes or reels. These tape recorders are capable of recording 10–30 sec of fluoroscopy time. If the unit is run continuously with fluoroscopy, only the previous 10–30 sec can be reviewed, since all previous information is erased from the tape.

Video Disc Recorders. The video disc recorder is similar in appearance to a large phonograph record. Instead of magnetic tape, a flat, rigid disc is used. One television frame or field is recorded on each revolution of the disc, depending on the speed of the recorder (i.e., 1800 or 3600 rpm). The recording heads must meet the same requirements as noted for a video tape player/recorder. Recording heads may be located above and/or below the magnetic disc. Video disc recorders are capable of recording frame by frame in a static mode or continuous fluoroscopy in a dynamic mode. Since only two or three minutes of fluoroscopy are recorded per procedure, multiple examinations can be recorded on one video disc. The great advantage of a video disc recorder is that single frame image displays are much higher in quality than single frame viewing with a video tape recorder.

The video disc recorder has not yet entered radiology departments on a large scale, although it has a number of very desirable features. One of the most potentially valuable features is the possibility of dose-reduction pulsed fluoroscopy. Recently, the video disc recorder has been generally accepted for fluoroscopy procedures in the operating room in conjunction with the mobile image intensifier fluoroscope. With this equipment a view may be recorded, stored, and reviewed during procedures such as hip-pinning. The video disc recorder is also used successfully in digital subtraction angiography. Reports indicate that dose reduction of over 90% can be realized with the use of the disc recorder, as compared to conventional continuous fluoroscopy. Other reports give dose reductions of 30 to 70% for gastrointestinal radiography and a similarly high dose reduction for angiographic procedures.

Some problems with the disc recorder itself, together with the requirement of careful interface with the total x-ray system, have contributed to the poor acceptance of the recorder in some main areas of radiology. For example, the generator has to pulse at high and low frame rates at reasonably low milliamperage values. The interface involves the automatic brightness control, the television camera, the x-ray tube, and many other components. In addition, most disc recorders have a limited bandwidth of 4.5–5 MHz, whereas a higher bandwidth of 8–10 MHz is desirable.

Although fraught with many problems at the outset, advances in solid-state technology have led to great advances in video disc recorders such that they are fast replacing video tape recorders, even in the commercial market. Reel-to-reel video tape recorders are fast catching dust in most x-ray departments and perhaps represent the most overbought and unused piece of accessory radiographic equipment ever sold.

Sometime in the near future, solid-state video discs will be replaced with laser optical discs. Toshiba Medical Systems[4] is currently offering a laser disc with their computed tomography (CT) scanner, which is capable of holding 10,000 CT images!

Resolution of the System

I was once contacted by a radiologist concerning the problems that he was having obtaining adequate resolution from an image intensifier system. The television pickup tubes had been replaced twice in a relatively short period of time, with only minor improvement in the resolving capability of the system. The reader should be reminded that a television or image intensifier system is composed of the x-ray tube, the image intensifier, the grid in front of the image intensifier, the spot-film device, the optical system connecting the image intensifier to the television camera, the television camera, and the television monitor. No system can be better than the weakest point in the system. The average resolving capability of a *good* cesium iodide image intensifier tube is about 4 line pairs per millimeter, and thus the total system cannot resolve anything greater than the 4 line pairs per millimeter on film.

As with radiographic film, the main factor that must be considered is the total resolving capability in the x-ray tube itself, which was the problem in the case noted above. The x-ray tube is limited in its resolving capability. To expect the image intensifier system to achieve the impossible is unfair to the manufacturer and is not realistic on the part of those who request such capabilities.

It should be pointed out that the image intensifier, particularly the zinc cadmium sulfide type, will degrade with time about 10 percent a year on the average. One manufacturer estimates that its image intensifier will last for approximately 3000 hours of fluoroscopic time. Translated into terms of average use, this means that an image inten-

[4]Toshiba Medical Systems, Tokyo, Japan.

sifier will last approximately three years. Eventually the image intensifier will degrade until resolution becomes unacceptable. In addition, mechanical and electronic adjustments in all the subsystems are very critical. For example, poor adjustment of the collimating lens between the output phosphor of the intensifier and television camera tube is so critical that it can be off just a few hundredths of a millimeter and the resolution of the entire system will be degraded to unacceptable levels. One of the most common problems with the image intensifier system is mishandling. It is not uncommon to see either a radiologist or a technologist slap an image intensifier into its holding bracket on the support system, without knowing that even these jarrings may throw the mechanical adjustments off to the point where the image is degraded.

Another factor to consider is the weight of the image system. The larger the image system, the more lead is required to protect the conventional image intensifier against extraneous radiation; the more weight involved in the system, the more counterbalance and thus moving effort are required. Any object that has weight has inertia, and any time movement occurs, an equal amount of inertia is required to stop that movement. When a big image intensifier is moved toward a patient, no type of electrical or mechanical control can stop movement of the intensifier as quickly as one would like.

As with almost any optical system, resolution is better in the center of the image intensifier than on the periphery. Even with the modern image intensifiers, there is still a 5–10% loss of resolution on the periphery as compared to the center of the field. As with any other radiographic system, the more closely the object to be radiographed is collimated, the better will be the resolution and the contrast. One should also be aware that there is a geometric magnification factor to be taken into consideration during fluoroscopic procedures. This geometric magnification varies according to how high or how far away the image intensifier is from the patient. The reader should also remember that the inverse square law has to be taken into consideration; the further away the image intensifier is from the source, the higher is the dose of radiation that must be given to maintain the brightness level in the intensifier. One must develop habits of keeping the image intensifier as close to the patient as is practical, in order to decrease radiation exposure to the patient and to improve image quality.

As far as a choice of size is concerned, the 6-inch image intensifier is probably the most versatile size to use from the standpoint of weight and total performance, in particular either for viewing only or in conjunction with the spot-film device. Dual-field or triple-field image intensifiers are more suitable if spot-film camera recording is desired, and they are a necessity if the photospot camera is to be the only recording modality in the fluoroscopic room. One should choose either a photospot camera or a spot-film device, but it is usually superfluous to have both.

The introduction of the cesium iodide image intensifier has been one of the greatest advances in radiology, second only to the discovery of x-rays and the introduction of computed tomography. With proper use it extends the diagnostic capability of radiologists to the limit only of their own ability to perceive the information displayed by the image intensifier system.

Retrofitting

There are methods of improving the fluoroscopic image without totally replacing an entire radiographic room. The updating of equipment by incorporating new designs is called *retrofitting*. Systems are available that can convert the old red-goggle fluoroscopy to an image intensification system. Retrofit packages are also available to convert spot-film devices to photospot cameras. New cesium iodide intensifier tubes are generally available as replacement items for zinc cadmium sulfide tubes, provided the existing electronics can be modified to accept the new intensifier. Grid-pulsed x-ray tubes can also be installed on a number of existing generators, which can provide a proper source of radiation for photospot and cinecameras.

In most instances it is wise to consult with a number of vendors to determine whether existing radiographic equipment may be adequately updated before purchasing new equipment.

References

1. Chamberlain, W. E. Fluoroscopes and fluoroscopy. *Radiology* 38:383, 1942.
2. Strum, R. E., and Morgan, R. H. Screen intensification systems and their limitations. *American Journal of Roentgenology and Radiation Therapy* 62:617, 1949.

Suggested Reading

Gebauer, A., Lissner, J., and Schott, O. *Roentgen Television*. New York: Grune & Stratton, 1967.

Kuhl, W. Image Intensifier Systems. *Medical X-ray Photo-Optical Systems Evaluation*, DHEW Publication (FDA) 76-8020. U.S. Department of Health, Education, and Welfare, Public Health Service, Food and Drug Administration, October 1975.

Rich, J. E. *Television in the Radiology Department*. Milwaukee: The General Electric Company, 1969.

Rich, J. E. *Intensified Fluoroscopy in the Radiology Department*. Milwaukee: The General Electric Company, 1970.

5   Automatic Exposure Controls (Phototimers)

Faced with a variety of patient sizes and varying tissue densities, radiologists and technologists have long hoped to have a device that would allow them to obtain a properly exposed radiograph regardless of the size of the patient. Furthermore, it has been hoped that exposures could be reproducible from one time to another and independent of the manual selection of technique factors. Such a device has been available for a number of years, but these devices were better known for their variability than for their reproducibility. A device that automatically determines the amount of radiation required to produce an acceptable level of film blackness is called a *phototimer*, or *automatic exposure control* (AEC). Although there are limitations in the use of phototimers, the equipment can be used much more effectively if one recognizes these limitations, and the goal of obtaining a properly exposed, reproducible radiograph can be achieved.

Types of Phototimers

There are two basic types of automatic exposure control mechanisms used in conjunction with modern x-ray generators: the *photomultiplier (PM) phototimer* and the *ionization chamber*. The main difference between the two types is that the ionization chamber measures an ionization current and the photomultiplier type measures light intensity. Phototimers may be located in front of the image receptor (entrance types) or behind the image receptor (exit types). In most cases, the image receptor is the film cassette. AECs used in conjunction with image intensifier fluorography utilize exclusively the photo-optical measuring principle. The term *photo-optical* means here that the light intensity output of the image intensifier is measured with a PM tube or silicon cell that sends exposure information to a comparitor circuit, which then signals for exposure termination to achieve the proper film density.

Photomultiplier Phototimers

From this point on we will discuss automatic exposure control as applied to film in cassettes. The detector of the most commonly used photo-optical type of AEC or phototimer is made from thin, segmented sheets of Lucite that have been covered with black paper, except in the region that will be used to detect radiation (Figure 5-1). One of the properties of Lucite is that it will transmit light in a manner similar to

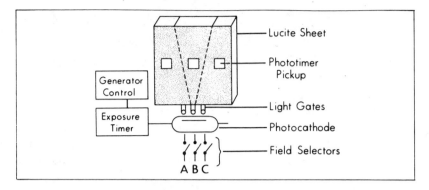

Fig. 5-1. Automatic exposure control of the photomultiplier type. Sheets of Lucite transmit light to the gates of the PM tube. The gates are controlled by switching circuits located on the generator console. A current that is proportional to the light transmitted to the PM tube is generated, amplified, and then used for terminating the exposure.

a fiberoptic bundle. The sheets of Lucite, called *Lucite paddles*, are coated with a phosphor, which, when struck with radiation, will produce light in proportion to the intensity of the radiation. This light is transmitted by the Lucite to an output area called a *light gate* and coupled to a PM tube. The photomultiplier converts the light photons into electrical signals, which can be amplified and used to control the termination of the exposure. Generally the exposure is forcibly terminated by *silicon-controlled rectifier (SCR) circuits* that use forced extinction techniques for maximum reproducibility at very short exposure times.

In an entrance-type PM phototimer, three apertures are normally available. The size of the aperture varies according to the manufacturer, but the average size is 100 square centimeters (cm²). Each aperture is connected via the Lucite sheet to a light gate. There may be one PM tube per gate, or there may be one tube for all three gates. Electrical or mechanical switches are provided on the generator console to allow the operator to choose any combination of fields or apertures up to a total of seven combinations.

There is an electronic advantage in having one PM tube that senses the lights from all three gates, because the selection of a PM tube is extremely critical with respect to its sensitivity. Moreover, the electrical supply to the tube is a crucial factor, since the tube is very sensitive to minute changes in the power supply. It is far easier electronically to adjust one PM tube than it is to balance three tubes. However, there are mechanical problems involved in connecting the light gates to the PM tube.

One of the major problems with PM-type phototimers has to do with the response of the phototimer to changes in kilovoltage. At low kilovoltage ranges (50–60 kV) there is proportionately more x-ray energy absorbed by the pickup paddle than at high kilovoltage levels, leaving less energy to expose the x-ray film; the result is that the exposure is terminated before the proper film density is obtained. Thus, more x-ray energy is at low kilovoltage levels to produce an acceptable film density than at higher kilovoltage levels. In technical jargon, this problem is referred to as "being oversensitive at low kilovoltage." The problem is overcome using what is known as *kilovoltage compensation,* a method in which the kilovoltage being used for a specific exposure is sensed, and an electrical signal is sent to the detector to raise or lower its sensitivity, thereby compensating for variability of absorption of the x-ray beam by the detector. At low kilovoltage levels, the phototimer detector is decreased in sensitivity, allowing more radiation to reach the film before the exposure is terminated.

Photomultiplier phototimers are the most commonly used exit-type detectors. Often a single aperture is used, with the size of the aperture varying according to different manufacturers. The aperture is generally larger in exit-type phototimers than in entrance-type phototimers. The main advantage of a single-aperture (single-field) phototimer is that it is generally much less expensive than a multiple-field type and is also less trouble to maintain. Another advantage is that thicker phosphors can be used on the pickup paddles, allowing a stronger signal to be generated. In addition, the same phosphor that is used in the film-screen combination can be readily used in the pickup paddle without the operator having to worry about absorption and scattering of the x-ray beam. The disadvantage of the single-field detector is not electromechanical but is primarily technical in nature: Its use requires far greater attention to be paid to the proper positioning of the patient over the detector sensor. Exit-type detectors have been de-emphasized by most manufacturers in recent years, primarily because of the need for special control of cassette opacity.

Ionization Chamber Phototimers

An ionization chamber (Figure 5-2) is composed of extremely thin sheets of foil, usually made of aluminum or lead, in which are enclosed radiation-sensitive ionization chambers. Each ionization chamber contains a gas, usually air under normal atmospheric pressure, that ionizes in direct proportion to the amount of radiation striking it. When the air is ionized, it forms a complete electrical circuit, and the current from this circuit can be amplified and used as a signal to terminate the exposure. Except for mammography, ionization chambers are almost universally used as entrance-type phototimer. AECs based on ion-

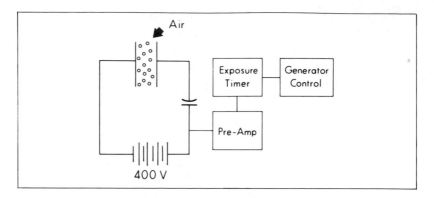

Fig. 5-2. Automatic exposure control of the ionization chamber type. A gas is ionized by radiation passing through the detector chamber, and the ionization produces a minute electrical current that can be amplified and used to terminate the exposure.

ization chambers are not phototimers in the strictest sense, but the term *phototimer* is almost universally used today for both classes of detector. The terms *autotiming device* and *automatic exposure control device* are beginning to be heard in place of *phototimer.*

Solid-State Automatic Exposure Controls

With the rapid incorporation of solid-state electronics into the design of radiographic equipment, changes are also taking place in AECs. Toshiba Medical Systems[1] was granted a patent on a solid-state AEC in 1980, and these units, or others like them, will gradually replace phototube and ionization chamber devices. The solid-state detector is made of an N-type silicon wafer with gold metal at one end and aluminum at the other (Figure 5-3). With the use of gold at one end of a silicon wafer and aluminum at the other, an area of decreased electrons (depletion region) and an area of increased electrons (diffusion region) appear. When the detector is struck by the x-ray beam, pairs of electrons and positive holes are produced inside of the silicon wafer, with movement of electrons towards the diffusion and depletion regions. This movement of electrons produces a minute electrical current that is amplified and used to control the x-ray exposure.

Solid-state AECs are extraordinarily thin, with the Toshiba unit measuring only 0.3 millimeters (mm) thick. This thinness provides minimal absorption of the x-ray beam and, additionally, allows the Bucky tray to be placed closer to the x-ray table top, thus decreasing magnification distortion. The response time of the AEC is approximately 3 milliseconds (ms) for three-phase equipment and is capable

[1]Toshiba Medical Systems, Tokyo, Japan.

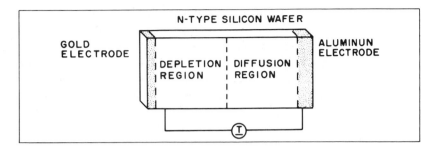

Fig. 5-3. A solid-state automatic exposure control. When the detector is struck by radiation, movement of electrons between the depletion and diffusion area occurs, producing a minute electrical current that ultimately is used to control the x-ray exposure. (Courtesy of Toshiba Medical Systems.)

of tracking the response to changes in kilovoltage exhibited by the various rare-earth film-screen systems.

Limitations of Phototimers

When phototimers are used it is necessary that the films, screens, and cassettes be standardized as far as speed and absorption are concerned. Inconsistency in the absorption characteristics of one screen or cassette as compared with another cannot be tolerated with the use of any type of phototimer, but particularly with exit-type phototimers. If either the film speed, screen speed, or absorption of the cassette varies from the standard set for a particular piece of equipment, the phototimer will not provide a correct film density. Phototimer cassettes are not absolutely essential for entrance-type phototimers, but film and screens must be standardized if one wants to avoid or eliminate operator error. Thus, one of the largest problems with phototimers is not with the phototimer itself, but with the variability of supplies used in conjunction with the device.

Another problem is related not to the automatic exposure device itself but to its size. When an exposure detector is placed between the table top and the Bucky tray, it takes up space. This additional need for space increases the object-to-film distance, thus increasing magnification distortion. The insertion of the exposure control device into the incident x-ray beam also produces scatter radiation, which decreases radiographic contrast. However, the loss in terms of unsharpness and diminished contrast is more than offset by the reproducibility offered by the exposure device and the automation afforded by the use of such devices.

A problem peculiar to both ionization chambers and PM-type automatic exposure controls is that of capacitor leakage. The signal developed by the AEC device is used to charge a capacitor. When the charge in the capacitor reaches a predetermined level that corresponds to the correct film density, the capacitor will discharge through a circuit terminating the exposure. Any capacitor will leak its charge over a period of time. As a general rule, the longer a charge is applied to a capacitor, the more leakage of the charge occurs. However, this leak can be sensed with modern electronic equipment, and with proper and carefully designed compensation circuits, the leakage problem can be overcome and ideal tracking can be provided. If the capacitor is not properly adjusted, the leakage may result in as high as 10% error in an exposure.

Performance of Automatic Exposure Controls

As indicated previously, the Bucky automatic exposure control is not an answer to all the exposure problems that face a machine operator. However, if the limitations of the equipment are understood, it is easier to obtain satisfactory results with almost any AEC mechanism rather than continually face the frustrations that commonly occur with this equipment. One should be aware that any given automatic exposure termination device is not perfect, even with the solid-state electronics now available to the industry. The automatic exposure control circuits must meet the regulations of the Department of Health and Human Services (HHS), which state that the generator must automatically terminate an exposure at 600 milliamperes-seconds (mAs) for 50 kV or greater, or at 2000 mAs for kilovoltage under 50. The performance standards also stipulate that the coefficient of variation in exposure reproducibility cannot exceed 0.1; that is, an exposure must be reproducible within 10% from exposure to exposure.

As far as exposure linearity variation is concerned, HHS restricts variation to the average ratio of exposure to the indicated milliamperage product (millirad/mAs) obtained at any two consecutive tube current settings to no more than 0.10 times their sum. This means that if 6 millirad/mAs is required for an exposure at 100 mA at 100 ms, and another exposure is made at 200 mA at 50 ms, the second exposure must be within 4.8–7.2 millirad/mAs. For practical use, the milliamperes-seconds cannot vary more than ± 10%.

Minimum Response Time

The automatic exposure timer is a component of a system the purpose of which is to provide an electrical signal to terminate an exposure. The time required for an automatic exposure control device to terminate an exposure after a proper amount of radiation has been sensed is called the *minimum reaction time*. There are a number of components that constitute this reaction time: The first of these is the reaction time of the AEC mechanism itself, which may be in the range of 2–3 ms. In addition, some time is required to operate mechanical relays in older equipment; this time can be stated to be about 8 ms. The third factor has to do with the interface with the generator. Without forced extinction, an additional variable of 8 ms must be added for the voltage waveform to cross zero potential. Repeatability of these older AECs is limited to 30 ms. With forced extinction and proper anticipation circuitry, the total system reaction time can be within 1 or 2 ms. Automatic exposure controls used with secondary contacting generators can provide density tracking down to 1 ms. However, 2 ms is a far more commonly used specification for those units without secondary contacting.

Checking the minimum reaction time of an AEC device usually requires the aid of an oscilloscope. If one calculates the nominal exposure factors that would be required for a routine radiographic exposure— taking into consideration screen speed, milliamperage station, and kilovoltage—an estimate of the exposure time can be made. If the exposure time is less than 20 ms, it is wise either to lower the kilovoltage or the milliamperage station or to use lower-speed screens to make certain that the exposure time will be longer than 20 ms (unless adequate controls, such as forced electronic termination, are available for shorter repeatable exposures).

Repeatability

Short-Time Exposures. An automatic exposure control should be capable of making repeated exposures at short times with no more than ± 10% variation in film density for an average of five exposures. To test this repeatability, one should place an aluminum step-wedge of the proper size (10 × 4 inches) over the detector aperture and manipulate the exposure factors (kilovoltage or milliamperage) until a short exposure time is reached (about 50 ms will do). Five exposures should be repeated using these technique factors; the density between the films should not vary more than ± 10%. For practical purposes, the film density should not vary more than one density step on the step-wedge.

Long-time Exposures. An AEC should meet the same requirements for long-time exposures as it does for short-time exposures. Long-time repeatability is tested in the same manner, except that technique factors are chosen so that the response time will be about 1 sec. Exposures longer than 1 sec are not generally used in radiography.

Response to Changes in Tissue Thickness

An AEC in good working condition should be able to compensate for varying thicknesses of the anatomical parts being radiographed. To check this response, one should place an aluminum step-wedge over the detector aperture and make an exposure at 70 kV. One should then repeat the procedure, adding a piece of old fluoroscopic apron, aluminum sheeting, or Lucite blocks between the x-ray beam and the automatic exposure control detector. The procedure should be repeated several times until the equivalent of five layers of apron are filtering the x-ray beam. Ideally, water should be used when the exposures are made; a large container should be placed with an inch or so of water in it, and water should be added gradually between exposures to simulate patient tissue. This procedure should be repeated for the 90- and 120-kV levels. Film density at all kilovoltage levels and for varying "patient thicknesses" should not exceed 10–15% over the entire range of thicknesses. For ideal comparison, a step-wedge can be placed outside the measuring field. Figure 5-4 illustrates an acceptable method for accurately checking phototiming tracking.

Fig. 5-4. Setup for testing automatic timer with water. Add water at 1-inch intervals; use the exposures of the step-wedge for comparison. The water serves as a scattering medium for the x-ray beam. The top illustration shows the placement of the water tank and the step-wedge over the automatic timer sensors.

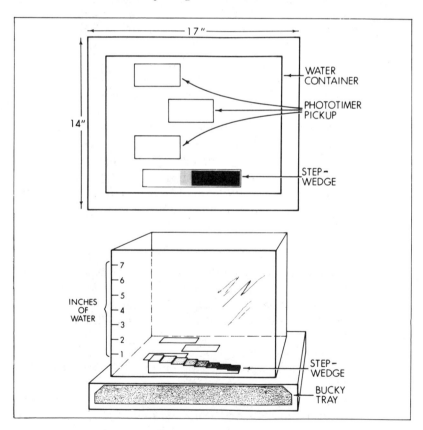

Technical Considerations

Assuming that an AEC meets the performance criteria noted previously, there are additional factors that must be considered (some of which have already been alluded to). One cannot overemphasize the importance of having matched screens, films, and cassettes for use with any type of AEC device. If the AEC is presented with variables other than those for which it has been calibrated, the expected film density will not be obtained. Electronic devices can be made to track evenly, but with the nonlinear responses of screens, even tracking becomes more difficult. Figure 5-5 illustrates this problem. The processing of the radiographic film is just as crucial a factor for films exposed by the AEC mechanism as it is anywhere else in the radiology department. Sensitometric control of the processing must occur before any adjustments are made in the x-ray equipment, particularly with the automatic exposure control.

The density control buttons or rotary knobs provided with the AEC can be used to increase or decrease the sensitivity of the detector. These controls normally allow a range of density controls from -50%, -25%, 0, $+25\%$, to $+50\%$. On some equipment these controls are labeled -2, -1, 0, $+1$, $+2$, but the effect is about the same. If a film is light or dark at different technique levels as demonstrated by the kilovoltage linearity test described previously, the operator may be able to use the density controls to bring the film density into a useful range while waiting for a service representative to adjust the automatic exposure control properly.

Fig. 5-5. Automatic exposure control tracking. An electronic device can be made to track linearly; however, when a nonuniform medium is added, such as x-ray intensifying screens, adjustments must be made to the electronic circuits to have them track with the screens. The linearity of screens, x-ray beam quality, and varying patient densities combine to make accurate phototiming tracking difficult.

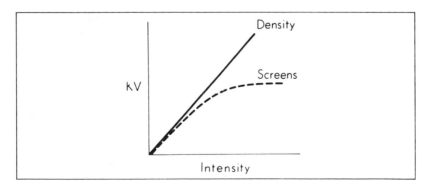

The use of AECs by student radiographers has long been the bane of training instructors since the students do not have to select appropriate technique factors. While the use of AECs may discourage students from learning correct exposure techniques, it is the experience of most manufacturers and users of automatic exposure controls that the positioning of the patient over the measuring field aperture is far more critical than when conventional techniques are used. The automatic exposure device cannot replace a well-trained technologist.

Selection of the aperture field is as important as correct positioning of the anatomical part over the aperture. For example, with a three-field pickup, choosing the middle aperture will allow a proper exposure of the mediastinum to be made on a chest radiograph, but the lung fields will be overexposed. Choosing the right field aperture may or may not allow a correct exposure of the lungs, depending on whether or not the cardiac shadow will interfere with the sensing of the correct film density. The selection of the left field aperture will allow a correct exposure of the lung fields, provided there is no gross pathology in the lung. Unless one is using very high kilovoltage techniques (125–150 kV), a chest radiograph performed with an AEC will not correctly expose both mediastinum and lung parenchyma. The same holds true for the lateral chest radiograph. Positioning the lateral thoracic spine over the aperture will allow a correct exposure to be made of the thoracic spine but will overexpose the lung fields; selection of an aperture for the lung fields will allow underexposure of the upper thoracic spine. Placement of the anatomical part over the aperture in such a way that only a portion of the anatomical part is sensed will result in underexposure or overexposure, depending on how much of the anatomical part is sensed by the detector.

Collimation is just as important with AEC devices as it is with standard exposures. Scattered radiation resulting from widely collimated areas will reach the detector and terminate the exposure before a proper film density is detected.

As far as the selection of an automatic exposure control is concerned, as with any radiographic equipment, the application should dictate the selection. For routine bone radiography, assuming that film screens and cassettes are standardized, a single-field, exit-type photomultiplier AEC is all that is required. For a mixture of radiographic examinations, a three-field pickup, entrance-type automatic exposure control is desirable; it may be either the ionization chamber or the photomultiplier type.

Although the capabilities of automatic exposure controls will vary widely from manufacturer to manufacturer, if the operator knows the capability of the equipment, it is possible to make and reproduce acceptable exposures. In the meantime, manufacturers are continuing to perfect a better and more reproducible AEC mechanism.

Suggested Reading

Exposure initiation/termination mechanisms and automatic exposure timers in diagnostic radiology. In *Proceedings of the Society of Photo-Optical Instrumentation Engineers*, Vol. 70, pp. 165–200. Atlanta, Georgia, September 1975.

Olson, A. P., and High, M. Performance of automatic exposure controls when used with rare-earth intensifying screens. *Radiology* 140:491, 1981.

6

Angiographic Injectors

In the recent past, when angiography was a relatively new procedure and injectors consisted mainly of a strong arm and a glass syringe, large bolus injections were fraught with difficulties and uncertainties. As angiography became more common, the need was perceived for an easier and more reliable method of injecting the contrast material. This need was first answered by mechanical devices that enabled the operator to hold the syringe more confidently and also to exert more pressure, but there was still a great need for more precise control of the injection as well as for greater consistency.

Further development of the injector led to the application of mechanics and pneumatics as well as hydraulics to the syringe to achieve greater injection pressures. The new equipment was in general much improved over the simple mechanical devices and was simple to use. These injectors were mostly volume injectors that delivered a preselected amount of contrast medium at a preselected pressure during a certain span of time. The time of the injection, however, was unpredictable. This shortcoming was a result of the many variables involved, such as the viscosity of the contrast medium, the length of the catheter, the inside diameter of the catheter, and the injection pressure. If the operator did not time the injection, he would not even be aware of how long it took. In addition, due to many variables the operator would set the pressure either too high or too low, or in many cases the injection pressures needed for a proper bolus were beyond the limits of the injector. These problems with injection pressure resulted in insufficient opacification and required that the examination be repeated, thus subjecting the patient to unnecessary radiation exposure as well as to an additional dose of contrast medium.

The modern angiographic injector resembles its counterpart of the past in very few ways. Improvements have been made in safety features, controls, programming, aesthetics, and uniformity of delivery.

Types of Injectors

Generally there are two methods of injection, with two of three parameters controlled by the operator. The third parameter can be measured after the injection has been completed, or it can be calculated and predicted. The three parameters are volume, pressure, and time.

One method, referred to as the *pressure-controlled injection*, involves controlling both the pressure applied to the syringe and the volume of contrast medium to be delivered, with the time of injection unknown.

With this system the time of injection can be predicted if sufficient data are available relative to the catheter diameter, the catheter length, and the viscosity of the contrast medium. The problem with pressure-controlled injections is that the injector may not be capable of the pressures that are required to deliver the volume desired to achieve a proper bolus of contrast medium within the time allotted. If the operator does not check his data, he may not even be aware of this occurrence, and additional injections and radiation exposures may be necessary to complete the examination satisfactorily.

The second type of injection is called a *flow-rate injection*. Because of variations in catheter length, inside diameter of the catheter, and contrast-medium viscosity, the pressure may vary to maintain a constant delivery rate per unit of time. However, the actual flow rate is monitored, and pressure is maintained to keep it constant.

All the units discussed in this chapter are flow-rate injectors and consist of a motor-driven power ram that drives a jacketed disposable plastic syringe. Each unit is mounted on a mobile pedestal at a convenient working height from the floor. All the units are sold as a basic package, with optional additional modules available for initial add-on or for retrofit to achieve extra versatility and convenience according to individual needs. The modular feature also simplifies service calls and reduces "down time," because the service representative can substitute a new electronics board or module for the defective one.

Electrical Safety

The injectors reviewed in this chapter have electrically isolated syringes to protect the patient from electrical shock hazards. It is wise to make sure that the injector unit is installed by a competent service representative, to ensure not only proper operation of the unit but also electrical and mechanical safety.

The use of electronic instruments, electrical appliances, and x-ray equipment in hospitals creates electrical shock hazards that can be subtle, if not entirely unexpected. Severe or mild shocks to the patient may occur even though equipment appears to be grounded, due to the fact that ground connections may have a high resistance or may be broken, allowing leakage or fault potentials, which normally are shorted to ground, to be present on the external surfaces of equipment.

Electrically sensitive patients may be found in special procedures rooms, intensive care units, angiographic laboratories, and other critical care areas. However, they can also appear in emergency rooms or surgical recovery rooms. These patients may have electrical conductors inserted into the heart or near the heart and thus no longer have a high resistance to external skin contact; the current path is now reduced to a much lower resistance. In this case, a pressure transducer or a catheter filled with a conductive solution, a pacemaker, or an electrocardiogram (ECG) lead can produce a large current flow at the heart sufficient to cause fibrillation. As little as 20 microamperes (μamp) can cause fibrillation in dogs; therefore, as a conservative limit, alternating currents passing directly to the human heart should not exceed 10 μamp [1]. This level is considered by medical authorities to be a safe one for electrically sensitive patients.

If the electrical equipment near the patient has high leakage potentials, it can introduce microampere currents directly to the heart. Moreover, because of the milliampere threshold of perception, an attendant can unknowingly be the electrical connection for these currents.

The conventional grounding of x-ray equipment may not be adequate to lower leakage potentials to safe levels. A #20 AWG grounding lead may be sufficient in one case and not in others. Inspections have found a water pipe to be the reference ground at some institutions, whereas the neutral lead and conduit may be the reference in others. Frequently, the ground connections in wall receptacles are found to be at different potentials with respect to each other and the other grounds in the same room. Although this condition may be safe and acceptable in some applications, it cannot be tolerated in special procedures rooms or other areas where electrically sensitive patients are present. However, it may be impractical to bring the leakage current on every piece of equipment down to zero. Reducing the leakage to zero is not necessary as long as each unit is at the same potential so that no current can flow between the surfaces of different pieces of equipment.

A ground system that allows all the equipment in a room to assume the same potential is called an *equipotential ground*. It is readily obtained by connecting all the equipment together through a common ground lead. *An equipotential ground system is recommended as an additional ground network and is not to be considered as a replacement for the conventional third or fourth wire to ground provided with the equipment.*

The power distribution system in special procedure rooms or any radiology room must employ a separate grounding lead that is distinct from the grounded neutral lead. This separate grounding lead (green) is to be connected to the building structural frame, the ground rod reference system, or the *cold* water plumbing. The grounded neutral lead (white) is a current-carrying conductor, which is not to be used as the grounding conductor. Although the neutral lead is grounded at the distribution transformer, the green lead must not be connected to the white lead as a short cut to ground. The neutral lead should not be grounded in the room and should not be connected to the equipotential system. The same principle applies to a three-phase system: The neutral lead should not be used as a grounding lead. A separate, non–

current-carrying conductor should always be used to connect equipment to ground.[1]

In reality, there is no one method of ensuring patient safety, but equipotential grounding coupled with equipment inspection and maintenance assures a high level of safety. The preceding information has been presented only to emphasize the importance of proper installation and maintenance of equipment. It is suggested that the manufacturer's instructions be understood and followed carefully before any patient examination is undertaken.

Components of Angiographic Injectors

Each of the injectors reviewed in this chapter has a basic configuration or model to which the purchaser can add or delete modules to achieve the types of features and versatility that may be required in a particular radiology department. Each injector basically consists of the following components:

1. Electromechanical power ram
2. Operator controls
3. Safety circuitry devices
4. Mobile stand

Electromechanical Power Ram

The power ram, which is the heart of the injector, is simply a motor-driven screw that translates electrical energy into mechanical energy in the form of a push or pull. The motor-driven ram in turn pushes or pulls the plunger of an injection syringe to achieve injection or filling according to the operator's wishes. The Medrad Mark IV[2] and the Angiomat injectors have removable power rams that make it convenient to place the syringe near the patient with the control panel and pedestal at a remote site.

Operator Controls

Control over the motor-driven ram is determined by circuitry and switches on the operator's control panel. "Start," "stop," "forward," and "reverse" translate to "inject," "fill," and "stop" relative to the power ram. Additional circuitry refines the operator's control over motor speed to include pressure, volume per second, and rate rise. Rate rise is the acceleration of the power ram per unit of time (second). This

[1]Refer to National Electrical Code 1975, article 517 for grounding requirements in electrically susceptible patient areas.
[2]Medrad, Inc. Pittsburgh, Pa.

factor is variable and permits the operator to control catheter whip and to maintain finer control over the injection.

Since it takes a specified time for the injector motor to reach nominal speed, it is possible to control this speed, which in turn affords the operator more precise control over the bolus of contrast material as it is injected. With this control, the operator can reduce the whipping effect of the catheter and still maintain sufficient injection pressure to ensure proper radiographic contrast.

Safety Circuitry Devices

The control panel also contains automatic circuitry for detecting an over-rate injection by sensing the motor speed and comparing it with the pressure and volume per second as set by the operator. This circuitry will terminate the injection and indicate a malfunction to the operator. If a malfunction occurs, a service representative should be summoned immediately, and use of the injector should cease until repairs can be made.

Mobile Stand

As the name implies, the mobile stand provides free movement of the injector. This is accomplished by using casters. Most angiographic injectors have provisions for raising or lowering the height of the injector, a useful feature—particularly when angiographic magnification is being done.

Additional Components and Controls

Syringe Heater. Since all injections are ultimately affected by the contrast medium and its viscosity, it is worthwhile to consider the syringe heater. The contrast medium is generally taken for granted because we have no direct control over its manufacture, and it is delivered in convenient, sterile bottles, ready for use. One of the factors affecting the viscosity of the medium is its temperature. The material to be injected should be at body temperature to provide an ideal injection. If the pressure tolerances of the catheter are exceeded, the viscosity of contrast media is increased when it is chilled, and the catheter may rupture when the injector is called on to provide a specific rate of injection if the pressure tolerances of the catheter are exceeded. All the injectors reviewed in this chapter have syringe heaters. The heater of the Cordis injector[3] is built into the syringe holder. The Mark IV and Angiomat injectors have flexible, detachable heater blankets that clip directly to

[3]Cordis Corporation, Miami, Fla.

the syringe or the syringe pressure jacket. All the syringe heaters are automatically controlled and are preset to maintain the syringe at 38° C. The heaters are not designed to raise the cold contrast medium to this temperature but to maintain the temperature of the syringe at 38° C. The contrast medium should be placed in a warm-water bath before the syringe is filled.

Trigger Control. Injection triggers are provided on the main operator's control as well as at a remote control device. If the remote trigger is not provided in the original purchase, it would be a wise investment because it allows the operator to set the injection parameters on the operator's console and to control the injection from a radiation-protected area. Injection triggers are of the "dead-man" type, meaning that when the "inject" button is pressed and held, the injection will be initiated and will continue until all of the contrast material has been injected, unless the button is released. If the button is released, the injection will be stopped. This beneficial safety feature provides the operator instant control in case of malfunction of the equipment or motion by the patient during the injection. If the injector is interfaced with a film-changer/programmer that controls the injection, the dead-man feature may not be applicable.

Injection Volume Control. The injection volume control permits selection of the volume of each injection. The injection volume can be preset for a small injection of 2 milliliters (ml) to check catheter placement. Additional 2-ml injections can then be made at will up to the capacity of the syringe, or the volume control can be reset to the volume required for the actual injection.

Interfacing. The injectors reviewed in this chapter can be interfaced with film changers, x-ray generators, and cinecameras or spot-film cameras to permit automatic control of all events in an injection program. Before purchasing an injector, the user should know the type of equipment with which it is to be used and also how it is to be used with other equipment, such as film changers. Each manufacturer of injectors has specific kits available that will permit interfacing of the injector to the x-ray apparatus. Generally, it is desirable to interconnect the injector with a film changer/programmer and x-ray generator in order to synchronize all the injection events; in this way one can obtain the best diagnostic results and reduce the human error factor to a minimum.

Delay Control. The injectors reviewed in this chapter also have a delay control, which permits the operator to select either an x-ray delay or an injection delay. If the operator chooses an injection delay, the x-ray film changer will start the x-ray exposures and film changing operation when the injector hand switch is pressed; the injection will not take place until the delay time, as set by the operator, has elapsed. When the x-ray delay mode is selected, the injection will occur when the hand switch is pressed. The film changer will initiate x-ray exposure after the delay time, as set on the injector control, has elapsed.

General Operation Mode. An injector can have a single mode or a combination of operational modes. These modes include *fill*, *standby*, and/ or *inject*. When the *fill mode* is selected, the power head and the syringe can be operated with the appropriate push button to fill the syringe with contrast medium. When an injector is in this mode, an injection cannot be made even if the remote hand control is activated. In those units with *standby modes*, the power ram cannot be operated for filling the syringe. When the *inject*, or *armed*, *mode* is selected, the injector is ready for an injection and will deliver the contrast medium when the remote control hand switch, the foot switch, or the auxiliary start control is activated. The Medrad Mark IV is constructed so that it is in either an armed or a disarmed mode. When the injector is armed, one can load or inject; if one starts to load, it disarms. When the injector is disarmed, one can only load, not inject.

Mechanical Stop. The mechanical stop is really a safety device and should be set for each injection. It is used, as its name implies, to limit the travel of the power ram to the desired volume of the injection. The mechanical stop should be set to the maximum volume of contrast material that can be tolerated by the patient or to 1–2 ml more than the desired volume to be delivered.

Remote Control. A remote control hand switch allows the operator to initiate the injection from a distance, out of the immediate area of radiation exposure. Full control of the injection is always possible because the hand switch is a dead-man type. When the button is pressed, the injection starts, and when the button is released, the injection is stopped. It should be pointed out, however, that when the injector is controlled by a programmer, there is no control over the injection once it is initiated.

Removable Power Ram. Two of the injectors have a removable power ram/syringe assembly. In these units, the power ram is placed on an adjustable arm remote from the control panel, which permits more flexibility when the operator is positioning the syringe near the patient. The power ram/syringe assembly is also removable from the arm and can be placed in a bracket on the x-ray table or hand held, if desired, for better approximation to the patient. The Medrad Mark IV has an optional mobile pedestal for the power ram/syringe assembly, making the assembly separate from the injector console for increased versatility. The Cordis injector has the power ram/syringe and the control circuitry as an integral unit. Although this is a satisfactory arrangement for most applications, it does not provide the positioning versatility of the Medrad or Angiomat injectors, and when the syringe is angled, the operator controls also angulate accordingly.

Disposable versus Nondisposable Syringes

All of the injectors reviewed in this chapter are made to accept either disposable or nondisposable syringes. The disposable syringe, which is made of plastic to facilitate removal of air bubbles, is quite similar in appearance to the nondisposable one; however, the similarity ends at this point. The nondisposable syringe requires meticulous cleaning and sterilization by gas or chemical means, and in addition, it requires a certain amount of maintenance. The syringe must be handled carefully so that it will not be damaged. The "O" rings on the plunger should be inspected and replaced regularly. The syringe must also be lubricated, as directed in the instruction manual, prior to sterilization. In addition to the syringe, other connecting parts associated with the injector power ram may need sterilization prior to operation.

The disposable syringe eliminates all the cleaning and maintenance steps outlined above. Each syringe comes complete in its own sterile, nonpyrogenic package. The disposable syringe is used inside a pressure jacket made of heavy, transparent plastic; this jacket keeps the syringe from bursting during an injection. A supply of disposable syringes should be kept on hand in the department ready for immediate use, because it is never wise to clean and reuse a disposable product. Reuse can result in failure due to structural damage to the syringe that has occurred during cleaning and resterilization.

Commercially Available Injectors

The Cordis Injector

The Cordis flow rate injector is a compact, mobile unit with all the features described in the previous sections. The unit is shipped with a supply of 100-ml disposable syringes and a clear plastic pressure jacket for their use. Also part of the basic unit is the rate/pressure computer; a serial programmer, consisting of 20 modules; a remote relay box; a ground cable; a foot switch for scout and programmed injections; a remote hand-triggering cable; and a dust cover. The Cordis injector can be supplied for use on 120 volts (v)/60 hertz (Hz), 220–240 v/50 Hz, or 100 v/50 Hz electrical supplies. An optional ECG programmer (20-module) is also available for addition to this unit.

The rate/pressure computer is used before an injection to predict the injection flow rate at the desired pressure or the injection pressure required to achieve the selected rate. The pressure limit is computed when catheter specifications such as length and internal lumen diameter, viscosity settings of the contrast medium, and the desired flow rate are entered by positioning the appropriate knobs on the computer. If the catheter specifications and flow rate combination desired require a pressure greater than that selected, an amber lamp flashes to warn the operator that the desired flow rate will not be achieved, although the pressure delivered will be equal to the pressure selected. The operator then must choose a higher pressure limit or a lower flow rate, within the capabilities of the injector.

Pressure injections can be made by selecting a flow rate that requires a pressure greater than the desired pressure, as indicated by the flashing amber lamp. The injection will be delivered at the selected pressure limits. Also supplied with the Cordis injector as part of the operator's manual is a set of nomograms for determining flow rates for applicable contrast media.

An optional ECG programmer can be installed in the control panel to synchronize the angiographic injection with the R wave from the cardiac trace, after a preselected delay of 20–1000 ms. There is also a provision for single or multiple injections. An adjustable gain control adjusts the ECG input in such a way that the R wave triggers the injection and distinguishes the R from the P and T waves. If multiple injections are selected, injections will occur with each succeeding R wave after the injection is initiated and will continue until the syringe is empty or until the trigger switch is released. If the ECG trigger is not to be used, normal injector operation can be regained by switching the module to "Manual."

The Angiomat 3000 Injector

Standard Features. The standard Angiomat 3000 flow rate injector[4] consists of a control cabinet with a removable power ram assembly mounted on a mobile pedestal base. The control panel contains a volume/flow rate selector, standby/ready switches, an x-ray inject selector switch with a delay selector, and a power switch. The power ram (syringe head) assembly has separate load, fast, and unload switches, mechanical stop adjustment knob, volume available and adjustable stop scales, manual load/unload knob, syringe locking mechanism, syringe pressure sleeve, disposable 100-ml syringe, and syringe heater. The radiographic film changer and other connection cords are stored on cable retractors to facilitate cord handling.

---

[4]Liebel-Flarsheim, Division of Sybron Corporation, Cincinnati, Ohio.

In addition to the normal over-rate monitor, the Angiomat 3000 has a secondary backup system that takes over control if the primary system malfunctions. In this case the injection will continue until completed at the proper flow rate. An indicator light on the control console will then flash to indicate the malfunction.

The Angiomat 3000 has a well-designed and easy-to-operate console arrangement with illuminated, color-coded, push-button switches and digital selector switches for accurate selection of flow rate, volume, and x-ray/injector delay. The combination fill/standby/ready switches are not only illuminated in color but are connected in such a way as to prevent accidental misfiring, because only one mode can be engaged at a time. The Angiomat automatically switches to the standby mode when the power is turned on, and then the equipment is energized to reduce accidental triggering. A transparent plastic shield is placed over the control to prevent contamination of the electrical components by spilled liquids.

The mechanical stop is set with a convenient knob and can be adjusted readily anywhere between 1 and 100 ml. A moving pointer provides quick readout of the volume remaining as well as the mechanical stop indication.

The power ram is removable for versatility in placement near the patient or for attachment to the x-ray table or patient rotator. Load and unload switches are grouped on the power ram and include a "fast" selection to increase the speed of the power ram before filling or after injection for rapid removal of the syringe.

The syringe heater is an automatically controlled detachable blanket type that clips directly onto the syringe pressure jacket. A lamp on the power ram housing glows to indicate the "heater-on" cycle.

A ready light on the power ram flashes in unison with the ready light on the main control to indicate when all injection parameters have been selected. There is also a manual control knob to permit precise movements of the syringe plunger for evacuation of air, thus minimizing the possibility of air entrapment during catheter hookup. This manual control knob can also be used to check for a trapped air bubble after hookup by drawing a small amount of blood.

Optional Features. For increased versatility, one or all of the options discussed below can be incorporated into the Angiomat injector.

The *pressure limit control* permits pressure-controlled injections. A rotary selector is used to select the maximum pressure desired, in 100-ml increments up to the maximum pressure set by the operator. The level of maximum pressure to be generated can be preselected to permit the use of low-pressure catheters.

The *rate rise control* (seconds) permits selection of the acceleration rate of the power ram syringe plunger to the selected flow rate. Selections are from "normal" or "2.0 sec," in order to reduce catheter whip and to permit the catheter to stay in the selected vessel during the injection. This option is especially useful during selective studies.

The *ECG trigger module* is used to synchronize the angiographic injection with the R wave from a cardiac trace, after a preselected delay. This module can be switched off and used in the monitor mode with the trace displayed on an oscilloscope (optional extra) with no injection, or it can be used to synchronize the injection with the R wave after a preselected delay. Other features of this module include an R-wave indicator lamp that flashes when the ECG module receives a signal from the patient that is sufficient to trigger the injector; a sensitivity control, lead selector switch that will accept up to three leads directly from the patient or from additional monitoring equipment on the rear panel of the Angiomat; and a variable inject/delay control with delays from 25 ms to 1 sec. Immediately beneath the R-wave indicator is a representative cardiac cycle waveform. A negative spike represents the start of injection. Injection delay is represented by a space of time between the R wave and the "Start" marker (negative).

Another available option is the *multiple-injection ECG trigger.* This option, which can be chosen in place of the single-injection module, is used to start and stop the injection in synchronism with the R wave after a preselected delay and also to provide multiple injections during one to nine consecutive cardiac cycles. In the repetitive R-wave termination position, the injection is terminated by the R wave if the heart rate is increased and the succeeding R wave occurs before the stop marker. An optional monitoring oscilloscope is available for displaying the cardiac trace. This unit is useful because the operator can monitor the patient's cardiac rhythm while selecting the beginning and terminating points in the cycle. A beeper with volume control provides audio indication of the R-wave spike.

The *dual rate control module* provides a dual flow rate control for filling the peripheral arterial tree at a slow rate and then automatically switching to a higher rate for retrograde filling of the iliac and aortic vessels.

The Medrad Mark IV Angiographic Injector

The Medrad[5] Mark IV flow rate injector is one of the most versatile and sophisticated injectors available commercially.[6] In addition to all the functions described for the other injectors, the Medrad injector features

---

[5]An injection pressure of 600 pounds per square inch (psi) cannot be used with the 65-ml syringe. Medrad, Inc., Pittsburg, Pa.
[6]Because this injector can be used with syringes of 65 or 130 ml capacity, care must be exercised when the 65-ml syringe is being used, since flow rate and pressure are calibrated for the 130-ml syringe. Actual flow rate with the 65-ml syringe will be one-half the set flow rate. Actual injection pressure is twice the indicated pressure.

automatic pressure monitoring and extended flow rate capacities to permit lymphangiography and venography.

The standard Mark IV flow rate injector consists of a streamlined control console with a movable and removable power ram/syringe assembly, all mounted on a mobile pedestal base. The remote control trigger switch and accessory cables are mounted on retractable cord reels. Digital push-button switches control volume, flow, and delay time settings.

A unique "volume set too high" indicator light on the control console warns that insufficient contrast medium remains in the syringe to complete the desired injection and disarms the injector.

A system monitor guards against over-rate and over-volume injections. The system monitor circuit constantly guards against discontinuity between the control console and the power ram/syringe and disarms the injector until corrections are made. During an injection, the system monitor compares the selected flow rate to the actual injection rate. If the injection rate exceeds the selected rate by a significant amount, the system monitor will light and the injector will be disarmed. The injector can be rearmed, but if the problem persists the injector will disarm repeatedly until the problem is corrected.

A pressure monitor indicates the maximum pressure obtained during an injection. Pressure limiting can be selected by push button up to 1200 psi. Two "arm" buttons are provided, for single or multiple injections. The single-injection button arms the injector for one single-volume injection or one complete ECG program, and the injector will then disarm. In the multiple-injection mode, the injector will not disarm after delivery. Several volume-limited injections are possible up to the capacity of the syringe.

All the following steps must be completed before the injector can be armed for an injection:

1. One of the flow scale buttons must be pressed.
2. One of the pressure limit/monitor buttons must be pressed.
3. One of the delay timer control buttons must be pressed.
4. Loading must be completed (not in progress).
5. The system monitor light must be extinguished.
6. The syringe plunger must not be at the forward limit of the syringe.
7. The mechanical stop must be locked.
8. The double-syringe assembly must be in "delivery" position.

An automatic disarm circuit disables the injector if the above injection parameters are not set.

A linear acceleration control permits selection of injection rise times from 50 ms to 9.9 sec to control catheter whip and improve cardiac synchronization. A universal flow module permits precise control of flow rates to permit biphasic aortic bifurcation and backflow studies, allowing the operator to select two flow rates and two acceleration rates. A delivered flow rate indicator indicates the flow rate in milliliters per second that was achieved during the injection. The digital injection timer shows the actual injection duration after each injection. The times range from 0 to 99.9 sec in the milliliters-per-second flow mode, and 0 to 99.9 minutes in the milliliters-per-minute and milliliters-per-hour modes.

ECG Trigger Modules I, II, and III. Three optional ECG trigger modules are available with the Medrad Mark IV injector. Module I synchronizes a single injection to the R wave with a preset delay up to 0.8 sec. Module II synchronizes single and repetitive injections to the R wave and can be set to terminate the injection during some portion of the ECG. Up to nine repetitive injections can be made. When repeat injections are selected, the program can be set to stop the injections as each successive R wave occurs. Additional features of Module II include the following:

1. A beep switch indicates each R wave audibly.
2. A repeat-injection switch sets the number of R waves that will pass before injection termination.
3. An R-wave termination switch determines whether the injection will stop in the R wave or be delayed past the R wave, as set on the injection delay control.
4. The R–R interval displays the time between R waves (up to 9.99 sec) and resets with each R wave for an instantaneous reading.

Module III has all the features of Modules I and II plus an injection profile programmer. Optional Modules I and II use the flow pattern set on the flow module; Module III allows either standard flow module settings or the eight profile controls. When the profile-type injection is selected, each R wave will trigger an injection. Every 0.1 sec, another control will determine flow rate from zero to 40 ml per second. If all eight controls are to be used, the R-R interval must be at least 0.8 sec with injection delay at zero. Shorter R-R intervals will restrict the number of profile controls that may be used.

Another feature of Module III is the program preview button. When this button is pressed, the flow pattern is displayed on the optional Mark IV oscilloscope module and the remote monitor/recorder, if incorporated. This feature permits the operator to preview a complete injection profile before an injection takes place and to make corrections at this time. When the program preview button is pressed, the injector is disarmed and no injection can take place.

Electrical Hazard Monitor Alarm. The electrical hazard alarm is a safety option that can be used to check the leakage potential of any object within a 15-foot radius of the injector. An alarm will sound if the leakage potential as measured exceeds a preset level.

Pulsed ECG Film Changer Delay. The ECG film changer synchronizes the film changer triggering with the injection each time an R wave is received to ensure that the injection will appear on the film. The pulse is delayed according to the time set on the digital switches; the switch closure can be set from 0.5 sec up to 1 sec.

Oscilloscope Module. The oscilloscope module features three sweep speeds. It monitors and stores the patient's ECG, injection flow/volume, or other patient variables. Additional features of the oscilloscope include the following:

1. Channel 1 input normally displays the patient's ECG, but it can also be used for auxiliary inputs. The sweep will be synchronized to the R wave.
2. The module has 1-millivolt (mV) calibration. The amplitude of the patient's signal can be determined by comparing it with the 1-mV signal. This capability is useful for evaluating electrode placement.
3. Channel 2 input normally displays the injection profile, but a switch allows other inputs to be fed into this channel in place of the injection profile. The injection flow/volume will be plotted as an injection takes place and will be stored and viewed after the termination of the injection.
4. The module has 20-ml calibration for determining actual values of the flow and volume, and gain controls to determine the amplitude on the scope graticule. Each channel has its own position control.

Three operational modes are possible with the oscilloscope:

1. Automatic: The sweep begins when the start switch is pressed and stops 0.25 sec after completion of the injection program. The injection begins on the second R wave.
2. Stop: The scope will store the most recent information but will not run when an injection takes place.
3. Run: The scope will continue sweeping regardless of injection triggering.

Brightness, focus, and sweep speed are also included to maximize the scope display.

The oscilloscope also has a nonfading storage display, which operates in a manner similar to a chart recorder in that information passes across the screen to the left and new information is added simultaneously. The information is stored for review at the end of the injection profile. There is no need for a chart recorder unless a permanent record is deemed necessary.

Reference
1. Bigger, J. T., Jr. Clinical Exposure of Patients to Electrical Hazards. In *Electrical Hazards in Hospitals*, Part 2, Section B. Washington, D.C.: National Academy of Sciences, 1970.

Suggested Reading
Fischer, H. W. Discussion on injectors for use in angiography. *Modern Concepts of Cardiovascular Disease*, May 1968.

Shea, F. J. Development and operation of the angiographic injector. *Radiologic Technology* 42:23, 1970.

7

Serial Film Changers

It is not within the scope of this book to trace the evolution of film changers and their applications to radiology. Amplatz [1] gave a historical account of the development of film changers and included a good description of their mechanics. For the reader who is interested in the forerunners of the present devices, Scott's Carmen Lecture [2] is a chronicle not only of the development of film changers but also of the history of angiography.

The selection of a film changer, like that of any other piece of radiographic equipment, is based on many considerations. The intent of this chapter is to present a description of the basic characteristics of a film changer and to describe some various types that are commercially available. Each type of film changer has unique electrical or mechanical features designed to meet specific applications.

General Considerations

Simply stated, a rapid-sequence film changer is a device that moves a film from a stored location within the device into an exposure area and, after the exposure is completed, into another area of the device where it is stored and protected against additional radiation exposure until it is removed. In all film changers the speed at which the film or cassette is moved from storage area to exposure area to receiving area is held constant for the exposure or film transport cycle. This means that once an exposure is initiated, the film or cassette will be transported throughout that cycle at a constant speed. The speed at which the film is transported is determined by the selection of the exposure rate (i.e., the number of exposures per unit time). The higher the exposure rate, the faster the film or cassette will be transported, and correspondingly, the shorter the amount of time there will be available to make an exposure. Other factors in the film transport cycle that affect the time available for exposure will be discussed later in the chapter, under Radiographic Considerations.

The very early film changers were hand operated and required an operator to exchange cassettes manually between exposures. A well-trained team could make exposures about every 2 seconds (sec), and thus angiographers were quite limited in the examinations they could perform, particularly in the cardiac and pulmonary areas. In order to meet the demands for higher filming rates, film changers were motorized. Today they are complex devices that allow exposures in the range of 12 exposures per second while preserving all of the geometrics affecting film quality.

Modern film changers are installed with devices for programming the exposures that will be made during the angiographic injection. The operator is able to select how many exposures will be made per unit time, to program a delay between a series of exposures, and even to trigger an exposure by a physiological event, such as diastole or systole during cardioventriculography. Thus, one may choose an exposure rate of six exposures per second and program three series of exposures: three exposures during the first second, followed by a delay of 2 sec; four exposures during the next second, followed by another pause; and then a third series at a different exposure rate, say one exposure every other second for 5 sec. The newer film changers are extremely versatile, and the operator must be aware of the physiological events that occur so that the programmer may be accurately set in order to record these events correctly.

Older programmers were usually wall-mounted devices containing a number of selector switches that were interconnected with clock motors to establish the desired sequence of events. Modern programmers use solid-state electronics controlled by either punched holes or mark sensors on an IBM-type card.

Types of Film Changers

Rapid film changers can be classified functionally as cassette units and cassetteless units.

Cassette Units

Cassette units generally have a storage compartment for unexposed film holders. The film holders, which may be either conventional rigid cassettes or plastic pouches, are moved to the "ready" position by springs. On exposure command, the cassettes are powered into the exposure compartment, and after exposure, into the receiving compartment. The Picker[1] CP-420 and CP-500 and the Sanchez-Perez[2] Universal Automatic Seriograph are current examples of commercially available cassette-type film changers. Cassette devices have a slower exposure rate primarily because of the weight of a casette. Increasing the exposure rate beyond four exposures per second requires a complicated design for vibration dampening.

Cassetteless Units

Cassetteless film changers transfer film either as individual sheets (cut film) or as a strip from a roll of film. Because of the reduced mass of

[1]Picker Medical Systems, Cleveland, Ohio.
[2]Automatic Serialograph, Division of Litton Industries, Des Plaines, Ill.

the film, cassetteless film changers transport films faster and consequently allow a greater exposure rate. Prior to the use of antistatic coatings on intensifying screens, cut-film changers had a tendency to jam. Cut films, unless loaded directly from the original package, curl when stored on end in an opened package, particularly if ambient humidity is high. Curled film can cause havoc in a cut-film changer. Needless to say, cut film should not be stored in the film changer.

Roll-film transport systems do not require the intricate mechanisms that are used to pick up and transport sheet films and are inherently more reliable. Cut-film magazines must be loaded in a darkroom, whereas roll films are supplied in cartridges that permit daylight loading of the magazines. During an examination that requires more exposures than can be made with a single magazine, there is less delay due to reloading with roll film unless, of course, extra magazines for cut film are available and preloaded.

Except for systems that provide sequential numbering of exposed films, sorting and arranging the films in proper sequence for viewing is necessary when a cut-film changer is used. In the case of roll film, sequential viewing of exposures is inherent. However, even though there is the possibility of losing an important single film in a series when cut film is used, the films can be filed in standard envelopes. In the case of roll film, individual frames must be cut to permit filing in standard envelopes. By taping the films in sequence before filing, one can make subsequent viewing of the examination series less troublesome. Processing is straightforward for cut film and more troublesome for roll film, although automatic processors have made the processing of roll film less awkward. Ideally, the automatic processor should use the receiving magazine from the roll-film changer as a feed magazine to the processor, with a spool take-up at the processor exit. The method of having one darkroom technician to feed the rolled film and another technician to collect the processed film is inefficient.

Two methods have been applied in the design of roll-film changers for handling the film. The first method uses a roll-film feed and a roll-film take-up magazine; this method provides the inherent reliability of a simple transport system plus ease of loading but requires added effort for processing and filing. The second method uses a roll-film feed with each exposure cut from the roll as it enters a receiving magazine; this method retains the advantages of reliability and ease of loading, and at the same time has the advantages of efficient processing and ease of filing in standard envelopes. However, depending on atmospheric conditions, and particularly when the film supply is near the end of the roll, the cut sections may form into severe curls, which, besides being awkward to handle, require extra care when they are fed into the processor.

The Siemens-Elema Puck,[3] the AOT-S,[4] and the Xonics film changer[5] are current examples of commercial devices of the cassetteless, cut-film type of film changers. The Franklin[6] variable-speed film changer is a current example of a "roll-film in–roll-film out" type of film changer. The CGR 8-15-35[7] is a current example of a "roll-film in–cut-film out" type of film changer.

Peripheral Changers

Film changers for peripheral angiography are not considered to be rapid-sequence devices. Many changers for peripheral vascular studies, using cassettes and films of special lengths, have been built and have had limited commercial success, like other equipment peculiar to the needs of radiology. With floor space in most radiology departments at a premium, the advantage of using large film in a peripheral examination was waived in favor of the conventional serial changer with its smaller film size. To use the conventional film changers, special procedures tables were designed to move the table top longitudinally in programmed increments at fixed speeds. Generators were designed to adjust kilovoltage according to the thickness of the patient at each increment of shift.

The early peripheral changers were designed for 36-inch cassettes and films, since those accessories were readily available. A current changer uses 14 × 51–inch cassettes, which are fitted with three pairs of intensifying screens of different speeds and three sheets of 14 × 17–inch film. The different screen speeds are required to compensate for the difference in tissue thickness at the lower abdomen and at the distal extremity. (DuPont markets a tapered screen for this purpose.) Precut film in the 14 × 51–inch size is also available from some manufacturers. Stationary grids are built into the supporting framework of the changer. The cassettes in the early units were stacked in the exposure area, with sheets of lead between the cassettes. At the end of each exposure the cassette was shifted into a protected area of the changer by counterweights. The object-film distance increased with each successive exposure. However, because of the long focal-film distance required to expose the long films, magnification between successive radiographs was not offensive. The radiographic definition was good, again because the long focal-film distance offset the effect of penumbra of the focal spot.

The current device for use with 51-inch cassettes moves the cassettes to the same plane for successive exposures. Because of the mass of the

[3]Siemens-Elema AB, Solna, Sweden.
[4]Siemens-Elema AB, Solna, Sweden.
[5]Xonics Corporation, Van Nuys, Calif.
[6]Liebel-Flarsheim, Division of Sybron Corporation, Cincinnati, Ohio.
[7]Compagnie Générale de Radiologie (CGR), Paris, France.

cassettes, the exposure rate is limited to one per 1.5 sec, with delays between exposures programmed over a range from 0 to 99 sec.

The BCM-500 and BCM-700[8] and the Siemens[9] motorized cassette changer are examples of commercially available cassette-type peripheral changers. Roll-film devices for peripheral arteriography are no longer commercially available.

The advantage of using large film (as opposed to the method of shifting the patient sequentially over a conventional rapid-film changer) lies in the ability to capture the contrast medium from the aortic bifurcation to the popliteal level without the need for a second injection, which may be necessary with small-format film changers if the hemodynamics have not been previously ascertained. During bilateral femoral arteriograms, the large film provides comparison of the blood flow in both lower extremities on a given time base. On the other hand, the rapid sequencing rate of the conventional film changer, when used with a shifting top, offers the possibility of studying blood vessel changes during the excursion of the contrast medium.

The x-ray tube requires a special cone, particularly if one wants to take advantage of the heel effect of the x-ray tube anode. A wedge filter is available when the units are not provided with intensifying screens with compensating speeds. The filter, while providing a uniform film density, produces a long-contrast scale, which is not always desirable in an angiographic study.

A focal-film distance of 76 inches is required for a 51-inch cassette. Conventional ceiling-mounted x-ray tube stands in typical radiology rooms are generally not able to provide sufficient focal-film distance to expose radiographically the entire film length used in peripheral angiographic film changers. Specially constructed mounting devices are required to mount the x-ray tube in a fixed location in the ceiling.

Although they are not in the generally accepted category of a rapid-sequence film changer, 105-millimeter (mm) and 100-mm photofluorographic cameras are indeed rapid-film sequencing devices (see Chapter 4). Current image intensifiers are capable of 3–4 line pairs per millimeter of resolution on the output phosphor of a 9-inch image intensifier. By comparison, a conventional nonscreen radiograph made with a 1.2-mm focal spot also presents approximately 3–4 line pairs per millimeter of resolution of an organ in the body. By appropriate modification of the table top and the fluoroscopic carriage movements, the entire torso and extremities can be recorded with such cameras at repetition rates up to 12 frames per second.

Cameras have not found much appeal in peripheral angiography, although they are occasionally used in peripheral venography. The field

[8]B. C. Medical Engineering, Ltd., Montreal, Canada.
[9]Siemens (Gorla-Siamia), Erlangen, West Germany.

coverage of image intensifiers does not permit bilateral exposures, and the image intensifier is too heavy and bulky to follow the excursion of the contrast medium, particularly as it approaches the distal areas of an extremity. In aortography the field coverage of the image intensifier does not permit one to record opacification in the peripheral branches of the arterial system without the contrast medium getting ahead of the imaging field. Roy [3] reported similar limitations with cineradiography in abdominal and peripheral angiography. On the other hand, Grollman et al. [4] found appeal in the rapid-sequencing capability of spot-film cameras in cardiac angiography. For cardiac angiography, the 9-inch image intensifier size covers an adequate area, and the capability of observing the catheter and the injection during filming offers appeal for this modality, particularly since imaging systems are constantly being improved in resolution capability. The introduction of the large field-of-view image intensifiers will probably change this concept, especially as digital radiography is further developed.

The range-of-field size coverage for rapid-sequence changers varies with the manufacturer, from 10 × 12 inches (24 × 30 centimeters [cm]) to 14 × 14 inches (35 × 35 cm), as follows: The Sanchez-Perez Universal Automatic Seriograph, the Siemens-Elema Puck, and the AOT-S film changers can use 10 × 12–inch (24 × 30–cm) cut film or 14 × 14–inch (35 × 35–cm) cut film; the Picker CP-420 and CP-500 and the Memco changer[10] use 14 × 14–inch (35 × 35–cm) cut film; the CGR 8-15-35 changer uses a 14 × 14–inch area (35 × 35 cm) from a roll; and the Liebel-Flarsheim Franklin changer uses an 11 × 14–inch (28 × 35–cm) area from a roll.

The 14 × 14–inch (35 × 35–cm) size is the popular one for universal applications. The 10 × 12–inch (24 × 30–cm) size is generally used in cerebral angiography and cardioangiography; with care in positioning, this size can also be used for abdominal angiography on average-size patients. The risk of missing the periphery of the vascular bed in large patients is generally not favorable to the smaller format.

Radiographic Considerations

Aortography and peripheral angiography are almost always performed with the patient recumbent on a table top that is extended over the film changers, except in the case of the cassette-type long film devices. By necessity, the table top is relatively thick to provide the support required to allow minimum sag as the patient is cantilevered over the film changers. In order to position the patient without interference because of sag, even with the most rigid table tops, patient-film geometry

[10]Memco Inc., Des Plaines, Ill.

is sacrificed. It is important, therefore, that the film plane be as close as possible to the top surface of the film changer to take advantage of the available geometry. During cerebral angiography it is good practice to place the patient directly on the exposure area of the changer in order to optimize the patient-film geometry. For optimum contrast the exposure area should have the least possible amount of attenuation of the x-ray beam and should be sharply collimated. At the same time, adequate support of the patient should be provided to avoid interference with film transport.

In the case of long-film changers, the patient is supported by the changer top. However, because of the need to provide clearance between the table top and the changer, the patient-to-film distance (up to 3 inches) is of concern, by normal radiographic standards. As stated previously, the long focal-film distance required with these changers compensates for the relatively poor patient-to-film distance.

Determination of Film Exposure Times

The objective in radiography of the opacified vascular tree is to obtain the optimum contrast and maximum definition of the radiographic detail. Since kilovoltage is adjusted according to thickness of the body part (keeping in mind the absorption characteristics of the contrast medium), radiographic exposure time is dictated by the film changer. The minimum radiographic exposure time in angiography is dependent primarily on the generator capacity and the loading characteristics of the x-ray tube. The maximum exposure time depends on whether the film changer is a cassette type or a cassetteless type. Since a cassette film changer operates generally at a lower rate, it allows a relatively longer exposure time. A cassetteless film changer has more electrical and mechanical interlocking actions and generally operates at higher rates, which influences the maximum exposure time that can be used.

A film changer cycle is made up of a number of incidents that must occur before an exposure can be made, and the operator must be aware of these occurrences in order to obtain the maximum utilization of the equipment. We are concerned primarily with the *total film cycle time*, the *film transport time*, the *film stationary time*, the *maximum exposure time*, and the *maximum available exposure time*. The maximum time available for exposure is the total available time per film cycle minus the transport time, which includes the time required for compression of the screens (in the case of a cassetteless changer) or the vibration dampening time (in the case of cassette changers), minus inherent time delays in the x-ray generator and the available time selections on the generator console (Figure 7-1).

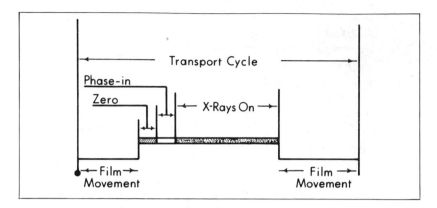

Fig. 7-1. Time graph showing x-rays–on time for a film changer. Although the transport cycle may be of considerable duration, zero and phase-in time must be subtracted, leaving little time available for the x-ray exposure.

Total Film Cycle Time

The total film cycle time is the time available for the film to be transported, the intensifying screens to open and close, and the time the film or cassette is stationary. To determine the total film cycle time, simply divide 1000 milliseconds (ms) by the exposure rate in seconds. For example, if a film changer is programmed at five exposures per second, 1000 ms divided by five films per second would allow 200 ms for the total film cycle. The total time available for transporting the film and for exposure depends on the maximum rate of exposures per second. At a higher rate, less time is available for exposure, and vice versa.

Film Transport Time

The film transport time is the time during which the screens are opened, film is removed from the exposure area and another film brought into place, and the screens are compressed again. As a general rule, the film transport time comprises approximately 60% of the total film cycle time. In the example just given in which the total film cycle time was 200 ms, 60% (or 120 ms) would be devoted to the film transport time, with the remaining 40% (or 80 ms) devoted to the film stationary time.

In a particular film changer, the speed of transport is fairly uniform. The time for moving the film into and out of the exposure area is constant, regardless of the exposure rate. The time available for exposure, on the other hand, depends on how long the film is stationary in the exposure area and on the rate of exposure.

In cassetteless-type film changers the intensifying screens are spaced away from the film surface during transfer between exposures. Therefore, the stationary time available for exposure must include the time needed to compress the screen against the film and, after the exposure is completed, to open the screen-film assembly for transferring the film to the receiving compartment. The time interval for this mechanical action is constant for a specific film changer and can be calibrated out during installation.

Film Stationary Time

The time during which the film is stationary determines the time in which an exposure can be made. During this period of time one must also take into consideration the inherent time delays of the film changer/generator interface and the technique factors available for selection on the generator console.

Generator Anticipation Time. Two time delays determine the time available for exposure in rapid-sequence film changers. The first time delay is the interval between the electrical command from the film changer to start the exposure and the conditioning time required by the generator to start the actual production of x-rays. The interval, which is inherent, is usually a constant value for a particular combination of x-ray generator and film changer and can be calibrated out by the installer. This time delay is known as *zero time* or *anticipation time*. In generators using relays and electromechanical contactors for energizing the x-ray transformer, the zero time would be an important percentage of the total available time for exposure if it could not be calibrated out in anticipation of the actual moment when x-rays are produced. In modern generators with solid-state circuitry, zero time is negligible.

Generator Interrogation Time. The second delay that affects the available exposure time is known as *phasing-in time*. In modern jargon this delay is also known as *interrogation time*, as noted in Chapter 2. This delay is the interval required by the generator to determine whether or not it is electrically ready to begin an exposure.

An x-ray transformer—or any transformer, for that matter—operates on the principle of mutual inductance, with alternating changes in polarity of a magnetic field at a rate whose frequency corresponds to the power line frequency. The polarity of the magnetic field must change in order for transformer action to occur. If the x-ray exposure is terminated while the magnetic field of the transformer is in a given polarity, a second exposure must not be made until the applied voltage has reversed its polarity and until the polarity of the magnetic field in the transformer core has also been reversed. In a 60-hertz (Hz) system, the voltage rises from zero to a peak and falls back to a zero value in 1/120 sec (8.33 ms). A change in polarity occurs when the current reaches a zero baseline on a sine curve. A complete cycle of alternating polarities occurs in 1/60 sec (16.7 ms).

For quite a number of years, x-ray generators have been supplied with exposure times that operate in synchronism with the electrical frequency of the main power lines. The timers in solid-state generators start and stop the exposure (unless the generator is equipped with a forced extinction device) at the zero points during the electrical cycle. Therefore, if the x-ray exposures were terminated at the end of 8.0 ms, a second exposure could begin almost instantaneously—theoretically 0.33 ms after the termination of the first exposure, assuming that a film could be transported into the exposure area in that time. However, if the first exposure were terminated at the end of 10.0 ms, a second exposure could not begin until 6.33 ms later, when a reverse in polarity of the power supply could begin at the zero baseline of the frequency curve.

In actual practice, a finite time is required to move the first exposed film out of the exposure area and to move a second film into the exposure area in a rapid-sequence film changer. In the example given above with an exposure time of 10.0 ms, if the time required to move a second film into the exposure area were also 10.0 ms, a second exposure would not take place in a single-phase x-ray generator until 23.3 ms later because the polarity of the power supply would be of the same sign as when the first exposure was terminated. Thus the second exposure could not begin until the phase-in of the current coincided with the required polarity for the transformer function. (The above example is simplified in order to explain the relationship between the requirement for proper polarity and the zero start point for exposure timing.)

A single-phase generator, unless it were equipped with a forced extinction device, would terminate the exposure at times that are multiples of 8.33 ms, for example, 1/120 sec (8.33 ms) or 1/60 sec (16.7 ms). For this reason, the longest interrogation time on a single-phase generator would be 16.7 ms.

In most installations, three-phase x-ray generators are used with rapid-sequence film changers. In a three-phase generator, the polarity in one of the phases changes every 2.77 ms from the phase ahead of it, in terms of time. At any given moment, two phases of the transformer have the same polarity, while the remaining phase has the opposite polarity.

All modern three-phase generators start and end exposures with solid-state switches, such as silicon-controlled rectifiers (SCRs). A characteristic of an SCR is that it cannot be turned on until the voltage across it is positive nor switched off until the current flowing through it reaches a zero value. Therefore, 2.77 ms are required to complete a circuit with a combination of two phases having the proper polarities to allow the transformer to function properly. Figure 7-2 illustrates these phase relationships.

The example given in Figure 7-2 is an elementary and simplified description of the phasing-in requirement of a three-phase generator. In actual practice, the production of usable radiation occurs about 1 ms later than the theoretical beginning of the exposure because of the rise time required for the voltage to build up across the x-ray tube.

Manufacturers have begun to produce generators that include transformer core memory circuitry to ensure that the exposure will not begin until the proper phase relationships have been established. In the case of the example in Figure 7-2, the kilovoltage across the x-ray tube would rise from zero without the usual spikes in the voltage wave that can occur when the exposure begins at a voltage greater than zero on the sine curve.

Another approach toward reducing the phase-in time is to start the exposure without regard to the sign of polarity of the transformer when the exposure begins. With such an arrangement, phase-in time can be as short as 1 ms. This method involves the use of a so-called *soft start* on the first impulse to protect the transformer. The soft start has a negligible effect so far as usable radiation is concerned.

The two time delays—zero time and interrogation time—occur after the film cassette has come to rest or after the screens are compressed, and they must be subtracted from the film stationary time. In a synchronous generator the interrogation time cannot be anticipated, and one must allow a full 16.6 ms for the interrogation time to occur. With a nonsynchronous generator the interrogation time can be anticipated and zeroed out during installation.

Fig. 7-2. Phase relationships in respect to film changers. Phase *A* begins at zero (*1*) when *A* is positive, but the exposure cannot begin until 2.77 ms later when *B* is at zero (*2*) and negative. A second exposure cannot begin until 5.54 ms later when *B* is at zero (*3*) and positive, and *A* is at zero (*4*) and negative. Without additional electronic circuits, the interval between exposures is 5.54 ms.

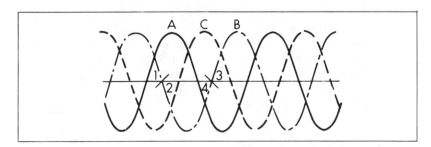

Maximum Permissible Exposure Time

Looking again at the example of five exposures per second, we note that the film stationary time is 40% of the total film cycle time, or 80 ms. From this 80 ms we must subtract 16.6 ms for interrogation time with a synchronous generator, leaving a total of 63 ms as the film exposure time. This figure, then, is the *maximum permissible exposure time.*

There are two other factors that must be considered in conjunction with the maximum permissible exposure time. The first of these has to do with *air dissipation time*, which is peculiar only to cassetteless film changers. As noted in Chapter 1, film cassettes will trap air and produce a transitory poor film-screen contact. When cassetteless film changer screens compress the film, they also compress a small amount of air, producing the same transitory poor film-screen contact. In older changers this time is computed to be about 16 ms, which again must be subtracted from the film stationary time. In the newer film changers, the screens have slight curves in them, with the result that this air dissipation time has been reduced to 2–4 ms.

The second factor to consider is the *maximum available exposure time.* If one needs 56 ms for an adequate exposure as calculated from the preceding data, the generator console may not be able to allow the operator to choose exactly 56 ms, and he must choose another station. In such cases the operator must choose the lower time station in order to prevent the screens from opening during the time of the exposure. The available time station then has the effect of again lowering the permissible exposure time.

The importance of calculating the proper exposure time for any film changer cannot be overemphasized. The quality of film obtained during an angiographic procedure is directly dependent on making an exposure exactly during the time when the cassette is motionless or the screens fully compressed.

Phototimer Applications

The problems involved with phototimed exposures made on rapid-sequence film changers have not been solved satisfactorily by most manufacturers. Efforts to grapple with attenuation averaging of a moving organ with varying opacity, as opposed to a relatively stationary organ with uniform tissue density, have produced an unappealing order of complexity in exposure control. The demand for repeatable short-time exposures, always within the stationary time limit of the film-changing cycle, did not ensure uniform exposures. In contrast to the possibility of calibrating out the zero start time of the generator, it has not been possible to anticipate random turn-off times in phototimers without comparator electronics.

Biplane Applications

Rapid-sequence film changers are available for biplane examinations, with two changers on individual stands or with a single changer on a stand that allows the changer to be tilted upright for lateral exposures. The more popular arrangement is with two changers on separate stands. Two exposures are made either simultaneously or alternately without disturbing the patient's position. Simultaneous exposures are desirable when congenital anomalies are being investigated. All current film changers, except the long film changers, can be programmed to synchronize the transport systems electrically for simultaneous or alternate biplane exposures.

Simultaneous biplane exposures require meticulous care in collimating the x-ray beam and orientation of the radiographic grid to avoid crossfogging the films with scattered radiation. The best results are obtained when the x-ray beams are collimated to expose only the area of interest and thus to reduce the amount of scatter. Some angiographers use profile masks at the x-ray tube and at the film side of the patient to confine the x-ray beam to its ultimate outer limits. Many installations make the radiographic exposures alternately to overcome the problem of crossfogging films where synchronous relationships of vessel dynamics are not predominant requirements. Crossfogging can be eliminated in biplane installations by alternate loading of films.

The so-called universal stands are feasible only when time relationships in opacification of the vessels can be forfeited and when the risk of a second injection of the contrast medium can be tolerated.

One of the important specifications for film changers in a biplane configuration is that the planes of the adjacent edges of the films in the two changers intersect. For mechanical reasons it is not possible to move the film in the vertical plane over the film which is in the horizontal plane, a feature that would be desirable in biplane cerebral angiography in nonmagnification techniques. Contoured table tops are required to optimize the patient-film geometry. With floating tabletops the need to reposition the patient between fluoroscopy and angiography is minimized.

Future Applications

With the rapid changes in the technology of image recording, the future for film-changing techniques is unpredictable. In just over a decade computed tomography has replaced a significant number of angiographic procedures. Real-time ultrasound and Doppler applications have had a similar impact. Larger input areas of image intensifiers with steadily improving resolution and electronic image enhancement and recording may eventually replace film changers. The increasing use of rare-earth intensifying screens will reduce, if not eliminate, the need for high capacity x-ray generators and high heat-loading x-ray tubes. In the near future developing technologies for transmitting and recording images of anatomical structures may totally replace all film recording systems. The x-ray beam as *the* source for producing images of anatomy and the conventional radiographic film as it exists today are, indeed, at the plateau of their evolution. Changes in the state of the art will affect the need for and the use of film changers.

References

1. Amplatz, K. Rapid Film Changers. In H. L. Abrams (Ed.), *Angiography* (3rd ed.). Boston: Little, Brown, 1983. Vol. 1, Chap. 6.
2. Scott, W. Carmen Lecture. *Radiology* 56:485, 1951.
3. Roy, P. Peripheral angiography in ischemic arterial diseases of the limbs. *Radiologic Clinics of North America* 5:3, 1967.
4. Grollman, J., Gray, R. K., Speigler, P., MacAlpin, R. N., and Olmsted, W. W. 105-mm serial photofluorography of the coronary arteries. *Radiology* 108:577, 1973.

Suggested Reading

Amplatz, K. A new rapid roll film changer. *Radiology* 90:130, 1968.
Fredzell, G., Lind, J., Ohlson, E., and Weglius, C. Direct serial roentgenography in two planes simultaneously at 0.08 second intervals. *American Journal of Roentgenology* 63:548, 1950.

8

Computed Tomography

Martin Trefler

In a short space of time following its introduction in 1972 [1], computed tomography (CT) became a premier diagnostic imaging modality, despite its formidable cost and the initial inconvenience of its use. The first scanners, produced by the British company EMI,[1] required 4 minutes to complete a scan and 5 minutes to reconstruct a single image slice consisting of an 80 × 80–square array of points. The patient was immobilized during the scan period with his head in a "water bag." Furthermore, the image sharpness was markedly inferior to that of radiographic techniques.

The reason for the immediate acceptance of CT can be appreciated when it is understood that in standard radiographic techniques, even large objects cannot be visualized unless they present a contrast difference of 10–15% with respect to their surroundings. In CT the *contrast resolution* is better than 0.25%—almost 100 times better than with a film-screen technique. This allows clear visualization of pathologic conditions that would otherwise require highly invasive methods.

Computed tomography involves the mathematical reconstruction, by a high-speed computer, of an image of a *tomographic plane*, or "slice," of the anatomy. The input data for the *reconstruction* process consists of a series of x-ray transmission measurements made in many different directions around the subject but all located in the tomographic plane of interest.

The mathematical techniques employed in the computer reconstruction process were developed in 1917 by Radon [2] and first applied by Bracewell [3] in the field of radioastronomy. The concept of medical reconstructions was introduced in the early 1960s by Oldendorf [4] and Kuhl and Edwards [5], but because the reconstruction techniques were slow and the radionuclide sources used were weak, the methods did not gain appreciable acceptance. It was not until 1967, when Hounsfield adapted x-ray sources and the high-speed modern computer to the problem, that a clinically useful system was developed [6].

Data Acquisition in CT Scanners

As previously mentioned, the data used in the reconstruction of a tomographic image, or slice, is a series of x-ray transmission measurements obtained in many different directions around the patient. In the

[1]EMI Ltd., Schoenberg House, Trevor Rd., Hayes, Middlesex UB3 1HH, England.

169

first clinical machines, a stationary anode x-ray tube was used as a source. This tube was collimated both in front of and behind the patient so that only a small, round beam entered the patient. The transmitted beam was incident upon a single NaI scintillation crystal–photomultiplier tube director system. This geometry virtually precludes any scattered radiation from being detected, and this scatter-free signal is one of the main reasons for the improved contrast resolution of CT. In this "first-generation" machine, an entire view was collected by moving the x-ray tube detector system in a straight line across the anatomical area of interest. The system was then rotated by 1 degree, and another traverse followed. To obtain the data for a single slice, 180 translate-rotate scans were required (Figure 8-1).

Within a short time of the introduction of these scanners, it was realized that the scan time could be reduced significantly if several detectors, positioned in a straight line in the plane of the tomographic slice, were used simultaneously (Figure 8-2). These second-generation machines allowed the rotation between traverses to be increased to

Fig. 8-1. First generation of CT scanner. X-ray tube/detector system traverses linearly across patient. The system rotates 1 degree and repeats the motion. This process continues until the system has rotated 180 degrees.

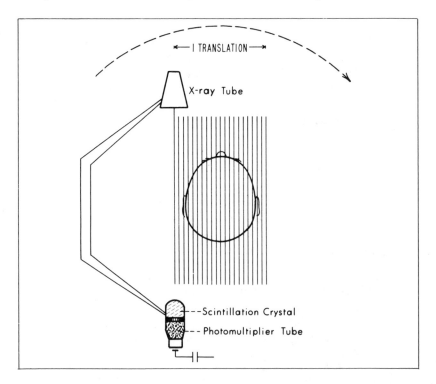

about 10 degrees, greatly reducing the number of traverses required. Interdetector *septa* made of highly absorbent material reduced the possibility of the detection of scattered radiation. These septa act in a manner similar to a grid in radiography, except that because they are located between the detectors they do not absorb primary radiation. Another innovation introduced with these scanners was the addition of a second row of detectors located parallel to the first and moving along with the first row. Thus, two adjacent planes could be scanned simultaneously—hence reducing the scan time by an additional factor of two. Typical scan times for this generation range from 20 seconds to 2 minutes.

In 1976, a major advancement in technology occurred when General Electric introduced a scanner in which the translate motion was completely eliminated. This was accomplished by greatly increasing the number of detectors from about 10 to approximately 280. Instead of being placed in a straight line perpendicular to the central ray of the x-ray tube, these detectors were aligned along the arc of a circle whose center was the x-ray focal spot (Figure 8-3). This was called a fan-beam

Fig. 8-2. Second-generation scanner operates similarly to the first generation except it utilizes multiple detectors. This allows the rotational increment to be increased and speeds up scanning time considerably.

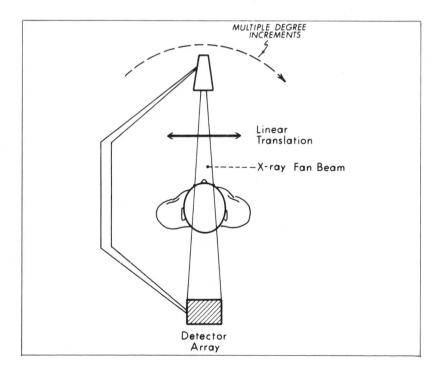

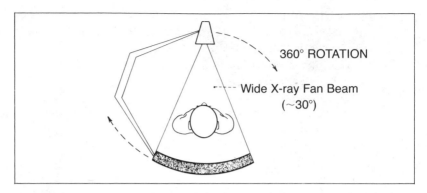

Fig. 8-3. Third-generation scanner. This scanner utilizes a wide angle fan beam and a large number (500) of small detectors. The traverse motion is completely eliminated allowing scans to be completed in only a few seconds.

geometry (third generation). In this geometry, the only motion is one of rotation. The detector fan and the x-ray tube rotate simultaneously in a circle whose center coincides with the approximate center of the patient. The x-ray tube is pulsed rapidly, similar to cine techniques, and in each pulse of the tube one "view"—a number of transmission measurements equal to the number of detectors—is collected. The first of these scanners had two operating modes, collecting 288 views in 4.9 seconds (sec) or 576 views in 9.8 sec. These short scan times allowed body sections to be imaged well for the first time. The detectors for this prototype scanner were a series of small ionization cells filled with xenon gas. X-rays entering one of the chambers ionize a xenon atom, and the ions are collected at miniature electrodes. The current gener-ated is proportional to the incident x-ray flux. This generation of scan-ners appeared to have several intrinsic design problems. First, the pres-surized gas detectors were more susceptable to the detection of scattered x-rays because of the extreme thinness of the interdetector septa. Second, the pressure plate needed to contain the gas absorbed significant numbers of x-ray photons. Third, the gas detectors were subject to *microphonics*—sound waves in the gas caused by vibration of the rotating system. The solutions to these problems involved con-siderable engineering innovation, but modern third-generation scan-ners continue to produce excellent images in scan times as short as 1 sec. These scanners use as many as 700 detectors, and although many systems now use solid-state, x-ray–sensitive detectors, xenon detectors are still in the preponderance.

The fourth distinct method of data collection is similar to third-gen-eration scanners in that the motion is one of rotation only. However, in the fourth-generation scanners, the detectors form a stationary ring while the x-ray tube rotates in a circle inside the detector ring (Figure

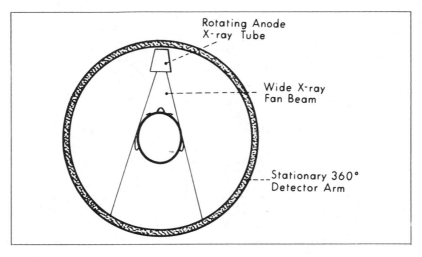

Fig. 8-4. Fourth-generation scanner. This scanner utilizes a stationary ring of detectors with the x-ray tube rotating inside the ring.

8-4). A fan of detectors is always in the x-ray beam, and the transmission information collected by all "on" detectors during an arbitrary time interval as the beam passes past them constitutes each view. The detectors in the fourth-generation systems were originally scintillation crystals—either NaI or CsI—with associated photomultiplier tubes, but as the number of detectors has increased (as many as 2400 have been proposed) and the size of each detector decreased, scintillation crystals have been exchanged for solid-state detectors. It was originally believed that fourth-generation designs would replace third-generation scanners, but experience with the more recent systems uncovered several design limitations—for instance, the large number of detectors needed. Thus, there is no clear mandate for either third- or fourth-generation systems, and both types of scanners are being manufactured. In fact, second-generation systems still enjoy a significant market share due to their low cost and durability.

Image Generation

The image produced by conventional radiographic techniques is usually a continuous range of changing gray levels. Because of the discrete way in which data is collected and the mathematical methods used in reconstruction, the CT image must be thought of as an *array* of individual, separate image points. These arrays are generally referred to by their mathematical descriptive, *matrix*. A *matrix* is any mathematical array and can be rectangular (unequal numbers of columns and rows) or square and can have more than two dimensions. In CT, all matrices

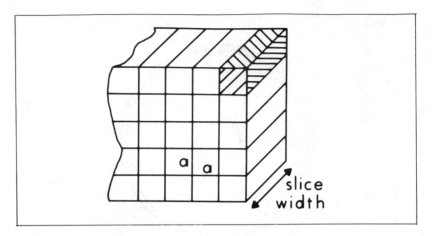

Fig. 8-5. Geometry of the reconstructed image. A pixel is an area of uniform brightness (a × a) and is the two-dimensional projection of a voxel (*shaded area*).

are square and two-dimensional. The image matrix is displayed on a television monitor, or cathode ray tube (CRT), and is therefore usually thought of as being composed of two-dimensional, square boxes, or *pixels* (short for picture elements). It must be remembered, however, that each of these squares really has a third dimension corresponding to the slice thickness (Figure 8-5) and therefore should be considered as a *voxel*, or volume element.

There are four different reconstruction methods that have been proposed for producing the final CT image from the collected transmission measurements. These methods are usually subdivided into two broad categories: *iterative techniques* and *analytic techniques*. The iterative techniques are fairly simple to implement mathematically and involve a succession of approximations. The analytic techniques require the implementation of fairly esoteric mathematical procedures that cannot be done without a high-speed digital computer. A short description will be given for each of these methods.

Iterative Techniques

Back Projection. The back-projection method of image reconstruction is one that can be implemented without any mathematics. In the early experiments of Oldendorf and Kuhl, the transmitted intensities of the radionuclide beams were converted into a density pattern on photographic film. Essentially, a separate pattern was produced for each traverse of the source past the object to be scanned. Two such patterns for a uniform square object are shown in Figure 8-6. Also shown in

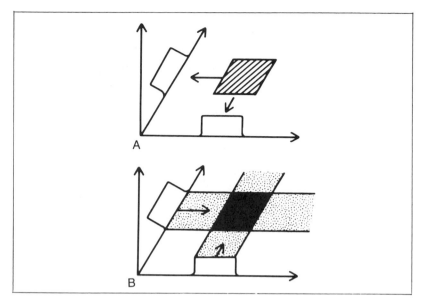

Fig. 8-6. Back projection. (A) Production of two perpendicular projections from a uniform square object. (B) Back-projected image. Note production of star artifact.

Figure 8-6 is the intensity distribution obtained when a visible light beam is passed through the resulting photographic films. The transmitted pattern of this light beam is recorded on a final photographic film and is the reconstructed image. It can be seen that for the fairly simple object in our example, the film would show a maximum density corresponding to the uniform square object. Also evident in the image would be obtrusive streaks caused by each back-projected beam. In a more complex object, these streaks would have the appearance of stars and are known as *star artifacts*. The simple back-projection method described here can be extended easily for implementation on a digital computer, but the images are characterized by the star artifacts.

Iteration. The method of reconstruction known as iteration is best implemented on a digital computer because of the large volume of data involved. However, the mathematics used does not go beyond simple arithmetic. To obtain an image by iteration, the first step is to establish a matrix on which the object densities will be superimposed. The size of each square box in this matrix and the number of such boxes are determined by the characteristics of the scan. It is easiest to picture this process in terms of a first-generation scanner. Here, it is easy to see that the smallest object that can be visualized is determined by the width of the x-ray beam. This width is determined by the smallest of the three significant geometrical parameters: the focal spot, the detector, and the collimator aperture. This size, often referred to as the *resolution aperture*, corresponds to the size of the pixel. The number of pixels is determined by the length of each traverse and by the angle of rotation between subsequent traverses.

Once the matrix size has been determined, a first approximation to the final image is made by placing a guessed number in each pixel. This guess is made "educated" by considering a single traverse of the tube detector system. This traverse results in a series of measurements of transmission along one side of the object. These measurements are known as *ray sums* since they are the result of the absorption produced by tissue in all of the object boxes along the path of the ray (Figure 8-7). The first guess at the pixel values is then equal to the ray sum divided by the number of pixels in the ray. All of the pixels along each ray will have the same value in this guess. The guess is then corrected by considering the set of ray sums collected in the following traverse. This correction process is illustrated in Figure 8-8 for the simple case of a 2×2 matrix of pixels.

In theory, the iterative method can give an exactly accurate presentation of the object densities. In practice, however, there are many limitations to the method. The most important of these limitations are 1) the large amount of computer time required, 2) the corruption of the ray-sum data by noise (either electrical or quantum), 3) the fact that the ray-sum analogy requires that the x-ray beam be monoenergetic, but in fact it contains a broad spectrum of energies. Thus, although Hounsfield's first scanner used this technique [1], it is almost never used today. The analytic techniques described next are more difficult to implement but allow the correction of many effects caused by the nonideal data collection processes.

Fig. 8-7. Production of "ray sums." Measured signal is given by $\ln \frac{1_{out}}{I_{in}} \alpha \mu_1 + \mu_2 \ldots + \mu_n$ where μ is the attenuation coefficient of the ith box in the patient.

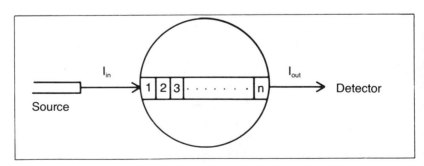

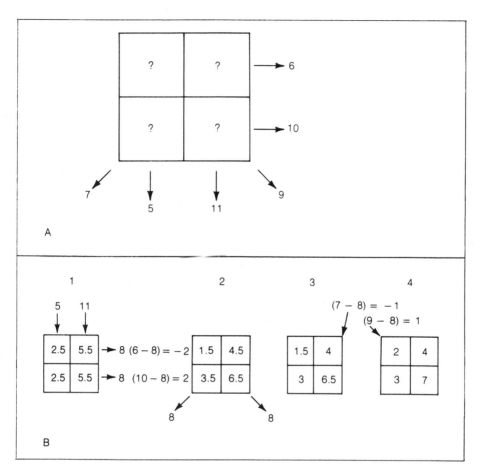

Fig. 8-8. A simple illustration of iterative reconstruction for a 2 × 2 matrix. (A) A sample object represented by boxes (pixels). Horizontal, vertical, and diagonal ray sums are indicated. The problem is to assign Hounsfield numbers to each of the pixels through the iterative (repetitive) method of reconstruction. (B) First iteration (*1*): Vertical sums are divided evenly between each box and summed horizontally. Second iteration (*2*): The first iteration horizontal ray sums are subtracted from the original horizontal ray sums (A) and the correction added evenly to each box. Third iteration (*3*): As with (*2*) but with corrections along the first diagonal. Fourth iteration (*4*): the same as the third iteration, but along the opposite diagonal. In situations where pixel measurements are corrupted by noise, the true attenuation coefficients for each pixel will never be obtained and the process must be repeated until no further improvements are generated.

Analytic Methods

Fourier Transforms. The method of Fourier transforms involves a complicated mathematical discussion that is beyond the scope of this text. Nevertheless, mention should be made of the reasons that much interest has been paid to this reconstruction technique. First, if the apertures of the instrumentation are properly chosen to be in accordance with the smallest object to be reconstructed, it has been shown that an exact reconstruction can be obtained [3]. Second, scientists studying the problems of artifact production in CT scanners have found that the Fourier techniques allow them to intensively investigate the reconstruction process. This had led to the utilization of this form of analysis in most of the scientific investigations of the subject. Nevertheless, the final method discussed below is faster, and commercial scanners seldom use Fourier analysis to perform high-speed reconstructions.

Filtered Back Projection. The method of filtered back projection is an elaboration of the iterative back-projection method described above. As mentioned previously, the main drawback of the back-projection technique is the production of star artifacts. The illustration in Figure 8-6 showed that a rectangular object produced a back-projected profile with rectangular edges, and it was this profile that resulted in the star artifact. It was found that the addition of small negative side lobes, symmetrically placed on each side of the back-projected profile of each ray, could, if judiciously chosen, cancel the star patterns. This negative contribution is the *filtering effect.* This process is somewhat more complicated than simple back-projection and requires a computer for implementation. However, since the reconstruction process for the data collected in any traverse can be begun at the immediate conclusion of that traverse, the total reconstruction time can be reduced considerably. The filter function that produces the best result for each scanner under any particular imaging task must be found by trial and error. In fact, some scanner manufacturers have given the operator the opportunity to choose among several filter functions. This method is the preferred reconstruction technique for all current commercial scanners.

Image Display Parameters

The reconstruction process yields a matrix of densities that correspond to the attenuation coefficients of the tissue in the scanned slice. These densities are converted into a gray scale for display on the CRT. The image displayed is therefore not a continuously varying one, as in a radiograph, but rather a series of discrete points. If the pixel size is sufficiently small, this discrete nature will be masked by the raster pattern of the CRT display.

In order to convert the calculated numbers to a scale that can be conveniently displayed, a new set of gray levels is defined by the relationship

$$H = \frac{\mu_t - \mu_w}{\mu_w} \times 1000$$

where μ_t stands for the linear attenuation coefficient of tissue and μ_w represents the linear attenuation coefficient of water. The symbol H is conventionally used for the units of display, called *CT numbers* or *Hounsfield units*, after the inventor of the CT scanner.

It is evident from the above definition that air has a value of -1000, and water a value of 0. Bone will have a large positive CT number. The range of CT display numbers contained in a typical slice will greatly exceed the density range normally found in a radiograph. For instance, the average radiograph contains gray levels ranging from about 0.2 (base + fog) to about 2.5. Since this is a logarithmic scale, it involves a factor of about 100 in gray-scale differences. The CT gray scale, however, can easily incorporate factors of 2000 between different levels. In order to be able to view or photograph this image, one either would have to compress it drastically (this would result in the display of an image that contained almost nothing but blacks and whites) or would have to view only part of the available range. The second alternative is the one usually chosen. This technique, known as *windowing*, involves the simultaneous adjustment of two controls. The first sets the *level*, or value, in Hounsfield units, of the center of the range of CT values to be displayed. The second sets the *window width*, or the total range of CT values to be displayed. Windowing corresponds closely to choosing a latitude or high-contrast film-screen combination in radiography.

There are many other display options available to the viewer. An exhaustive description is beyond the scope of this presentation but a list of the more common ones will be presented. When a sheet of x-ray film is the storage medium for recording an image, there are very few processing options available to the technologist or physician. However, when (as with a CT scan) the image is stored as digital information in a high-speed computer, a great deal of flexibility is available. One of the most common and useful processing functions is *magnification*. There is, however, a subtlety in this process that is seldom recognized. Digital magnification is implemented by having the user specify a region of interest in the image. The computer then expands that part of the image to fill the total display matrix. To do this, the computer performs a simple calculation to determine what values to put in each pixel. For example, if one wishes to magnify the image by a factor of two, the new image will contain half of the old image displayed as shown on the display matrix. Thus, each pixel will correspond to half of the size—as measured in the anatomy of the patient—of each pixel in the old image.

The simplest way to calculate the pixel values is to use an average value determined by the values of the pixels surrounding each of the new pixels. This process is known as *interpolation* and is the method used for magnification in many scanners, the GE 8800 for example.[2] This method can be implemented very quickly, but, as can be seen, the new image contains no more information than the old one; it is just spread out more. With some scanners, the ray sums contain resolution information finer than the manufacturer wishes to display in the normal display matrix. This is usually done to shorten reconstruction times but, as will be discussed later with respect to contrast resolution, it also smooths the image. When this choice has been made, magnification can also be accomplished by performing the reconstruction process over again, but with a finer display matrix. This method has the advantage in that the magnified image will contain more information than the original.

Another commonly used processing technique is that of calculating image statistics in a region. Here, the user specifies a region of interest, and the computer calculates the average pixel value in that region as well as the standard deviation in the region. The average value has significance in determining the type of material in the pixel; that is, it may be of significance to know whether a pixel contains mostly blood or mostly tissue. Because of random variations in pixel values, the value of one pixel may not be an accurate representation of density, and an average of several pixels within a particular anatomical landmark must be obtained. These random variations are caused by several factors, the most important of which are electrical noise in the data-acquisition system, random variations in x-ray output or detector sensitivity, and the random nature of the interaction processes between x-rays and matter. A measure of the magnitude of these fluctuations is the *standard deviation*—the average magnitude of the variations from the mean pixel value. Measurements of standard deviation are useful when made on a water phantom, since increases in this parameter signify deterioration of the imaging system.

The final processing function to be discussed here is the off-axis reconstruction program found on many of the newer scanners. Since the collection of projection data is performed in an axial plane, the most logical reconstruction geometry is a matrix of pixels in an axial plane. However, in a normal scanning sequence, several contiguous axial planes are scanned and reconstructed in sequence. Thus a three-dimensional block of voxels is produced. It is a fairly easy computational task to produce an image in a plane perpendicular to the original scan

[2]General Electric Co., Milwaukee, Wis.

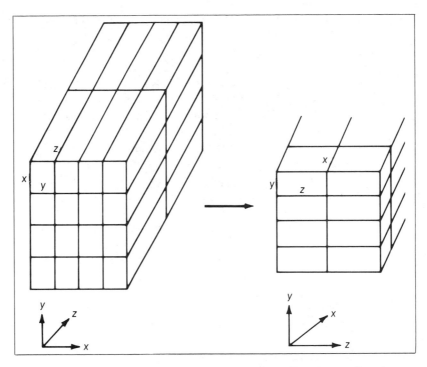

Fig. 8-9. Image re-registration. The use of contiguous slices in one direction can be used to produce a different slice in a perpendicular plane. In practice any nonaxial slice can be generated if interpolation methods are used. In this illustration two slices taken from a X, Y plane with a depth of Z are used to produce another slice in a Y, Z plane with a depth of X. Note that the resulting reconstructed pixels will be asymmetrical in respect to the original pixel.

plane (Figure 8-9) simply by assembling the pixels in a new order. The only drawback with this procedure is that the size of the pixels is usually on the order of a millimeter or less, while the depth of the voxel (slice thickness) is several millimeters. Thus, the image produced in this manner will have a resolution that is different in different directions. This factor will be less important when the slice thickness is reduced to millimeter size, as is possible with some scanners. It is also fairly easy to visualize that this process can be extended to allow reconstructions in any nonaxial plane. The computations simply become somewhat more complex and, hence, more time-consuming.

Image Quality in CT Scanners

Image quality will be discussed briefly under two subheadings: spatial resolution and contrast resolution. It will then be shown that these topics are not independent but must be considered together with the radiation dose delivered to the patient.

Spatial Resolution

The spatial resolution of an image is the ability of the image to display, as separate images, two objects located very close together. In a CT scanner this ability is determined by the combination of a number of factors, each of which combines to blur out the edges of objects in the patient. Some of the more important of these factors are the x-ray focal spot, the detector size, the display matrix, and motion. In most ways these factors are analogous to similar effects in radiography, and a detailed discussion here is unnecessary. It suffices to remark that the focal spot and detector size combine with geometrical magnification to produce image blur in the same way that the focal spot and film-screen resolving power cause image blur in radiography. The display matrix is chosen by the manufacturer and can produce pixels either smaller, larger, or comparable to the resolution limit allowed by the focal spot and detector combination. Depending on which of the above is chosen, this parameter may or may not be significant. The most efficient system will occur when the pixel size is almost the same as the resolving power of x-ray tube and detector. Typical spatial resolution of CT scanners is of the order of 5 line pairs per centimeter. This is almost a factor of 10 worse than radiographic spatial resolutions, which are of the order of 5 line pairs per millimeter. Some manufacturers specify resolution in terms of the minimum size of a hole that can be visualized. These specifications should be viewed carefully, since a single, isolated object can always be visualized if it absorbs enough x-rays.

Contrast Resolution

Contrast resolution is the ability to visualize, as distinct objects, two moderately large areas that differ in density by a small amount. In radiography, density differences of 10–15% are generally required in order to resolve such objects. In CT, this difference is a fraction of a percent. The three main reasons for this large difference are 1) almost complete elimination of scatter in CT scanners, 2) elimination from the image of all anatomical objects except those in the plane of the slice, and 3) an increased dynamic range of detectors. Because of these factors, the ability to distinguish two objects as separate finally becomes a factor of the random fluctuations in pixel values, or the image standard deviation (Figure 8-10). In most scanners, the predominant factor in the production of image noise is the random fluctuations in the absorption of x-rays. This can only be reduced by increasing the number of x-rays absorbed; that is, by increasing the patient dose.

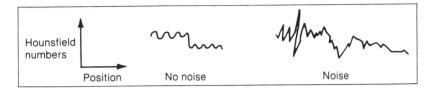

Fig. 8-10. Effect of noise on accuracy of Hounsfield numbers. With no noise there is a clear delineation of the signal generated from two adjacent anatomical parts. In the presence of significant noise from any source, the Hounsfield units are distorted and no clear demarcation is present.

Patient Dose

The dose delivered to the patient in any radiographic procedure is a measure of the number of photons absorbed in the patient. The random difference between the values of pixels when the anatomy contains the same absorbing medium is inversely proportional to the number of photons absorbed per pixel. The following conclusions can thus be made: 1) Improvements in contrast resolution made by increasing the number of photons absorbed leads to increased patient dose. 2) Improvements in spatial resolution are made by decreasing the x-ray focal spot and the detector, which allows the pixel size to be reduced. If the dose to the patient is kept constant, this results in fewer photons per pixel and, hence, reduced contrast resolution. In fact, one way to obtain improved contrast resolution is to average the values of adjacent pixels. The standard deviation is thus reduced (the number of photons per pixel is increased) at the expense of larger pixel size (decreased spatial resolution). It is a well-known paradox of CT imaging that simultaneous improvements in both contrast and spatial resolution can only be obtained at the price of increased patient dose.

One final point concerning patient dose in CT scanning must be made. In conventional tomography and in radiography, the entire thickness of the patient is exposed for each image obtained. The patient dose is linearly proportional to the number of images taken. In CT, ideally only the slice that is imaged is exposed during each scan. Thus, the total dose should be equal to the dose for a single slice and be independent of the number of slices imaged. In reality, this ideal situation is never achieved because of the sum of two effects: *interslice scatter*—scattered radiation from the slice being imaged into the adjacent slices, and *penumbra*—the shadow of the prepatient collimator causing some primary radiation to extend beyond the border of the slice. These effects should be small for slice thicknesses on the order of a centimeter, but they increase as the slice thickness decreases. By convention, these effects are measured by placing a dosimeter on the surface of a phantom in the center of a slice and measuring the single slice dose. A series of 7 contiguous slices is then scanned, with a dosimeter in the center of the center slice, and the cumulative dose for the series is then obtained. The ratio of the dose for multiple slices to the dose for a single slice reflects the collimator design for the scanner and should be on the order of 1.5.

Image Artifacts

The previous discussion of image production in CT scanners leads to the conclusion that, except for the corruption caused by image noise, the reconstructed image is an accurate representation of the density variations in the patient's anatomy. Unfortunately, this discussion has incorporated several simplifications that conceal the possibility of the existence of image structure that bears no relation to patient anatomy. Such image structure has become known as *artifact*. There are many sources of artifact in CT scanners, and we will attempt a short discussion of several of the more-common artifacts so that their effect on image quality can be recognized.

Beam-Hardening Effects

Our discussion of CT image production implicitly assumed that the x-ray beam contained components of only one energy. In fact, the beam is the standard broad spectrum produced by an x-ray tube. Although this beam is normally characterized by the peak energy, it contains a broad spectrum extending from very low energies to the peak (as well as the characteristic energy of the tungsten anode). Thus, as the beam enters the patient, the lower energy components are preferentially absorbed so that the exit beam is considerably harder than the entering beam. This is particularly true in head imaging because of the high calcium content in the skull. If one considers the beam just after it passes through the skull on the entrance side of the head, this beam will be considerably harder than the computer might expect it to be. The computer attributes the removal of the soft radiation to an increased density just inside the skull. This light ring was originally mistaken for cerebral cortex. In the original EMI head scanner, this artifact was partially removed by surrounding the head in a water bag. With the advent of body scanners, this was no longer feasible. Two corrective methods are currently employed. The first involves the use of a beam-hardening filter, usually copper. This method minimizes the artifact. The second method is to use a software correction to remove most of the residual artifact in head scans. This is done by calculating the effects of beam hardening on a typical skull. A correction table is then stored in the computer and each pixel is multiplied by the appropriate factor to remove the effects of the hardening.

Motion Artifacts

Patient motion produces a second major type of artifact. Whenever part of the object in the field of view moves during the course of a scan, the reconstruction program cannot adequately treat the scan information. The image will therefore contain a streak that appears as though the object's position is averaged over the region of motion. The intensity of these streaks is dependent on the density differences between the moving anatomy and its surroundings. The motion of bowel gases is particularly troublesome because of the large density difference between gas and the surrounding tissue. The only corrective action for these artifacts is to employ shorter scan times, and this has been the major impetus behind the development of scanners that complete an entire scan in times on the order of 1 sec. Several of these scanners are currently available.

Ring Artifacts

With the advent of third-generation scanners, ring artifacts became a serious problem. When the calibration of one detector becomes significantly different from the others in the detector array, that detector appears to receive more (or less) intensity during the entire rotation of the gantry. Thus, in each view, one ray sum is inconsistent with the patient information, and the error occurs in the same ray sum for each view. The reconstructed image will contain a ring whose diameter will depend on the position of the faulty detector. A similar effect can occur in fourth-generation scanners if the anode of the x-ray tube wobbles during the scan. These artifacts have been largely eliminated through improved detector and x-ray tube design.

Streak Artifacts

The last artifact to be discussed here is the streak artifact that is observed at sharp edges of objects that have high contrast. Bone and metal clips are the most common absorbers that produce streaks. These artifacts are the result of the reconstruction technique's inability to handle information at sharp edges. One of the implicit assumptions in all analytic reconstruction techniques is that the information is "band limited." This means that there are no objects in the patient that are smaller than some minimum size. If this is true, then the process of taking discrete digital values as representing the continuous signal that the detector receives provides an exact representation of the detector's signal. If this is not true, several mathematical approximations in the reconstruction techniques are violated. The resulting artifacts are known mathematically as "aliasing" and "Gibb's phenomena." None of the data obtained in a CT scan is truly band limited, but usually the signal strength, or contrast, in such edges is very low, for example at blood vessel boundaries. At the edges of bone, however, there can be enough contrast in the very sharp edges to make the sampling errors significant, and the result is the occurrence of streaks in the image.

Purchasing a CT Scanner

A modern CT scanner costs at least $600,000 and can easily be more than $1 million. Clearly the purchase of so expensive a piece of equipment involves a very difficult decision. The complexities of this decision process are such that we cannot begin to exhaust the possibilities. Nevertheless, it can be instructive to examine some of the trade-offs.

In the area of image quality, there exists a wide spectrum of possibilities. A medical school research and teaching environment usually demands the highest possible image quality to cover all possible imaging tasks. In a busy clinic, it is conceivable that the specialization implied by the type of patient population might allow some simplification and an associated cost reduction. For instance, an application that is mainly oriented toward brain scanning might have little use for a scanner capable of the very thin slices and high resolution over a limited field of view that is required for the best spine imaging. An application that involves mostly body scanning would benefit from a scanner that is capable of very fast scans, but high speed would be superfluous in brain scanning. Thus, the first task in choosing a scanner is to define clearly the application and investigate whether any specialization in imaging tasks is associated with a particular company's product.

A second priority in the selection process might be to define which optional configurations are best suited to the application. A particularly expensive option available for many scanners is a second diagnostic console. The purpose of this console is to allow the physician access to the images on the display screen without interrupting the scanning and archiving process. In an environment where the physician prefers to diagnose from hard copies, this console is unnecessary. Even if the decision is made to purchase such a console, a wide variety of processing options is available that can increase the cost considerably. The application can call for a minimum configuration that acts as a single display console, giving the operator the ability to only change window widths and levels, or it may be desired to acquire the most sophisticated configuration, which allows reconstruction from raw data, and even an independent multiformat camera. Another important hardware option that must be considered is the *archiving system*. The two most common methods of archiving scans are floppy discs and magnetic tape. Floppy discs are convenient for the storage of a single patient's images on an individual memory device. They allow random access to any slice taken on a patient. Magnetic tape is somewhat slower but allows the storage of many patient's images on a single tape and is somewhat more convenient from a storage standpoint. In an environment where it is necessary to store only the final images, a multiformat camera may replace both of these systems.

Most CT scanners are continually being updated with developed software programs from the manufacturer. These programs are generally offered at no cost or as part of a service agreement. However, the type of programs available may influence the choice of scanner. For instance, the ease of reconstructing nonaxial views from a series of axial slices is an important consideration in an application involving a large volume of spine imaging. The ability to vary the reconstruction matrix may be important in a scanner that allows the resolution to be varied by either changing the collimator (second-generation scanners) or varying the patient-detector distance (some third-generation scanners). The resolution determined by the reconstruction matrix should be consistent with the other factors that determine resolution, or else the operator will be continually forced to geometrically magnify the images in order to obtain high resolution.

In conclusion, the choice of a CT scanner involves serious consideration of many competing factors, a very important one being cost. Once these factors have been analyzed carefully, a professional consultant may be required to examine the various trade-offs and help maximize the utility of an installation while preventing unnecessary expenditures.

Inside the Multiformat Camera

The multiformat camera provides the radiologist with the means of viewing nonconventional images in a standard, conventional manner—a film and view box. By nonconventional images, one refers to two classes of radiological images: analog and digital. Analog images are obtained from devices such as video fluorographs, ultrasound imagers, and nuclear scanners. These images are presented either as light images at the output of image intensifiers or as electrical signals at the output of a scanning device. The images may be either static—the final result of a relatively long scanning time—or dynamic—rapidly changing visual information. The second class of images are presented in a digital format. These are the images produced by the ever broadening array of modern digital imaging modalities such as computed tomography, magnetic resonance imaging, ultrasound, nuclear medicine, digital subtraction angiography, and digital radiography. While most of these images are static—the information does not change in the time required to record it on film—the increasing speed of some digital modalities will soon require the ability to record dynamic digital images.

The main components of the multiformat camera are the interface to the imaging device, an image producing system, and the recording system. Each of these subsystems will be discussed in order to develop understanding of the workings and problems of the multiformat camera.

The Interface

The interface is the system that relays information from the imaging device to the camera. In analog imaging systems, the interface is normally a simple electrical connection that carries image information in the form of an electrical signal. Variations in signal levels directly represent the variations in image intensities. In contrast, the interface between a digital imaging device and a multiformat camera is one of the most controversial elements of these systems. Over 100 different standards exist throughout the world for the transmission of digital information. Some vendors have been known to employ a number of different methods throughout their product line. This proliferation of standards makes it extremely difficult to purchase a multiformat camera from one vendor and simply connect it to a digital device produced by another vendor or even by the same vendor at a different time. A discussion of these systems beyond the scope of this article is impossible but the reader should understand that the problem exists. Standardization is mandatory if progress is to be made in digital imaging.

The Image-Producing System

The central component of the multiformat camera is a system that produces an image for recording. Although the first cameras were simply attached to the outputs of oscilloscopes, modern multiformat cameras use a sophisticated television monitor to produce a visible image. This type of monitor is familiar as it has been in use for many years in fluoroscopic systems. The monitor employs a raster scanning technique to record on its output phosphor a visible picture that corresponds to the electrical or digital input signal. The design of the television-type monitor is extremely critical if the resultant image is to be of high quality.

Consider the problem of the raster system. The most commonly used standard (known as RS 170) uses a 525 line raster with half of the lines scanned in successive 1/60 second intervals. The successive half-images are interlaced. This is known as a 2 : 1 interlace system. This system was devised in order that the eye would not perceive any flicker. The eye, in a dynamic mode, however, is very forgiving. When these images are viewed statically on film, raster lines can be seen as well as any variations in intensity caused by electrical drift. The drift problem can be corrected by accurate monitoring and control of the electron beam intensity. The raster lines can be removed by a system which very accurately displaces successive video fields in the vertical direction (up and down the film) in order to fill in the spaces between the lines. To do this successfully, at least eight frames must be recorded for each image.

Another problem in the video monitor is the synchronization of the scans with the recording process. If a small number of video frames are recorded per image, as is done in dynamic imaging, the image might contain several complete scans plus a partial scan. This will result in an abrupt change in image density that will be very visible in the static film image. This problem is avoided by careful synchronization of the display and the camera shutter.

Inexpensive video monitors have two significant problems. The first is the inherent curve of the face of the monitor on which the phosphor is deposited. This curvature results in a distorted image. This is corrected in modern multiformat cameras by using flat monitors which are relatively expensive. Secondly, any variations in the thickness of the phosphor result in intensity variations which, while not important in viewing the video image with the eye, are very distracting in a static image on film. The composition of the phosphor itself is important. Monitors used in older multiformat cameras contained a phosphor that was made up of two components. These components emitted light of different wavelengths. When coupled to a film that is not uniformly sensitive over the wavelength regions presented by the phosphor, objectionable image density variations can result.

Image Recording

The image on the video monitor is conveyed to the recording system (usually a film) by a lens. With a multiformat camera several images can be recorded on one film. Furthermore, one may desire to alter the number and size of the images on one film depending on the examination being performed. To do this, the lens system and/or the film must be moved since degradations in image quality will occur if the lens is used off-axis, i.e., the line joining the center of the monitor and center of the film do not pass along the axis of the lens. Thus, to change the magnification (size of the image), the lens is moved along the line between the monitor and the film. The lens and film are moved together to change the image position. These motions must be done smoothly and accurately in order that images are faithfully reproduced.

The film itself must be chosen with care. Four different phosphors are presently in use in video monitors and a match between the film and the phosphor can be critical. The contrast (gamma) of the film must be chosen to display adequately the information contained in the images. Major film manufacturers produce CRT films with gammas between 1.9 and 2.2 with a range of contrasts available. The film should be evaluated for use with the particular multiformat camera for which it is intended. Although spatial resolution is not a problem— the resolution of all films is several orders of magnitude greater than the resolution of the monitors—contrast resolution can be severely affected if the film is not correctly matched to the multiformat camera.

Modern Developments

Two impending developments are worth mentioning as digital radiology moves towards ever increasing information capacities. First, digital radiography systems are available in 1024 × 1024 matrix formats. This requires a 1050 line monitor for image presentation. These systems are available but are expensive and difficult to maintain. Another alternative, especially with a view to even larger image matrices, is a system that uses a laser to directly record the image on film. Such systems are currently under development by several companies but will be very expensive and are several years from clinical implementation.

References

1. Hounsfield, G. N. British Patent No. 1283915, 1972.
2. Radon, J. On the determination of functions from their integrals along certain manifolds. Berichte über die Verhandlungen der königlich Sächsischen Akademie der Wissenschaften zu Leipzig, *Mathematisch-Physische Klasse* 69:262, 1917.
3. Bracewell, R. N. Strip integration in radio astronomy. *Australian Journal of Physics* 9:198, 1956.
4. Oldendorf, W. H. Isolated flying spot detection of radiodensity discontinuities—displaying the internal structural pattern of a complex object. *IEEE Transactions on Bio-Medical Electronics* BMW 8:68, 1961.
5. Kuhl, D. E., and Edwards, R. Q. Image separation radioisotope scanning. *Radiology* 80:653, 1963.
6. Hounsfield, G. N. Computerized transverse axial scanning (tomography). Part I. Description of system. *British Journal of Radiology* 46:1016, 1973.

9   Special Considerations for Angiographic Suites

One of the basic reasons for writing this text was to present solutions to the problem of acquiring the proper radiographic equipment for a general or specific application. It was assumed that better-informed personnel would make wiser decisions both in the acquisition and in the operation of radiographic equipment. The approach to this educational problem has been primarily conceptual rather than specific in nature, and the presentation has by necessity been superficial at times. Another reason for the text is to make the reader aware of the complexity of radiographic equipment and the importance of regarding all the parameters that must be considered, particularly with respect to angiographic suites.

In this chapter, data will be presented that will attest to the complexity of radiographic equipment and from which the reader may derive useful information regarding the operation of an angiographic suite. An attempt will be made to correlate concepts with application. These concepts are applicable to any radiographic room, but they become more significant for angiographic suites because of the economic impact. A typical radiographic-fluoroscopic room today will cost about $200,000–$400,000, compared to $1,000,000 or so for a fully equipped special procedures room.

Speed Relationships in Angiography

It is the purpose of angiography to opacify vascular structures and to record this opacification in such a manner that structures may be minutely and critically examined. Detection of early vascular changes is critically important in diagnosing and subsequently planning the treatment of disease processes. If the physician is to examine an organ critically, detail must be presented as sharply as possible. This need for sharp detail is somewhat akin to the visualization of bone trabecular patterns, in which one is normally required to use a high-detail recording system such as detail-intensifying screens and fine focal spots.

In recording bone detail, no significant problems are encountered except for the problems of magnification distortion and excessive radiation exposure to the patient. In angiographic procedures, however, one encounters the additional problem of motion caused by vascular

Some of the information in this chapter was presented by Heinz Klosterman at the Symposium on Angiographic Equipment at Las Vegas, Nevada, in February, 1976.

pulsations. To obtain *stop motion*, radiographic exposures must be made within physiological time frames according to the capability of the generator to produce a sufficient amount of radiation during a particular time frame. To succeed in this task requires that the appropriate amount of radiation be given within a specified time frame; thus larger generators are required and more accurately timed electronic circuits become mandatory. An exposure timing error of 8.3 milliseconds (ms) can be tolerated if a total exposure of 100 ms is to be made, but an error of 8.3 ms in a total exposure of 20 ms cannot be tolerated. The need for more accurate, shorter exposure times and greater radiation output makes it mandatory that three-phase 12-pulse equipment be used for modern angiographic procedures.

Stop Motion

There are three types of motion that must be taken into consideration with respect to angiography: peristaltic, respiratory, and vascular. Respiratory motion is of little concern, since most patients can hold their breath for short periods of time during the radiographic exposures. Peristaltic motion is much slower in relationship to vascular motion, and if vascular motion is controlled, peristaltic motion is automatically controlled; that is, if stop motion is obtained for vascular pulsations, peristaltic motion, since it is slower, is automatically controlled.

As Figure 9-1 shows, the anatomical area dictates the angiographic recording method, and, in turn, the angiographic recording method is dictated by the exposure/film rate requirements. As a general rule, vascular motion within the head and neck occurs at 5–30 mm per second. Motion of the coronary arteries occurs at about 50 mm per second, but valvular motion occurs at 200–500 mm per second, particularly if there is any turbulent blood flow. Vascular pulsations in the abdomen are about the same as those in the head and neck and range from 10 to 50 mm per second. Object motion in the extremities is relatively quite slow and averages about 1 mm per second. The equipment used for recording events within these specified areas must therefore have remarkably different capabilities, with much higher filming rates required in the cardiac area than in the extremities. A high amount of object motion, such as that seen in the heart, requires that exposures be made by high-speed photospot cameras or cinecameras. Films of the extremities require equipment with slower capabilities, such as a film changer or peripheral film changer, since vascular motion is slower, and there is no need for a fast recording mechanism.

In order to obtain stop motion in a specific organ in the body, there must be a compromise between the recording speed, the kilowatt rating of the generator, and the kilowatt rating of the x-ray tube as af-

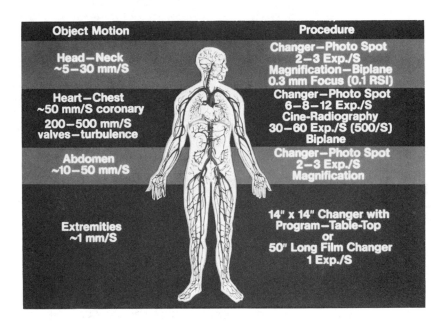

Object Motion	Procedure
Head—Neck ~5—30 mm/S	Changer—Photo Spot 2—3 Exp./S Magnification—Biplane 0.3 mm Focus (0.1 RSI)
Heart—Chest ~50 mm/S coronary 200—500 mm/S valves—turbulence	Changer—Photo Spot 6—8—12 Exp./S Cine-Radiography 30—60 Exp./S (500/S) Biplane
Abdomen ~10—50 mm/S	Changer—Photo Spot 2—3 Exp./S Magnification
Extremities ~1 mm/S	14" x 14" Changer with Program—Table-Top or 50" Long Film Changer 1 Exp./S

Fig. 9-1. The object motion produced by individual anatomical areas determines the procedure for recording the event. Anticipated object motion is listed in the left-hand column. For the corresponding anatomical area, the right-hand column shows the recording method that may be used for that area and the exposure limitations.

fected by factors such as target angle and focal-spot size. One also has to determine how much motion is acceptable. Table 9-1 depicts the exposure times that are required for different degrees of motion.

A glance at the table shows that some of these exposure times are not realistic. For example, if an exposure of 1 sec were made of the lower extremities, all of the contrast medium would be gone by the time the exposure was completed. In addition, cranial vessels are relatively fixed in comparison to the vessels in the neck, and hence exposure times would not be required at that frequency. It therefore follows that motion must be optimized, and it is generally accepted that exposure times of 100–200 ms are adequate for cranial studies; 50–80 ms for abdominal viscera; 20 ms for pulmonary angiography; 3–7 ms for coronary angiography; and 1–2 ms for valvular studies. Thus, if greater detail is desired, shorter exposure times are required. One must consider the effects of vascular motion in recording minute detail. On the other hand, the use of longer times of 100–200 ms for head and neck examinations allows one to make use of smaller focal-spot sizes and lower kilovoltages, thus providing higher contrast for maximum detail.

Table 9-1. Exposure Time Requirements for Various Degrees of Motion

Organ	Motion	Exposure Time to Limit Motion to:			
		1.0 mm	0.5 mm	0.1 mm	Optimal
Head and neck	1–5 mm/sec	1000 ms	500 ms	100.0 ms	50–100 ms
Heart	100 mm/sec	10 ms	5 ms	1.0 ms	3–7 ms
Lungs	100 mm/sec	10 ms	5 ms	1.0 ms	20 ms
Heart valves	500 mm/sec	2 ms	1 ms	0.2 ms	1–2 ms
Abdomen	50 mm/sec	20 ms	10 ms	2.0 ms	50–80 ms
Extremities	1 mm/sec	1000 ms	500 ms	100.0 ms	50 ms

Radiation Requirements

The next factor to consider is the radiation levels required for the different recording systems. Again, the reader should remember that speed is related to stop motion. Table 9-2 illustrates the radiation levels required for different recording systems. For comparison, a typical value for radiation exposure to the input screen during fluoroscopy in a 6-inch mode is 5 milliroentgens (mR) per minute, which equals about 2.5 microroentgens (μR) per television frame. On the 9-inch mode, the dose requirement would be approximately μR per television frame. Since this radiation level may present a rather noisy image, a special density setting may increase the radiation to a higher level. Proper television image presentation in this case will be provided by an iris, that is, an f-stop control, on the front of the television camera or by an automatic gain control, as described in Chapter 5.

In reference to Table 9-2, for a film changer at six exposures per second, 800 μR may be required for an exposure using high-speed calcium tungstate screens. A relative speed of 1 has been assigned to this system. If one uses rare-earth screens with a relative speed two times that of high-speed calcium tungstate screens, the required radiation exposure is cut in half to 400 μR; it is reduced to 200 μR for rare-earth screens with four times the speed of high-speed screens. In comparison, consider the radiation exposure requirements for a 9/6/4 image intensification system with a photospot camera and a 35-mm cinecamera. For a photospot camera with an $f11$ lens opening (speed), a 100-μR exposure is required using the 9-inch mode. On the 6-inch magnification mode, the radiation exposure is 220 μR, which is essentially equal to four times that required for rare-earth screen exposures. On the 4-inch mode, the radiation exposure requirement is 500 μR, about one-half the requirement of high-speed calcium tungstate screens. For a photospot camera using a faster $f8$ lens opening, radiation exposure requirements are even lower; for a cinecamera with an $f4$ lens opening, exposure requirements are lower than with all image intensifier modes.

Table 9-2. Systems Speed Relationship in Angiography (70- to 100-kV Exposure)

	Cut-Film Changer (6 exp/sec)			9/6/4 Image Intensifier			
	Screen Selection			Photospot Camera	35-mm Cinecamera		Mode
	High-speed Ca tungstate Relative speed = 1	Rare-earth ×2	×4	f11	f8	f4	
Input in micro-roentgens	800	400	200	100 220 500	50 120 250	13 30 52	9″ 6″ 4″
Exposure requirements	More	10 ms 1200 mA 1–1.4 mm focal spot 10–13° target angle	Less	5 ms 200–500 mA 0.3–0.6 mm focal spot 7–10° target angle			

However, the reader should remember that for cinecameras, we are referring to the exposure requirement per frame. The overall result is that less radiation exposure and shorter exposure times are required for exposures made with the image intensifier system. The disadvantage is that resolution with the image intensifier system is not quite as good as with the film changer, but this decreased resolution is offset by the fact that the shorter exposure times used by the image intensifier system afford less motion unsharpness.

Generator Capability

In order to obtain stop motion and therefore to improve the detail of angiograms, the generator must be able to produce enough power to allow the film to be exposed to the proper density in the required short time. Unfortunately, the generator is usually capable of producing much more power than the x-ray tube can handle. Furthermore, in order to decrease magnification distortion, the smallest possible focal spot must be used. Table 9-3 illustrates the kilowatt ratings of typical x-ray tubes with respect to focal-spot size. The larger the focal spot, the higher is the kilowatt rating of the x-ray tube. The angiographic procedure dictates the need for the kilowatt capacity and the focal-spot capability of the x-ray tube. For abdominal procedures, large focal-spot sizes and increased kilowatt ratings are necessary in order to obtain stop motion in a relatively thick anatomical part. Again, one must reach a compromise between the available generator and x-ray tube capabilities and the available screen-film and/or image intensifier recording system.

Table 9-3. Relationship of Kilowatt Tube Ratings, Focal Spots, and Recording Systems

Anatomical Area	kW Power	Focal-Spot Size	Recording System
Head and neck	100	1.2	
	30	0.6	Screen-film
	14	0.3	
	0.3	0.1(microfocus)	
Thorax	150	1.8	
	100	1.2	Screen-film
	30–70	0.6	Cinecamera
	14–20	0.3	Photospot camera
Abdomen	150	1.8	Screen-film
	100	1.2	or
	70	0.6	Photospot camera
Extremities	100	1.2	
	30	0.6	Screen-film

It was noted in Chapter 3 that target angle and focal-spot size are related to tube loading, with the smaller the target angle number, the higher the tube loading for a given focal-spot size. In order to obtain good detail, a small target angle number (e.g., 7–10 degrees) must be used with a small focal spot (e.g., 0.6 millimeters [mm]). The net result of using the smaller focal-spot size is a decrease in the kilowatt rating of the x-ray tube. Figure 9-2 illustrates this concept. The figure shows that a 13-degree target angle and a 1.8-mm focal spot are required if one is to use a 1500-milliampere (mA) station. At 1250 mA, one may use a 1.0 to 1.4-mm focal spot and a target angle of 10–13 degrees. A minimum focal spot of 1.0 mm is required in order to produce spot motion at a 10-ms exposure with a high-speed film-screen combination. For magnification work, a focal spot of 0.3 or 0.6 mm is required, but stop motion is not possible because of the limited kilowatt ratings of the x-ray tube (unless the anatomical part being recorded is of small size).

Magnification Angiography

In Chapter 3 the basic principles of magnification angiography were discussed. Figure 9-3 is another depiction of the basic principles of magnification of small vessels. The reader should recall the basic problems involved in photographic magnification in regard to magnification of the mottle present in the recording system. There is a direct benefit in geometric magnification, in that more image information is made available for critical study. There is some question of the validity of more than 2× magnification, but there are some angiographers who contend that greater than 2× magnification allows them to see smaller

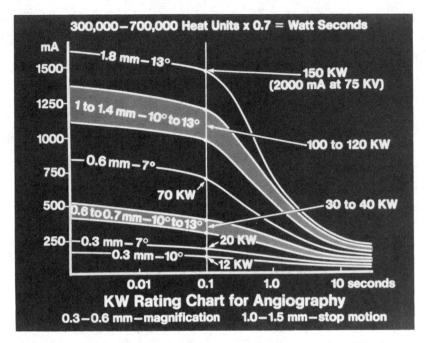

Fig. 9-2. Kilowatt rating chart for angiography, showing relationship of focal-spot sizes and target angles. A higher kilowatt rating is obtained by using large focal-spot sizes. These large focal spots are necessary to achieve the high power ratings that are required to afford stop motion. The smaller focal-spot sizes and higher target angles are used in magnification angiography; this combination allows sharp defintion at a compromise of having less than optimal stop motion.

vessels and thus to make earlier and more accurate diagnoses. Unfortunately, the higher the magnification, the smaller is the focal spot that is required to resolve the structures.

The problems involved in magnification angiography are related to the focal-spot size. Most magnification angiography is done with a 0.3-mm focal spot, a biased 0.3-mm focal spot, or a microfocal spot of 0.1 mm; occasionally a 0.6 focal spot is used. The use of these small focal spots allows less loading of the x-ray tube. The loading capability of a typical 0.6-mm focal spot is 70 kilowatts (kW); for a 0.3-mm focal spot, 20 kW; and 0.3 kW for a microfocal spot. These low loadings require the use of steep target angles ranging from 7 to 10 degrees, and the compromise here is less field coverage. Because of the low loading, optimal stop motion cannot generally be obtained. The trade-off, then, is the gain of more image information by geometry, but it is achieved at the expense of loss of sharpness (detail) due to motion.

Table 9-4 illustrates some basic x-ray tube characteristics needed for magnification angiography. As one can see from the table, there must

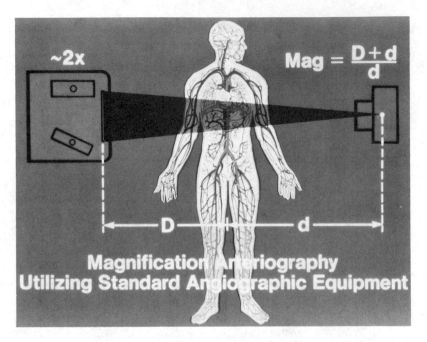

Fig. 9-3. Basic principle of magnification arteriography. The anatomical part is placed at distance *D* from the film changer and at distance *d* from the x-ray tube focal spot. The formula for the magnification factor is noted in the upper right-hand corner.

be a compromise between field size coverage and the size of the focal spot. A 9-inch field size will not allow adequate coverage of some organs (e.g., the liver) and will barely cover some kidneys. Therefore, in order to accomplish magnification angiography with large organs, less magnification has to be done so that larger focal spots and greater field coverage can be obtained.

The field size coverage must again be taken into consideration with respect to the recording mechanism. Figure 9-4 shows the relationship between film changers and image intensifier recording systems with respect to target angles, kilowatt ratings, available milliamperage, and the times required to obtain stop motion and magnification. Again, one should note that to obtain magnification, longer exposure times are required because of the limited radiation output from the small-focal-spot x-ray tubes.

Radiation Exposure

Much concern has been expressed regarding the amount of radiation exposure to the patient during magnification angiography. Many authorities feel that the skin radiation dose is largely increased. Figure 9-

Table 9-4. X-ray Tube Characteristics for Magnification

Focus (mm)	Anode Target Angle	Field Coverage S.I.D.		kW	Workable mA	Magnification Factor
		100 cm	80 cm			
0.6	7°	9″	7.2″	70	700	×1.6
	10°	14″	11″	40	400	
0.3	7°	9″	7.2″	20	200	×2
0.3 biased	10°	14″	11″	14	140	×3
0.1 microfocus	45° stationary			0.3	3	×5

5 illustrates the principle of magnification angiography in relationship to the skull. It is possible to perform 2 : 1 magnification without changing the source-image distance. For a skull measuring 10 inches the source-object distance is 30 inches for contact radiography; for 2 : 1 magnification angiography, the object-film distance is 15 inches. In contact radiography a 10 : 1 or 12 : 1 grid must be used, but in magnification angiography, due to the 15-inch object-film distance, the object-film distance serves as an air gap and hence the grid should not be used. The use of a 10 : 1 grid requires a dose rate increase of three to four times, depending on what grid is used. On the other hand, for a source-object distance of 15 inches versus 30 inches, the dose rate decreases by a factor of four due to the inverse square law. Thus, both grid-contact radiography and nongrid-magnification radiography require close to equal dose rates in head examinations. The total skin area radiation dose is equal, therefore, because the x-ray beam enters the patient at a shorter distance but in a proportionately smaller area.

Equipment Used in Magnification Angiography

Because there is a lot of interest in magnification angiography, it would be helpful at this point to illustrate some basic equipment concepts. Again, the reader should be reminded that equipment needs must be matched to the procedure being performed. Figure 9-6 illustrates the basic equipment that may be used in a combination cerebral-renal-visceral magnification setup. In this example, a 9/6/4 image intensifier is used with a 105-mm photospot camera. The x-ray tube to be used with the image intensifier system is a 0.3/0.6-mm focal spot, which allows 20 kW on the small focus and 70 kW on the large focal spot. The table being used is a raise/lower table with a floating table top. The film changer is a Puck[1] changer with lowering and tilting capabilities. A 0.3/1.2-mm radiographic tube is being used, which allows a

[1]Siemens-Elema AB, Solna, Sweden.

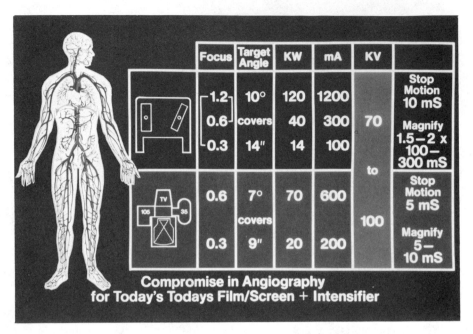

Focus	Target Angle	KW	mA	KV	
1.2 0.6 0.3	10° covers 14"	120 40 14	1200 300 100	70 to 100	Stop Motion 10 mS Magnify 1.5–2 x 100–300 mS
0.6 0.3	7° covers 9"	70 20	600 200		Stop Motion 5 mS Magnify 5–10 mS

Compromise in Angiography for Today's Todays Film/Screen + Intensifier

Fig. 9-4. A compromise must be made in considering equipment, focal spots, and target angles in relation to kilowatt ratings, available milliamperage, and kilovoltage. For film changers, a 1.2-mm focal spot with a 10-degree target angle provides the high power ratings needed for stop motion at 10-ms exposures. Small focal spots allow magnification at 1.5-2× but do not have the power capability for stop motion. Photospot cameras and cinecameras, by recording indirectly from the phosphor of the image intensifier, can provide both stop motion and magnification but are restricted due to limited field coverage.

kilowatt loading of 14 kW on the small focal spot and 120 kW on the larger focal spot. The use of equipment such as this allows a great deal of freedom in doing almost any type of angiography and routine magnification work. To do peripheral angiography, an additional tube must be installed in the ceiling and an appropriate changer obtained. To obtain 2× magnification, the anatomical part is placed halfway between the x-ray tube focal spot and the top of the Puck changer.

For coronary arteriography, a different setup is required. In this case, as shown in Figure 9-7, a film changer probably would not allow short enough exposure times to obtain stop motion, which is critical in coronary angiography. We are therefore relying on the image intensifier recording systems of a 105-mm photospot camera and a 35-mm cinecamera. A 9/6/4 image intensifier tube is being used as the input to the recording system. A standard angiographic table is used. Since the recording system is in reverse location to a film changer, magnification distances are measured in reverse, with the table top being lowered

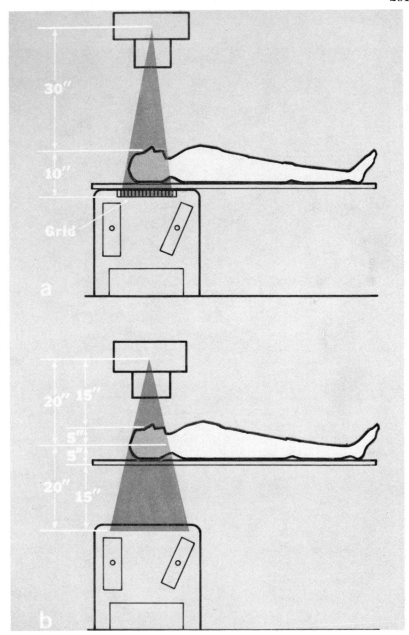

Fig. 9-5. Comparison between grid and nongrid techniques in angiography. In normal angiography the anatomical part is placed as closely as possible to the recording system, as noted in the upper illustration (a). A grid is necessary to absorb scattered radiation. In magnification techniques, the midportion of the skull is equidistant from the x-ray tube focal spot and the image receptor, as noted in the lower illustration (b). In this case no grid is necessary because the air gap provides cleanup from scattered radiation. Since the grid has been removed, radiation exposure to the patient can be reduced, but this reduction is offset by the fact that patient is closer to the x-ray tube (inverse square law).

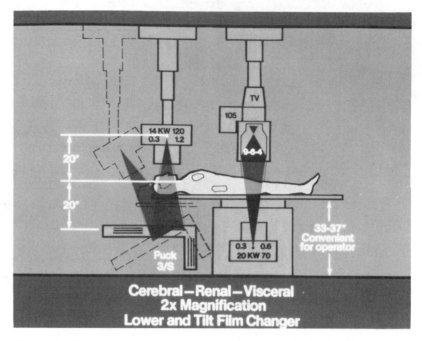

Fig. 9-6. Equipment used for general angiographic magnification. This equipment setup provides flexibility for most angiographic procedures. A peripheral film changer and ceiling isolated tube could be added for peripheral angiography. The Puck film changer can be raised or lowered to provide the magnification factor determined to meet the needs of the examination; this changer will also tilt, allowing cerebral magnification procedures.

closer to the x-ray tube and the image intensifier being moved further away from the patient. In this case we still retain the 0.3/0.6-mm focal spot, which provides the loading required in this application.

Digital Fluorography

The computerization and display of fluoroscopic and radiographic images is known by many terms including *digital radiography, digital fluoroscopy, digital video angiography,* and *digital subtraction angiography.* Regardless of the terminology, the process is one in which an image used as a mask is electronically subtracted from a subsequent image. In subtraction, the purpose of the process is to remove unwanted anatomy in order to visualize a contrast-filled vessel more effectively. For example, by subtracting bony structures, one may be able to see contrast-filled vessels far better than with the unwanted overlying bony structures obscuring detail. In order to obtain good subtrac-

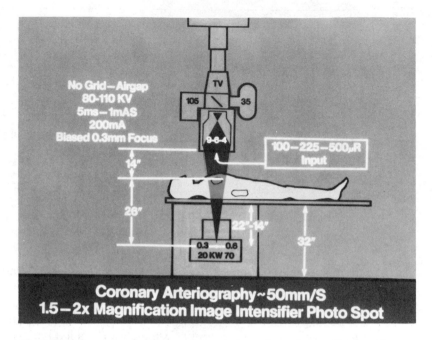

No Grid–Airgap
80-110 KV
5ms–1mAS
200mA
Biased 0.3mm Focus

TV

105   \   35

9-6-4

100–225–500μR
Input

14"

26"

22"-14"

32"

0.3   0.6
20 KW 70

Coronary Arteriography~50mm/S
1.5–2x Magnification Image Intensifier Photo Spot

Fig. 9-7. Equipment setup for coronary arteriography and magnification. A floating table top is used, although a cardiac cradle could be added for more flexibility. A biased focal spot is used for recording greater detail. Since the recording medium is above the patient, calculation of the magnification factor would be the reverse of that noted in Figure 9-3.

tion, almost perfect registration must be available. By *registration* it is meant that the two images must be perfectly superimposable and any movement between the two images will produce registration artifacts and degrade the image. Any involuntary movement such as breathing, swallowing, and vascular pulsations can produce registration artifacts sufficiently bad enough to completely ruin the examination.

In a digital subtraction system, the electronic signals from a television camera are fed into an *analog processor* with the analog signals converted to digital; from the *digital processor* the image may be stored or fed into another processor, where the image may be reprocessed and manipulated. Figure 9-8 is a block diagram of a typical digital subtraction system. In order to understand the basics of a digital system, it is important to understand the fundamentals of each of the components necessary for a digital subtraction system.

In digital subtraction angiography, a small 5 or 6 French catheter is normally inserted into the venous system, with the tip of the catheter in the superior vena cava. A bolus of 30–50 ml of contrast material is

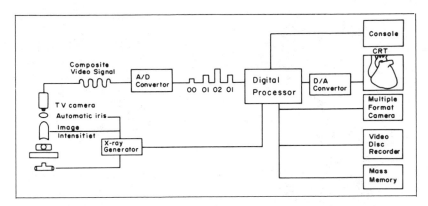

Fig. 9-8. Block diagram of a typical digital subtraction system.

then power-injected into the superior vena cava. After a suitable delay—depending upon the circulatory time of the patient, which can vary considerably—a fluoroscopic image is made of the anatomical area of interest with the television monitor presenting the digitized image in *real time.* Since only approximately 2 ml of iodinated contrast material can be given per kilogram to a patient, there is a limit to the number of contrast-material injections that can be given, and it is therefore important to correctly position the anatomical part to eliminate overlap of vascular structures that are to be visualized. Correct positioning is critical in these procedures.

X-ray Tube and Generator

Even though an x-ray generator may be supplied wth three-phase power, most fluoroscopic systems are in fact operated on single phase. Hence, many add-on digital units, that is, systems added to existing radiographic/fluoroscopic rooms, have been designed to operate on a single-phase input. Although in many instances the use of a single-phase power supply is cost effective, it must be remembered that with single-phase power there is a 100% ripple, as discussed in Chapter 4. This means that the power input to the image intensifier is not of a constant brightness level, although for routine use of the image intensifier this is of no concern. However, in digital subtraction units, it is important to control the brightness level to as near a constant as possible, and some electronic mechanism is necessary to maintain this near-constant brightness level. In some units, the radiographic mode (the mode used for spot films) is used for the digital application. As will be discussed later, in some instances it is necessary for the x-ray beam to be pulsed in short time frames to achieve adequate digital subtraction, and in these cases it is necessary for three-phase, pulsed

power input to be available. In some cases constant-potential generators are used because of their ultrashort exposure times. Sometimes it is also necessary to greatly increase the milliamperage of the exposure in order to increase the photon flux to the image intensifier, again to maintain a uniform image for subtraction. It is therefore important to be aware of the exact generator performance specifications to be certain of a good match between existing equipment and add-on unit.

Image Intensifier

The design and quality of the image intensifier tube itself are important because this piece of equipment may be the most important part of the digital imaging chain. The old cadmium image intensifier tubes, with the separation of the photon cathode from the entrance window, allow distortion of the image and are not useable for digital fluorography. The important factors to consider for reasonably new cesium iodide tubes are the contrast ratio, resolution, veiling glare, and pincushion distortion. The newer metal entrance windows, along with the relatively new fiberoptic output windows, provide higher resolution, better *quantum detection efficiency* (QDE), and better contrast than a standard cesium iodide image tube and may represent the tube to be used for digital fluorography. Moreover, as an image intensifier tube is used, the image degrades, losing both resolution and contrast at a rate of approximately 10% per year. Hence, in five years of routine use, a standard cesium iodide tube will have lost half of its initial resolution and contrast. Replacement image tubes are expensive, ranging from $25,000 to $45,000; it is extremely important to completely evaluate the image intensifier tube in terms of performance specifications before an add-on unit is obtained.

Television Cameras

In a standard fluoroscopic application, the aperture opening of the television camera is fixed, with the f-stop of the camera being set at the time of installation. The signal-to-noise ratio (SNR) of the standard camera is approximately 200 to 1. The standard vidicon has a very small lag, with the lag being less perceptible with the lead oxide vidicon. In a pulsed digital fluorographic procedure the brightness output of the image intensifier is considerably higher than when the instrument is in the standard fluoroscopic mode. Since the television camera cannot accommodate the massive changes in brightness levels, there must be a means for the aperture of the television camera to be changed automatically according to the brightness entering the television camera lens. Hence, with pulsed digital fluorography there must be automatic control of the aperture of the television camera. Also, the SNR of the television camera must be considerably higher, perhaps in the range of 1000 to 1.

In a standard fluoroscopic mode there is a continuous exposure to the patient while the television camera is reading the output of the image intensifier tube. The normal presentation is a 2 : 1 interlace system, as previously described, in which there is one-half of a picture interposed between another half, with one-half picture called a *field* and the two composite fields making up a *frame*. In the standard system, there is a finite period of time for the television camera to stabilize after an x-ray exposure has begun and a finite time for the television camera to return to zero potential after x-ray exposure has stopped. This means that the first few fields acquired after x-ray exposure has begun are of little use.

Continuous fluoroscopy, especially for angiocardiography, is necessary in order to obtain 30 frames per second and to eliminate bothersome flicker. There are obviously many applications in digital fluorography in which 30 frames per second exposure are not necessary. In those cases, a pulsed x-ray beam is used, with an interlaced camera reading out the output of the image intensifier after each pulse. Again the buildup time of the television camera wastes the first few fields. There is a necessity to keep the pulse as short as possible in order to have sharp, frozen images, but at the same time the pulse must be long enough to offset wasted frames during buildup time. In order to obtain extremely short exposures, a pulsed progressive scan is used. In some applications a very short exposure is used (approximately 1/30 of a second), and a single progressive scan of the exposure is made. Following the acquisition of this single frame, there is what is called a *scrub frame*. The purpose of the scrub frame is to remove any charge remaining on the target prior to the next expose-and-read sequence. Figure 9-9 compares the different types of acquisitions of the television field. Again, for applications that require a high frame rate, continuous x-ray exposures are necessary. In some applications this continuous exposure is at the fluoroscopic level, and in others the exposure is at the radiographic level. It appears that the pulsed x-ray exposure is necessary in order to obtain high-quality freeze frames and in order to decrease the radiation exposure to the patient when low frame rates are being used.

Analog-to-Digital Converters

The output signal from a television camera is analog, which means that the signal output is continuously and rapidly changing. This output analog signal is easily stored on numerous types of storage devices such as magnetic discs. In fact, more analog signals can be stored on a magnetic disc than digital signals. However, analog signals are difficult to manipulate as compared to digital signals, so some mechanism must be available to convert the analog signal from the television camera to a binary computer-usable digital signal. The purpose of an analog-to-

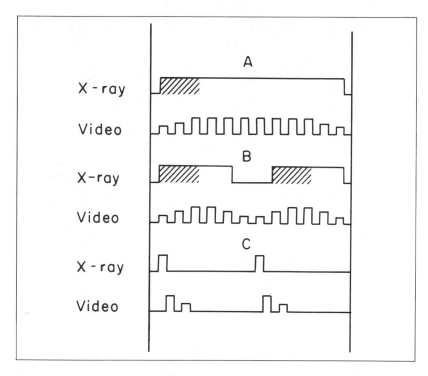

Fig. 9-9. A comparison of the different types of acquisitions of the television field. (Courtesy of General Electric Medical Systems.)

digital (A/D) converter is to take the amplified television video signal from the television camera and convert this voltage to a series of numbers that can be manipulated by a binary digital computer. Figure 9-10 illustrates a conversion of an analog signal to a digital signal. Because the television signal is rapidly changing, the A/D converter must be fast and must accurately convert analog signals to the digital numbers that the image processor will work with. The television signal is related to brightness levels, with each pulse related to the bandwidth being used. Recall that bandwidth (or "bandpass," as it is sometimes called) relates to how many times the electron is turned off and on per unit time. One signal relates to one brightness level; with an "8-bit-deep system," this is assigned a brightness level of 256.

Image Processor

An image processor is another term for a digital computer that has been programmed to manipulate the data obtained from the A/D converter so as to eventually present that data to a television monitor that can be manipulated by the user. At this point it is important that the reader understand some basic terminology concerning computers. The

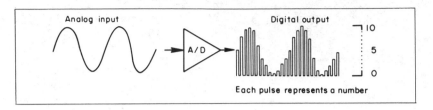

Fig. 9-10. Conversion of an analog signal to a digital signal.

term *bit* is commonly used and is an abbreviation for *binary digit*. A binary digit is a fundamental unit of the digital computer. Digital computers store information through a series of *flip-flops*, also called *logic elements*. A flip-flop is a simple circuit, the sole purpose of which is to be on or off. When the circuit is activated a "1" is sensed by the computer; zero is sensed when it is not activated. One bit represents one flip-flop circuit. By combining these flip-flop circuits, it is possible to store an astronomical amount of numbers. The bit is related to two to the nth power. For example, with a 2-bit system containing two flip-flops, there is the possibility of four combinations or two squared (2^2). On the other hand, an 8-bit system would be equivalent to the 2^8 power or 256 possible combinations of the logic elements. Thus, by using only a small number of flip-flop circuits, the computer has numerous available combinations. This is essentially how a binary computer operates. Another term used in computer terminology is *byte*, which represents 8 bits. This unit was originally chosen because 8 bits of memory are required to store the alphabet in lower and upper case, the numbers 0 to 9 and, all normally used punctuation marks found on a typewriter. Hence, in a computer printout, at least 1 byte of storage is required to print one number, letter, or punctuation mark normally found on a typewriter. In computer terminology, bits are rounded off such that a 64K (meaning 64,000) actually represents two to the sixteenth power, or 65,536 logic elements.

As in computed tomography, digital information is presented in the form of a matrix. With a 256-square matrix (or any other matrix, as far as that is concerned), each horizontal television line is divided into 256 data points called *pixels*. Since it requires approximately 6 ms for a horizontal television line to be pulsed, this means that each of the 256 data points has to be calculated in 23 nanoseconds (ns). For a 512-square matrix, the time becomes approximately 12 ns, and for a 1024 system, the time is approximately 6 ns.

In a 256-square matrix, therefore, data for 65,536 data points is necessary. For a 512-square matrix, the number of data points goes up by four times and is equal to 262,144. Hence, the larger the matrix size, the more data points that have to be calculated in the time of a single television field. It also stands to reason that the higher the matrix, the more resolution one would expect to obtain. Unfortunately, since the scanning time of a standard television system is fixed, the larger the matrix size, the larger the computer must be to calculate information for each data point or pixel.

Assuming again a 256 × 256 matrix, in order to have usable image, a brightness level must be assigned to each of the pixel units. This presentation of a brightness level can vary considerably. For example, at an 8-bit-deep brightness level there are 8 bits for 256 different brightness levels that can be assigned to each pixel, with zero representing minimum blackness and 256 representing maximum brightness. Since the human eye cannot possibly see 256 gradations in gray from black to white, different controls are available in the console. By manipulating these controls, the range of gray levels on the cathode ray tube (CRT) is manipulated. The range of brightest levels selected is called the *window* and the lowest number is called the *level*. For example, in a 512-square matrix with an 8-bit-deep brightness level, if a level of 80 and a window of 100 are selected, all the pixel values less than 80 are displayed as black on the television monitor and all pixel values between 80 and 180 are displayed proportionately between black and white with all pixel values of greater than 180 displayed as white. This capability of maneuvering window and level controls is probably more important with digital fluorography than with computed tomography.

To summarize, the larger the matrix size and the higher the brightness depth, the faster the computer must be. In addition, there must be provisions for storing a larger amount of information as the depth of the brightness level and the matrix size are increased. Accordingly, the faster the computer and the greater the storage capacity, the more expensive is the computer. For example, with a 512 × 512 matrix and an 8-bit-deep brightness level, for a 30-frames-per-second acquisition, a storage of 7.8 megabytes (Mbytes) of information is needed as a minimum for each 1-sec acquisition.

Computer Storage

In computer terminology, *core memory* refers to the storage capacity of a computer and is contained within the computer itself. Core memory storage in most computers is obtained through a series of *microprocessors*. Normally a computer is programmed to perform all calculations within the core memory of the computer. The larger the core memory, the greater is the number of computations that can be made. The core memory then is a limiting factor in the capability of a computer. Information in the core memory is not permanently stored and will be lost if the power supply to the computer is interrupted. In computer terminology, this is called *volatile information*. Core memory is used to provide high-speed random access to data.

At this point it would be helpful to describe the basic operation of the microprocessor. The fundamental building block of the microprocessor is a *silicon chip*. A chip is a small rectangular piece of silicon in which there is an integrated circuit of several electronic components. Figure 9-11 is a block diagram of a basic microprocessor.

Almost any computer system is composed of five basic functional units. These include the *input unit*, *central processing unit* (CPU), the *output unit*, the *memory unit*, and *bus lines*. The CPU is composed of the *control unit* and the *arithmetic-logical unit* (ALU). Connecting each of these units are electrical lines called buses. Connecting the input unit to the CPU is the *input bus*, and connecting the CPU to the output unit is the *output bus*, with the data bus connecting the memory unit to the CPU. A microprocessor is basically a CPU on a chip. The function of the input and output modules is communication with the outside world. The input module supplies information to the ALU and throughout the ALU to the memory unit. In the case of a digital fluorography unit, the input is the digitized analog signal coming from the television camera. The output module displays the manipulated data, and again in terms of digital fluorography, presents the data as an image on a CRT. A bus is used to transmit information in the form of signals from one unit to another. Input and output buses are similar to one-way streets, with data going only in one direction. A *data bus*, on the other hand, has to provide a mechanism for two-way communication between the memory unit and the CPU. Hence, a data bus normally has to transmit information in parallel fashion, since if only one bus line were available, information could not be transmitted at the same time as it was being received. This means that the data bus is bidirectional and can transmit and receive signals essentially simul-

Fig. 9-11. A basic microprocessor.

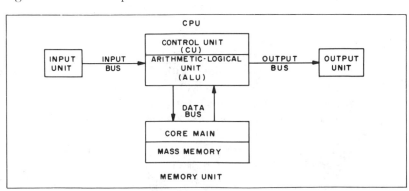

taneously. An *address bus* is used to route signals to different places within the chip. A *control bus* synchronizes the activity of the entire system, and the number of control buses available on a chip is related to the function of that chip. In the designing of chips, the design of the bus lines is crucial to the speed of the entire chip.

The memory module of a microprocessor is used to store information. Normally the memory may contain two types of information: programs and data. A *program* is a sequence of instructions that have been coded into binary form, which can reside in an electronic digital memory. The programs specify the sequence of steps executed by the computer. Under the supervision of the control unit of the CPU, each of the successive instructions of the program is obtained ("fetched," in computer language) and deposited into a special register of the control unit, where it is decoded and executed. For example, a typical instruction might add the contents of two registers and then deposit the results back into a third register. The data contained in the memory unit are processed by the ALU.

There are many different types of memory, and two types are normally used in any computer: main memory (core memory) and mass memory. The *main memory* is used to store the program currently being executed and the data required or generated by their execution. Retrievable information from the core memory must be rapidly transmitted to the CPU; hence, the speed in which information can be retrieved from core memory controls the overall speed of the central processing unit. A typical cycle time for a core memory unit is on the order of several hundred nanoseconds (1 ns $= 10^{-9}$ sec). Hence, the size of the main memory is limited by the number of transistors, its speed, and the speed with which the information from the core memory can be transmitted to the CPU. Likewise, core memory is expensive, and there must be a mechanism to store information that is not being used by the CPU during its current operation. This type of memory is known as *mass memory* and is used to store programs and data that are not immediately required by the CPU. In comparison with core memory, mass memory is slow, and, on demand, required information is transferred from the mass memory into the core memory for manipulation and then transferred to the CPU.

For mass memory that requires temporary storage, the storage device is normally a *disc*, with magnetic tapes being used for long-term storage. Transfer from a disc to the core memory is rather rapid in relationship to the time it takes to transfer information from a magnetic tape to core memory. Two major types of discs are available. The *Winchester disc-drive units* are electromechanical, hermetically sealed magnetic discs. They can store large amounts of digital information. The discs themselves are about the size and shape of a phonographic record. *Floppy discs*, also known as minifloppies or macrofloppies, are discs that are commonly used for word processors and personal computers. These are small, semiflexible discs that are inserted into slots in a disc-drive unit. Retrieval and access time of a floppy disc is slower than for a Winchester, but floppy discs have the advantage of being considerably cheaper.

Before moving on from the memory unit of a microprocessor, the terms *random-access memory* (RAM) and *read-only memory* (ROM) should be explained. A random-access memory is one in which the information stored on a chip can be accessed almost instantaneously and retrieved from anywhere on the disc. RAM memory can either be written or read. RAM memory is *volatile*, which means that all the information is lost when the power to the unit is lost. RAM memory is used in core memory because of its speed. Unfortunately, if a power interruption occurs during the acquisition of information, all the information is lost. With ROM memory, once its contents are written or fixed, it is permanent. These units are not affected by loss of power.

As noted previously, the CPU is composed of a control unit and an ALU. The control unit takes the program from main memory and acts as a "traffic cop" in determining what information is sent, where it goes, and in what sequence. As the name implies, the arithmetic-logical unit performs arithmetic and logical operations on the data passing through it, with the most common arithmetic functions including addition and subtraction. One should remember that division is nothing more than rapid subtraction; multiplication is nothing more than rapid addition. The logical operations of the ALU units include such things as "nor" and "or" operations. This means that under the direction of the program coming from the control unit, the logical unit could compare contents of one register to another to determine if the contents of the two registers could be added, subtracted, shifted, or what have you. Likewise, the logical unit could determine whether the contents in one register had reached a certain level, and whether the contents should be shifted to another register.

Now that we have briefly described the components of a microprocessor, a microprocessor can be defined as a large-scale integrated (LSI) mode of numerous microtransistors that implements the function of an arithmetic-logical unit plus its associated control on a single chip. Microprocessors may be composed of literally hundreds of chips representing millions of transistors. On the average an LSI chip contains from a few hundred to 10,000 or 20,000 transistors. A very large-scale integrated (VLSI) chip contains up to perhaps 100,000 transistors, and a super LSI (SLSI) chip contains more than 100,000 transistors. A typical microprocessor with a chip the size of this square □ has the capability of storing more information and at a more rapid speed than a whole room full of electromechanical computers available just 30 years ago. It should also be noted that many microprocessors may be used together to perform a task such as digital subtraction, in which one microprocessor output may be used to instruct another and so on. Units may be *hard-wired*, which means the program cannot be changed, or *programmable*, which means there is a capability by the computer or service personnel to alter the mode of the operation. The whole goal of digital fluoroscopic examinations is to examine low-contrast objects at a frame rate of approximately 30 frames per second, with some flexibility in the processing of the obtained data.

Types of Digital Subtraction

There are three basic types of digital subtraction: temporal, energy, and hybrid. Each of these methods appears to have some application in digital fluorography.

In *temporal subtraction*, a frame or the sum of several frames is sequentially subtracted from the succeeding frame in order to eliminate unwanted information and to allow the visualization of contrast material within vascular structures. As with standard subtraction techniques, the initial frame or frames are called a *mask*, with the mask containing the unwanted information that eventually is subtracted from the frames containing contrast material. Figure 9-12 is a diagram of temporal subtraction.

The primary limitation to digital fluorography using temporal subtraction mechanisms has to do with movement of the patient either by breathing, peristalsis, motion of the heart, vascular pulsations, or swallowing. The motion artifacts that result frequently require the examination to be repeated.

Fig. 9-12. Temporal digital subtraction. In this method a mask is recorded and sequentially subtracted from subsequent frames with the result that enhanced vascular patterns are visualized. (Courtesy Philips Medical Systems.)

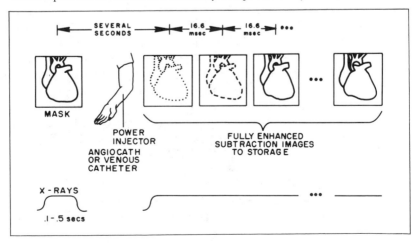

Energy subtraction is a mechanism that requires a very rapid switching of kilovoltage and milliamperage of the x-ray generator from a high energy, such as 130 kV, and to a low energy, such as 60 kV. Energy subtraction appears effective in subtracting soft-tissue densities from the image, with the residual image containing vessels and bony structures. Although the concept of digital energy subtraction eliminates some of the problems related to patient motion, there are problems that may limit its usefulness. With energy subtraction the effects of electronic and physical artifacts are far more severe than with temporal subtraction, with artifacts including scatter of x-ray and light and the inherent characteristics of the television imaging system. Hence, energy subtraction may be useful in removing soft-tissue artifacts caused by such things as swallowing, but there are a number of other factors that may severely limit its use in digital fluorography.

Hybrid subtraction is a technique developed by General Electric that is a combination of temporal and energy subtraction. In hybrid subtraction, a mask is obtained at low and high kilovoltages just prior to the arrival of the contrast material. These two energies obtained at high and low kilovoltages are combined to eliminate soft-tissue artifacts, especially swallowing, and the mask is then used as a subtraction mask. Energy-subtracted images are obtained as the contrast material goes through the anatomical area being examined, with each of these pairs again being processed to eliminate soft-tissue artifacts. What is accomplished with this method is that there remains a digitized image containing primarily bone and contrast material. These energy-subtracted images of the mask and post-contrast images are then temporarily subtracted, which effectively eliminates bone and soft-tissue images. Although still in the clinical evaluation stage, it appears that the use of hybrid subtraction may eliminate some of the problems produced by motion artifact. As with other imaging systems, there are also problems with hybrid subtraction. There is a reduction in the signal-to-noise ratio (as compared to energy or temporal subtraction). There is also an increase in radiation exposure to the patient when compared with temporal subtraction. However, in terms of the population as a group, the elimination of repeated examinations may be offset by the elimination of motion artifacts. Figure 9-13 illustrates the concept of hybrid subtraction. Hybrid subtraction, along with digital energy subtraction, requires a high-powered generator with the capability of very fast switching in (microseconds) between high and low kilovoltages, for example, between 60 kV and 130 kV.

Manipulating the Digital Image

Once a digitized image is presented on the CRT, it is necessary to manipulate the image for better visualization of the contrast-filled vessels.

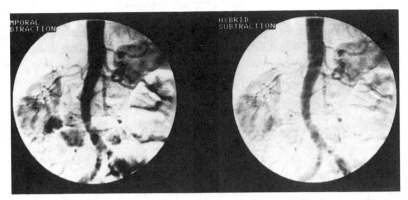

Fig. 9-13. Hybrid subtraction. (Courtesy General Electric Medical Systems.)

There are a number of post-processing procedures for image manipulation that can be used to enhance the digitized image.

The two common procedures for image manipulation are *remasking* and *pixel shifting.* In the remasking technique, a subtracted image is entered into the image processor, and another image containing the unwanted information is located from anywhere in the recording; these two images are then subtracted. Remasking is very helpful for eliminating minor motion artifacts or for eliminating bowel gas overlying a point of interest. Remasking is commonly performed and can salvage an otherwise useless examination. In pixel shifting, again the purpose of the procedure is to eliminate unwanted information, such as that coming from patient motion. In pixel shifting, one image is moved either horizontally or vertically in order to more or less align two different images before subtracting one from another. Both procedures are mechanisms to overcome minor misregistration between the mask and the subtracted image.

The answer to the question of whether one should buy an add-on unit or an integrated unit is a complex one. There are literally dozens of manufacturers now involved in supplying digital equipment to radiologists. In some cases it appears that digital units have been added on without a clear understanding of how the process works, with the result that there has been a tremendous financial loss simply because the add-on units sometimes will not work with the existing equipment. In addition to the image processor, the most important aspect is deciding whether the existing image intensification system will in fact accept a digital unit. It was noted previously that an image intensifier will degrade approximately 10% during each year of use. Additionally, the performance characteristics of the image intensifier may not have been good to start with. From an unbiased evaluation of the possibilities, it appears that the use of a continuous fluoroscopic mode with add-on digital process may only be capable of performing digital images of first- and second-order vascular structures at best. Provided the image intensification system is of proved and tested quality, an add-on unit may be sufficient to perform common digital procedures that do not require a high frame rate. An economic evaluation based on the number of procedures to be performed, the cost of the equipment, and the public relations benefits of having a unit must be made. For those departments that would be interested in the entire spectrum of digital applications, it appears that a totally integrated system designed especially for digital applications would be necessary. In between the two units are add-on units that have been designed to meet high-level performance specifications of existing imaging equipment.

The acquisition of a digital system is a very complex one; therefore, before acquiring any type of add-on unit, it would be well worth the time, effort, and financial expenditure to hire an independent consultant to assess the add-on unit, the projected work load, the projected applications, and the capability of the existing equipment. Acquiescence to the eagerness of an equipment salesman is no substitute for a knowledgeable, informed understanding of present and future capabilities of digital radiography. It should be understood from the preceding discussion on digital fluorography that this is a very complex subject, and indeed the presentation may be oversimplified. Hopefully, the discussion has emphasized the need to have in-depth information concerning digital fluoroscopy and the equipment involved before committing one's self to the addition of this new procedure.

A discussion of multiformat cameras is presented in Chapter 8. Multiformat cameras are used to provide static images for digital fluorography.

Equipment Considerations

The complexity of requirements for angiography has placed a burden on the industry to develop a mechanical and electronic design that is suitable for a wide range of vascular procedures accomplished today. The design should allow the operator full control of the procedure in an aseptic environment, and allowances should be made for easy and rapid completion of the procedure. The application of the equipment should be limited only by the knowledge and experience of the operator and not by the mechanical design of the equipment itself. In other words, angiographic equipment should have universal application today and also flexibility for the future.

No commercially available equipment has been designed for total application in angiography. The equipment requirements for neuroradiology, cardioangiography, and cardiovascular angiography are such that if a room of equipment were available for all applications, it would be too complex, too costly, and too massive to be used. A compromise might consist of equipment designed to operate on a C- or U-arm principle. The rotational benefit of this configuration, together with craniocaudad positioning of the x-ray tube/image receptor, can be utilized to perform most angiographic examinations, including 2 : 1 magnification angiography. Therefore, as with recording systems and generator capabilities, there has to be a compromise with equipment. With the feasibility of computed tomography (CT) scanning already having been demonstrated, pneumoencephalography is no longer a viable diagnostic entity in neuroradiography.

Figure 9-14 illustrates the basic components of a standard, conventional, universal angiographic laboratory. This equipment is capable of routine angiographic examinations when coupled with a 125-kV, 1000-mA, 12-pulse three-phase generator. The peripheral changer and its associated radiographic tube are not illustrated. The equipment shown in Figure 9-6 is also acceptable for most routine angiographic procedures.

Figure 9-15 shows the equipment requirements for biplane fluoroscopic/cine applications. This equipment is much more costly and complex than that in a standard angiographic room and is required

Fig. 9-14. Concept of a conventional angiographic laboratory utilizing a floating table top and serial film changer. Instead of the film changer shown here, one could substitute a see-through changer or a Puck changer. A triple-mode image intensifier is illustrated, and suggested focal spot/kilowatt ratings of x-ray tubes are noted.

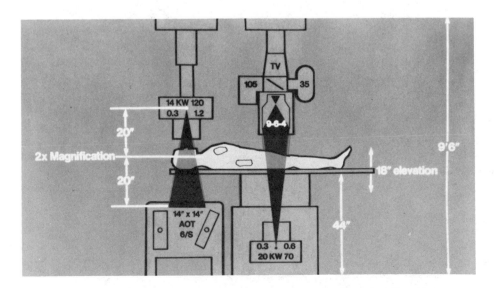

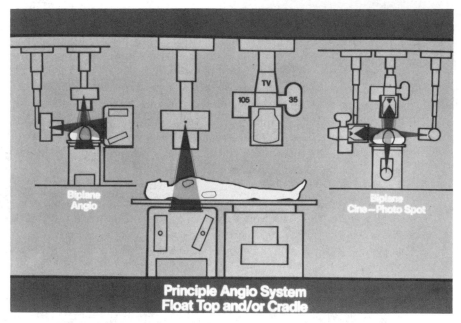

Fig. 9-15. Basic concept of a biplane angiographic setup for coronary arteriography. The film changer would not be necessary for pure angiocardiography, but it would be an ideal accessory for those installations having the work load to justify the addition.

Fig. 9-16. Basic principle of video recording in the angiographic suite, showing the applications of the video tape and video disc. Electronic recording with these parameters allows the potential of providing permanent film storage through electron beam or laser beam recorders.

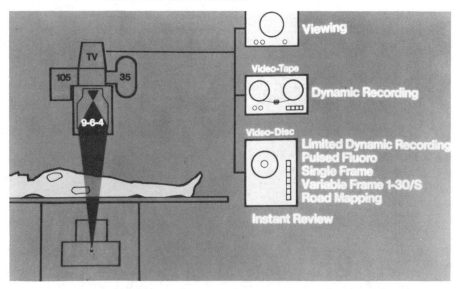

primarily for cardiac catheterization work and coronary arteriograms. The film changer is not necessary unless one has the work load to do routine angiography.

In the future we can look forward to more electronic recording of radiographic information. Video-disc recorders are becoming more popular every day. The advantages of video recording in angiography are illustrated in Figure 9-16. We can also look forward to devices such as electronic or laser beam recorders, which will either take the information recorded on video tape decks and transfer this information to permanent film records or will directly record the video information on film.

Progress in angiographic equipment has been rather remarkable in just a few short years. The advent of rare-earth screens and higher-resolution image intensifiers allows a greater utilization of current equipment. To gain the best results from this new technology, it is imperative that one understand the concepts of how the equipment operates and the compromises one must take into consideration.

Suggested Reading

Arnold, B., Eisenberg, H., and Borger, D. Digital video angiography system evaluation. *Applied Radiology* 10:81, Nov./Dec., 1981.

Buonocore, E., Pavlicek, W. A., and Meany, T. F. Viewpoint: digital imaging equipment—manufacturer's specifications. *RadioGraphics* 2:100, 1982.

Eskridge, J. M., Becker, G. J., Rabe, F. E., Holden, R. W., and Klatte, E. C. Digital vascular imaging: practical aspects. *Radiology* 148:703, 1983.

Harrington, D. P., Boxt, L. M., and Murray, P. D. Digital subtraction angiography: overview of technical principles. *American Journal of Roentgenology* 139:781, 1982.

Keyes, G. S., Pelc, N. J., Rieder, S. J., Sieb, L. E., and Enzmann, D. R. *Digital Fluorography: A Technology Update*. Milwaukee: The General Electric Company, 1981.

Riederer, S. J., Brody, W. R., Enzmann, D. R., Hall, A. L., and Maier, J. K. Application of temporal filtering techniques to hybrid subtraction in digital subtraction angiography. *Radiology* 147:859, 1983.

10

Instrumentation in Nuclear Medicine

Laurence P. Clarke

In the design of x-ray imaging equipment one is primarily concerned with measuring attenuation of a nonmonoenergetic beam of x-rays traversing a patient. The x-ray beam is of low photon energy (30–100 kiloelectrovolts [keV]), and since the photon flux is high, its presence can be detected with specially prepared x-ray film. It is therefore possible to obtain a high-resolution x-ray image of the patient. The image contrast obtained is related to the ability to differentiate photon attenuation in adjacent tissues.

The design considerations in nuclear medicine instrumentation are radically different. First, gamma rays of higher photon energy (80–511 keV) are being measured. Their increased penetrating properties make them more difficult to detect than x-rays. Second, the radioactive tracers are administered to the patient intravenously or by inhalation. The dose administered is kept to a minimum to reduce the radiation exposure to the patient. A very sensitive detector is therefore required, since the externally measured photon flux is very small in comparison to x-ray beams. Third, the radioactive tracer is generally of the form of a labeled compound, and hence it is unevenly distributed in vivo. Its spatial distribution or concentration may also vary with time. A specially designed collimated detector is therefore required to spatially resolve the distribution at different depths in the patient and for varying time frames. Finally, it is also important to minimize the detection of gamma photons of degraded energy that have undergone Compton scattering in the patient's tissues and lost their original direction. The latter can be achieved to a limited extent by determining the energy of the gamma rays externally and counting only those whose energy has not changed. Image contrast obtained by external detection is now related to the ability to differentiate the uptake of a radioactive tracer in the body organs or within an organ. However in x-ray imaging, photon attenuation degrades image contrast.

The design constraints on nuclear medicine instrumentation are therefore greater than those on conventional x-ray imaging devices. Specialized *scintillation crystals* are required that have the property of determining both the presence and energy of gamma rays with high sensitivity. The sodium iodide (NaI) crystal is the one most commonly used. This crystal has a high atomic number ($z = 32$) and a high density (3.67 gm/cm^3). Hence, the probability of interaction of a high-energy photon within the crystal is much greater than for x-ray film.

The presence of the gamma ray can be detected, since its interaction in the crystal results in a scintillation event or an instantaneous light flash. The scintillation photons are then converted to a short electrical pulse, which, in turn, is amplified and recorded. The amount of light output of the scintillation event is related to the energy of the gamma ray deposited in the crystal. Hence gamma rays can be measured as discrete voltage pulses, with pulse size related to their energy.

The detector systems used in nuclear medicine are essentially modifications of the system just described. The *scintillation camera* or *gamma camera* is the device most frequently employed. This camera is essentially designed to project a planar image of a radiotracer distribution in vivo onto the crystal. This is achieved through the use of a parallel-hole lead *collimator.* Gamma cameras of different crystal sizes and collimator types are employed, depending on the nature of the clinical study to be performed and on the energy of the gamma rays being imaged. Modifications to this camera include *single photon emission computed tomography* (SPECT) systems, where the camera head rotates around the patient in a similar manner as x-ray computed tomography (CT).

In this chapter the basic principles of these detector systems will be reviewed. Specific reference will be made as to their ability to determine the energy of gamma rays and to resolve their spatial distribution in vivo. Other more-specialized detector systems such as positron emission tomography (PET) devices, multicrystal cameras, and rectilinear scanners will not be discussed, as they are not as universally employed as the gamma camera.

In most investigational procedures in this field, emphasis is placed on functional imaging, for example, renal blood flow or blood clearance, brain flow or left ventricle wall motion, and ejection fraction. The gamma camera must therefore be interfaced to a minicomputer in order to perform dynamic imaging. The acquired data is available for subsequent calculation of functional parameters, such as those outlined previously. The important design factors and operation of these computers in the clinical area will be reviewed briefly.

It is assumed that the reader has read the earlier chapters on x-ray equipment, tomography, and computer data processing and has a basic knowledge of nuclear medicine or x-ray physics. A list of recommended literature on the topics described in this chapter is included.

Scintillation Counters

A schematic diagram of a scintillation counter is shown in Figure 10-1. It may be considered as having three basic components: the *crystal*, the *photomultiplier (PM) tube*, and the *pulse-processing system*. The interaction of a gamma ray of energy E_γ with the crystal is illustrated.

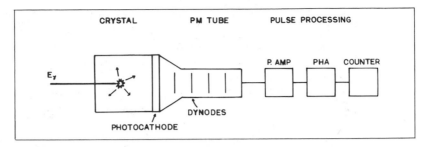

Fig. 10-1. Scintillation detector. The interaction of a gamma ray Eγ with the crystal and the resulting scintillation event is illustrated.

The energy of the gamma ray is absorbed by the crystal, and this energy is released in the form of light photons. The occurrence of a light flash of a very short time interval is referred to as a *scintillation event*. The light is collected by the PM tube and converted into a weak electrical pulse, which is amplified and processed before display. The *display system* therefore counts discrete electrical pulses, which correspond to each scintillation event in the crystal. Normally display systems are either *scaler counters* or *ratemeters* that register the number of electrical pulses (counts) detected for a given time interval, for example, in counts per second (cps) or counts per minute (cpm). In order for the reader to fully understand the principle of scintillation detection, the limitations of its application in imaging devices, and the calibrations necessary to measure the energy of the incident gamma ray, the properties of each detector component in Figure 10-1 will be described in more detail.

The Crystal

Materials that absorb energy from ionizing radiation such as gamma rays and release this energy in the form of a light flash are called *scintillators*. There are several different types of scintillators. Some are made of inorganic materials such as sodium iodide (NaI) or cesium iodide (CsI); others are made of organic material such as plastic. In nuclear medicine, NaI crystals are used because of their higher detection efficiency (greater light output) for gamma rays (80–511 keV). They are transparent to visible light, and hence light photons produced may be collected by the photocathode of the PM tube. When one inspects the scintillation counter, it is not possible to see the NaI crystal itself as it is enclosed in an aluminum case. This is necessary to exclude external light from the crystal, since the light from the scintillation event is very small compared to ambient room light. Furthermore, sodium iodide, like its chemical analog common salt, is hygroscopic and therefore must be hermetically sealed to protect it from moisture. The inner faces of the crystal are coated with a diffuse reflective material (usually powdered titanium dioxide), in order to direct the light to the photocathode. Finally, the side of the crystal facing the photocathode is unshielded. It is normally interfaced to a transparent light pipe, the latter optically coupled to the photocathode surface to ensure maximum light collection. The size and shape of the crystal will depend on the detector type. Cylindrically shaped crystals of 5 centimeters (cm) in diameter are normally used for radionuclide uptake probes, and crystals in the form of thin discs of 40 cm in diameter are used for gamma cameras.

The manner by which gamma rays interact with the crystal material is important in understanding how the energy of each gamma ray is measured. A schematic diagram of this interaction is illustrated in Figure 10-2 for a monoenergetic radiation source with a gamma ray of energy E_γ. The incident photon will primarily interact via Compton scattering and/or photoelectric events. It is initially assumed that the gamma ray energy is totally absorbed in the crystal. It interacts by one or more Compton scattering events, with the degraded photon energy finally absorbed by a photoelectric event in the crystal. Secondary electrons are produced by these interactions, which will cause further ionization within the crystal. Ion pairs become trapped or captured by imperfections in the crystal lattice. These imperfections are purposefully made by adding small traces of thallium (Tl) atoms to the crystal lattice (normally about 0.1–0.5% thallium). The energy absorbed by these imperfections is primarily released in the form of light energy (3 keV), corresponding to the wavelength of blue-green light. Two factors are important to remember with respect to the above scintillation event. First, the quantity of light generated within the crystal is directly proportional to the degree of ionization produced. This in turn is di-

Fig. 10-2. Gamma ray interaction with crystal via Compton scattering and photoelectric events. Case 1 illustrates total energy absorption in crystal; Case 2 - partial energy absorption. The corresponding pulse sizes produced and pulse height analyzer (PHA) settings to be discussed later are also schematically illustrated.

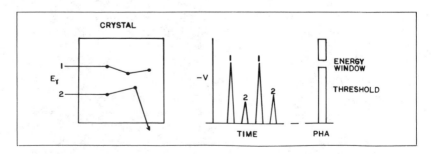

rectly related to the energy of the incident gamma ray. Second, the entire scintillation event is very short (0.3 microseconds [μs]), and hence discrete voltage pulses are measured.

The Photomultiplier Tube

Figure 10-1 also shows a schematic diagram of the PM tube. It consists of an evacuated glass envelope generally sealed within the same housing as the crystal. The front face contains the photocathode, the surface of which is coated with a thin, semitransparent coating of a photoemissive material (for example, cesium antimonide). The photocathode absorbs the light energy from scintillation events and emits a few low-energy electrons. The PM tube contains 10–13 charged plates or dynodes as indicated, each one having a progressively increasing positive potential (typically 100–1200 volts [v] for ten dynodes) with respect to the photocathode. The negative electrons emitted from the photocathode are therefore accelerated to the first dynode, their kinetic energy being sufficient to release more electrons from this plate. Each dynode is coated with a similar material to the coating of the photocathode. The electron amplification process is therefore increased from dynode to dynode until a measurable current is detected at the rear of the PM tube.

Important factors related to operation of the PM tube are as follows. The photocathode has a low quantum efficiency, where only 10–20% of the incident photons will result in emission of electrons. It is important that the photocathode surface is spectrally matched to the characteristic light output (wavelength) of the crystal for maximum detector efficiency. The gain of the PM tube is very high (10^6). Background pulses are therefore detected due to the release of photoelectrons from thermonic effects. However, they essentially can be excluded by pulse-height processing, to be discussed later. The PM tube is a linear gain device (for a constant applied voltage), and hence for this detector system the proportionality between the pulse size and the energy of the gamma ray absorbed in the crystal is maintained. However, it is important to realize that small changes in the high voltage applied to the PM tube will significantly change the size of the voltage pulse measured; this is because the gain of all dynodes are affected by this high voltage variation. Another important factor is the inability to design PM tubes with identical gain characteristics, and this will be discussed later with respect to the gamma camera.

The Energy Spectrum

Because of its finite size, the NaI (Tl) crystal does not always absorb the total energy of the incident gamma ray. In a sense it is an imperfect detector system. Instead of producing voltage pulses of the same size for monoenergetic gamma rays it yields a spectrum of pulse sizes, the shape of which is characteristic of the NaI–PM tube detector system. Hence the term *energy spectrum* for a scintillation counter. This concept is schematically illustrated in Figure 10-2 for two incident gamma rays of the same energy. In the first case indicated, the gamma ray undergoes one or more Compton interactions and is finally absorbed by a photoelectric interaction as previously discussed. This corresponds to a *total absorption event.* In the second case, it may undergo one or more Compton reactions, with its energy successively degraded but finally escaping the crystal. This corresponds to a *partial absorption event.* Hence, various amounts of energy will be deposited in the crystal depending on the number of Compton events occurring for each gamma ray. A range of pulse sizes will therefore be detected when measuring a radiation source. The pulse sizes for total absorption and partial absorption are illustrated in the same figure.

A convenient method for representing the spectrum of pulse sizes is illustrated in Figure 10-3. Consider a radiation source emitting monoenergic gamma rays being measured using the detector system just described. For a finite period of time, the number of pulses occurring at a given size is recorded. A frequency-type plot may then be drawn (with the number of pulses at a given size versus the pulse size as abscissa). This graph represents the energy spectrum for the detector, since the pulse size is equivalent to the energy of the gamma ray deposited in the crystal. This is true if the system is first calibrated with a gamma ray source of known energy (keV). The main peak of the spectrum primarily corresponds to the case of total absorption and is

Fig. 10-3. A gamma ray energy spectrum for a scintillation detector illustrating the photopeak position and energy window of pulse height analyzer (PHA). The intrinsic energy resolution index (FWHM of photopeak) to be discussed later is also shown.

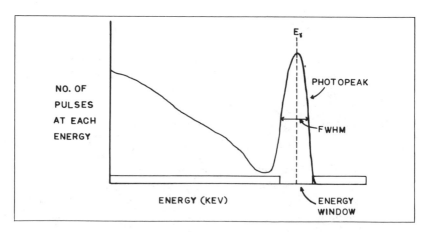

referred to as the *total absorption peak*, or *photopeak*. The smaller pulse sizes correspond to partial energy absorption in the crystal.

The Pulse Height Analyzer

As noted previously, nuclear medicine instrumentation is designed to measure the energy of gamma rays. In order to do so, it is necessary to discriminate between pulses corresponding to partial and total energy absorption as accurately as possible. This is achieved by discriminating between the corresponding pulse sizes as illustrated in Figures 10-2 and 10-3.

A *pulse height analyzer* (PHA), or *energy discriminator*, electronically measures each pulse height and accepts only those pulses lying within certain predefined limits. The energy limits are normally symmetrically positioned to enclose the photopeak, as illustrated in Figure 10-3. The energy window width is generally expressed as a percentage of the gamma ray energy. For example, for ^{99m}Tc (140 keV), an energy window of 15–20% is used for a modern gamma camera, centered on the photopeak.

Detector Response Characteristics

The important response characteristics of a scintillation detector are a) total detection efficiency, b) photopeak detector efficiency, c) intrinsic energy resolution, and d) pulse resolving time.

The *total detection efficiency* is a measure of the detector's sensitivity for gamma rays of a known energy. This response parameter refers to all photons that interact with the crystal and result in a current pulse of any size (that is, any pulse within the total energy spectrum). It is measured as counts per minute per microcurie cpm/μCi. However, for the gamma camera, one is interested only in those gamma rays whose energy has not been degraded by scattering in the patient and which are totally absorbed in the crystal. This parameter is now defined as the *photopeak fraction*.

$$\text{Photopeak fraction} = \frac{\text{number of pulses in photopeak}}{\text{number of pulses in energy spectrum}}$$

Photopeak detector efficiency = total detector efficiency $\times$ photofraction

The *photopeak detector efficiency* (cpm/μCi) is a function of the size of the crystal and the energy of the incident gamma ray. Table 10-1 shows typical values for this parameter for various crystal sizes used.

The *intrinsic energy resolution* is a measure of the ability to distinguish between gamma rays of similar energy. As previously mentioned, the scintillation counter is an imperfect system in that if monoenergic gamma rays interact with the crystal, a spectrum of pulse sizes results

*Table 10-1. Typical Values of Photopeak
Detector Efficiency for Different Crystal Sizes*

Crystal Size	Photon Energy (keV)	Photopeak Efficiency (% per gamma ray)
5 cm × 3 cm	140	90
(Probe)	364	30
40 cm × 9.5 mm	140	90
(Camera ⅜″ crystal)	364	30
40 cm × 6.35 mm	140	80
(Camera ¼″ crystal)	364	20

(Figure 10-1). More importantly, if all of the energy of the gamma rays is totally absorbed, it does not result in pulses of fixed size. Instead, a narrow range of pulse sizes is observed, resulting in a photopeak of finite width in kiloelectrovolts. The latter can be explained as being due to statistical variation in the amount of light produced (ionization) in the crystal, in the reproducibility in the light collection, and in the gain stability of the photomultiplier and electronics tube. An index is used to measure the spread in the width of the photopeak, as illustrated in Figure 10-3. It is defined as the *full width half maximum* (FWHM) of the photopeak. It is often calculated as a percentage of the photopeak position (keV).

$$\text{Intrinsic energy resolution (\%)} = \frac{\text{FWHM (keV)}}{\text{E}\gamma \text{ (keV)}} \times 100$$

This index is a function of several factors in the design of a scintillation counter, such as the crystal size, the PM tube characteristics, and the geometrical configuration and type of PM tubes employed in gamma cameras. The energy resolution increases as the gamma ray energy increases, being approximately proportional to $E^{-1/2}$. Note the smaller resolution index value, the greater the energy resolution. This parameter is normally measured for ^{99m}Tc (140 keV) or ^{137}Cs (622 keV). Typical values for various crystal types are shown in Table 10-2.

The Gamma Camera

A schematic diagram of the scintillation camera is shown in Figure 10-4. It essentially consists of a large disc shaped NaI (Tl) crystal, typically 40 cm in diameter and 9.5 millimeters (mm) thick. This crystal is generally optically coupled via a thin light pipe to 37–61 PM tubes arranged in a hexagonal array. When a gamma ray interacts with the crystal, photons are emitted from the discrete scintillation site and collected by the photocathodes of each PM tube. The amount of light

Table 10-2. Typical Values for Intrinsic Energy Resolution for Different Crystal Sizes

Crystal Size	Photon Energy (keV)	Energy Resolution (%)
5 cm × 3 cm	140	11
	662	8
40 cm × 9.5 mm (Camera ⅜″ crystal)	140	11–12
40 cm × 6.35 mm (Camera ¼″ crystal)	140	11–12

reaching a particular photocathode, and consequently the size of the current pulse measured, depends on the distance between the photocathode and the site of interaction. The signal from each PM tube is fed to a series of electronic circuits and register networks located in the head of the camera. These logic circuits analyze the relative strengths or timing of the incoming signals to determine the position at which the scintillation event took place in the crystal. Since the PM tubes are geometrically located in a symmetrical hexagonal array, it is possible to determine the X,Y coordinates of the event, assuming the crystal lies in an X,Y plane.

Voltage pulses whose height corresponds to these coordinates are used to display each event on a *storage (persistence) oscilloscope, image formatter,* or television monitor. In contrast, the sum of the signals from all of the PM tubes for each event reflects the total energy of the gamma ray absorbed in the crystal. PHAs are therefore employed and calibrated such that only total absorption events are detected. This is illustrated in the diagram where a PHA presents a logic on pulse (referred to as the *Z pulse*) for each accepted event to the display and counter systems. An image is therefore accumulated by the camera, where the detection of scattered photons with degraded energy from the patient is minimized. In practice, many modifications and refinements to the design of the camera as previously described are used. For example, they may include corrections to compensate for the partial absorption effects at the edge of the crystal, the enhanced sensitivity observed directly underneath the PM tube photocathode, or the maximum count rate capability.

Intrinsic Response Characteristics

The intrinsic response characteristics of the gamma camera refer to the properties of the detector head without the collimator. Therefore, they are primarily a measure of the overall response characteristics of the crystal, the light pipe design, the PM tube array, the electronic pulse processing, and the PHA and its energy settings. They reflect how well the camera head is designed for imaging photons of a given energy

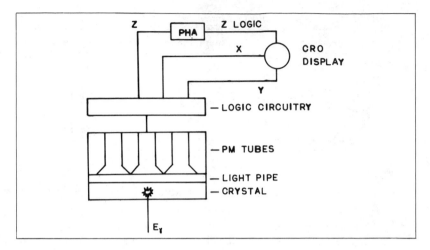

Fig. 10-4. Scintillation camera and electronics. Only three PM tubes are shown for simplicity.

(keV). The important parameters include the intrinsic energy resolution, spatial resolution, spatial linearity, flood field uniformity, and count-rate performance. Other parameters such as the system (overall) spatial resolution and sensitivity refer to the collimated camera system and will be discussed later.

The National Electrical Manufacturers Association (NEMA) has published a specific set of criteria for measuring these intrinsic response characteristics for the purpose of providing a reproducible method for comparison of gamma cameras from different vendors. The response characteristics are based on these recommendations. They are measured for both the *useful field of view* (UFOV) of the camera, which refers to the area of the crystal exposed by the collimator and used for clinical imaging, and the *central field of view* (CFOV), which is defined as 75% of the diameter of the UFOV. The latter field size is included, as the response characteristics of the camera are generally better in this more central region. This is because of the symmetrical array of the PM tubes and because of more efficient light collection. For details of the actual measurement methods, the reader is referred to the published 1980 NEMA specifications [1]. The definition of each response parameter, the related camera design factors, and their importance to clinical measurements will only be described here. It should be realized that many of these parameters are interrelated in terms of improved camera performance.

The *intrinsic energy resolution* of the gamma camera is its ability to distinguish between gamma rays of similar energy and is defined as the FWHM of the photopeak as described earlier for a single crystal–

PM tube system. It is measured using a distant point source of ^{99m}Tc (140 keV) that uniformly irradiates the UFOV and CFOV of the camera crystal. For the modern gamma camera, the intrinsic energy resolution is on the order of 11–12%. The magnitude of its value is important, particularly for imaging low-energy photons, where it is necessary to minimize the detection of small-angle scattered photons whose energy is only slightly degraded. The energy window of the PHA (Figure 10-3) is normally symmetrically centered on the photopeak. The size of this energy window is generally 20%. However, it may be reduced to 15% for better scatter reduction for cameras that have an intrinsic energy resolution of 11% or better.

The *intrinsic spatial resolution* of the gamma camera is its ability to accurately determine the origin or X,Y location of each scintillation event in the crystal. For every event in the crystal, the energy and position of the interacting photon is determined. It is measured by imaging a collimated narrow line source positioned in the X and Y direction and across the UFOV and CFOV of the camera. The count profile across the line source is obtained, defined as the line spread function, which is normally of a Gaussian shape. The FWHM of this profile is used as an index of spatial resolution. The FWHM is calculated in a similar manner as the intrinsic energy resolution index. The magnitude of the FWHM for a modern gamma camera is approximately 3–5 mm. Its magnitude is important as the intrinsic spatial resolution is related to the overall spatial resolution of the camera and may be a limiting factor if very high resolution collimators are used. An additional resolution index, the *full width tenth maximum* (FWTM) of the line spread function is also determined for the comparison of camera characteristics. Its magnitude is on the order of 8–10 mm. In reality, the NEMA method for measuring intrinsic spatial resolution employs a specialized parallel lead slit phantom for determining the average value of the previously discussed parameters across the full UFOV and CFOV.

The intrinsic spatial linearity is also measured by imaging the parallel lead slit phantom above. It is an index of the extent of X,Y positional distortion in the measured scintillation events across the camera crystal. The spatial linearity is measured as the deviation of the line images obtained from the actual linear dimensions of the parallel slit pattern. Two terms are used. The *differential linearity* (in mm) is the standard deviation of the measured displacement across the UFOV and CFOV. The *absolute linearity* (also in mm) is a measure of the full displacement across both fields of view. Typical values obtained are 0.45 mm and 0.4 mm, respectively, for the UFOV.

The *intrinsic flood field uniformity* is a measure of the uniformity of the response of the gamma camera when exposed to homogeneous photon flux across its crystal. It is generally measured by irradiating the crystal using a distant point source of ^{99m}Tc (140 keV) and acquiring the image matrix (64 × 64 pixels) into a computer. Again, two terms are used. The *differential uniformity* is related to the maximum rate of change of the camera sensitivity over a finite distance (for NEMA, 5 pixels or matrix points) within the UFOV or CFOV. The *integral uniformity* is in turn an index of the maximum change across each entire field of view. Typical values obtained for a modern gamma camera are ± 9.5% and ± 60%, respectively, for the UFOV. The importance of the uniformity and linearity of response of the camera is clear since an undistorted image of the radioisotope distribution in vivo is required for quantitative functional measurements.

The *intrinsic count rate performance* of the gamma camera is its ability to accurately record the rate of scintillation events in the crystal at count rates near the detector's maximum. This aspect of the camera's performance is related to the finite time required to fully process each scintillation event. For example, a finite time period (deadtime) is needed to determine the photon energy and its X,Y location and display the accepted event on the CRO (Figure 10-4). This parameter is measured by counting two sources of ^{99m}Tc, both together and separately, and determining the count loss when the greater activity of the two sources is detected. The deadtime is calculated from this data. The index normally used, however, is the calculated count rate at which the percentage of loss due to deadtime effects is 20%. The latter is a more meaningful parameter as it can be directly related to a particular clinical study where an estimate of the count rate can be made. A typical value for a modern camera is 140,000 cps.

It is important to realize that the measured deadtime is related to the size of the energy window used and the extent of the scattering medium present. The magnitude of the deadtime is important for clinical investigations where high count rates are observed, such as in first-pass cardiac studies in which a small bolus of 20 millicuries (mCi) of ^{99m}Tc is generally used. The deadtime will significantly reduce the maximum count rate observed and the statistical accuracy or value of the measured indices in this study. Furthermore, for some cameras other response characteristics are degraded, such as the spatial resolution and uniformity at high count rates. The NEMA specifications require that the camera's response parameters should also be measured at a count rate of 75,000 cps to determine related deadtime effects.

Design Factors That Influence Intrinsic Response Characteristics

The design factors that influence the intrinsic response characteristics are listed in Table 10-3. Many of these have been modified to improve the spatial resolution of modern gamma cameras. For example, if the scintillation event occurs closer to the array of PM tubes behind the light pipe, the ability to detect its location will be improved due to simple geometrical considerations. For this reason, cameras with thin-

Table 10-3. Examples of Design Factors That Influence
the Intrinsic Response Characteristics of the Gamma Camera

Crystal thickness ⅜″, ¼″

Light pipe design, thickness, or exclusion

PM tube design, number, shape, and packing density

Signal processing or microprocessor correction systems for intrinsic energy
   resolution, uniformity, and linearity

---

ner crystals (⅜ inch or ¼ inch) or designed with thin light pipes exhibit improved intrinsic spatial resolution. Similarly, if the number of PM tubes is increased (for instance, from 37 to 61), the number of spatial sampling points for the scintillation event is increased, and this in turn will result in improved spatial resolution. The reduction in light pipe thickness, however, introduces greater nonuniformity of response across the camera's crystal. This is because of the variability in response underneath the PM cathodes, now more immediately adjacent to the crystal surface. Modifications to the design of thinner light pipes have been necessary to achieve a more uniform response, such as the use of variable-density media that diffuses the light within the light pipe under the photocathode of each PM tube and reduces the nonuniformity of response across their faces.

Microprocessor Correction Systems

A recent development has been the introduction of microprocessor correction systems that are incorporated into the detector head. This microprocessor generally has a memory size of 16K, 8 bits deep. In essence, it subdivides the crystal area into a 128 × 128 matrix and uses this matrix as a correction table for each element of crystal associated with each matrix point. The earlier systems were designed to correct only for variations in crystal sensitivity across the crystal's surface. This was achieved by acquiring and storing an image of a uniform flood source in the memory to be later used as a correction matrix for the clinical images. However, the overall nonuniformity of response of the camera can also be attributed to other factors, such as the variation in intrinsic energy resolution across the crystal and the effects of the system's spatial nonlinearity. As a result, microprocessor correction systems have now been introduced that correct for variations in sensitivity, energy resolution, and linearity. The overall uniformity of response has been improved from typically ± 10% without correction to less than ± 4% with the above corrections applied. The success in the use of the microprocessor systems has in some instances led to the removal of the light pipe to further improve the intrinsic spatial resolution of the camera, since any greater nonuniformities in response can be corrected for.

Many hardware modifications to the above concept are employed by different camera manufacturers. These include real-time tuning of the PM tubes and more sophisticated electronics that correct for nonlinearities in the detector system. In some instances the camera electronics are more fully integrated into a digital computer, which permits the signal output of PM tubes to be more immediately processed and digitized than is possible with conventional analog camera circuitry. Digital systems of this type have one advantage in that the corrections discussed above are applied based on the current status of the camera's operation, intelligently monitored by the computer. The microprocessor-based systems assume the camera electronics remain stable for the time period after the correction matrix is loaded into memory, and this is generally the case for several days. A further advantage of the digital system is their higher count-rate ability, where count rates on the order of 200K are possible. However, the choice of camera system should be carefully judged on all aspects of its intrinsic performance characteristics, since in some instances they are degraded in the more digital systems due to incomplete sampling of the positional pulses (excessive pulse clipping) in the detector head. Significant development in digital camera technology is expected in the near future from several imaging companies, and the reader is advised to monitor this development before purchasing a camera.

Collimator Design and Types

The purpose of the collimator is to obtain a two-dimensional image of the radionuclide distribution in the patient, distributed in three dimensions in vivo. Its objective is therefore to project an image of the radionuclide distribution, such that there is approximately a 1 : 1 linear relationship between the latter and the image formed by the gamma camera. This is generally achieved by positioning a parallel multihole collimator in front of the crystal, as illustrated in Figure 10-5. It geometrically restricts the direction of the photons originating in the patient and interacting with the crystal to those that are directed parallel to the Z axis of the detector/collimator system. The collimator is normally made from lead such that photons that do not pass directly through the collimator holes, or apertures, are strongly attenuated or absorbed in the surrounding shielding, also called the collimator septa. The collimators are generally designed so that the septal thickness (seen as S in Figure 10-5) is sufficiently large, such that septal penetration of photons is very small (of the order of 2–8%).

Neglecting the effects of septal penetration, the response characteristics of the collimator can be approximately estimated on simple geometrical considerations. For example, the response characteristics for each aperture can be geometrically estimated, and the total characteristics of the collimator can be determined from the sum of the effects

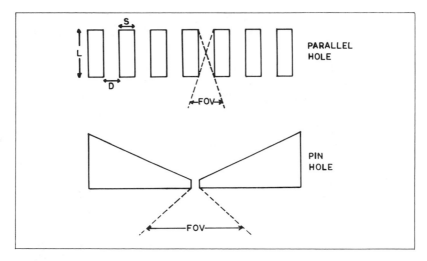

Fig. 10-5. Parallel hole and pinhole collimators. The field of view (FOV) of the collimator apertures are illustrated.

of all the apertures. Using this approach, the spatial resolution (Rg) is related to the field of view (FOV) of a given aperture and is therefore a function of its length (L) and diameter (D) (Figure 10-5). Hence, the smaller the value of D and the longer the collimator aperture L, the better is the spatial resolution obtained. If the aperture length L is long, the smaller the variation in spatial resolution is with distance from the collimator face. The sensitivity of the collimator, or more correctly its geometrical efficiency (G), is unfortunately inversely related to the spatial resolution. It is related to the total crystal area exposed by all N apertures:

$$G = \frac{\pi N\, r^4}{4\, L^2} \qquad r\ (radius)\ =\ D/2$$

Hence, the geometrical efficiency is sharply reduced with reduction in aperture diameter or if the length of the collimator is increased.

Many modifications to the design of parallel-hole collimators have been made with the objective of optimizing their geometrical efficiency, or sensitivity, for a desired spatial resolution. For example, triangular, square, and more recently hexagonally shaped holes have been introduced to increase the packing density of the apertures for a given septal thickness. Hence, the crystal area exposed by the collimator is increased. This has resulted in the design of more efficient collimators with better spatial resolution, in particular for those designed for low-photon energy (140 keV) where photon penetration is not a major problem. However, the design of higher-energy collimators with correspondingly thicker septa has not seen much improvement, as these collimators must be cast from lead or other high Z material, and photon septal penetration remains a limiting factor.

Collimator types designed for varying sensitivity and spatial resolution requirements and photon energy are tabulated in Table 10-4. Some important factors should be realized in the choice of collimator for a specific clinical study. For example, a high-sensitivity collimator may be used to improve the statistical accuracy of some measured functional parameters such as the cardiac ejection fraction. However it should be realized that its spatial resolution varies significantly with depth (Figure 10-5, Table 10-4), since the hole length L is made shorter or the hole diameter is increased. Conversely, the high-resolution or ultrafine-resolution collimator has lower sensitivity and will require the use of longer imaging times. Its choice will depend on the degree of patient motion to be expected during the study period.

The overall spatial resolution (R_o) of the camera collimator system is related by the following expression neglecting scattering effects:

$$R_o = \sqrt{R_i^2 + Rg^2}$$

where Rg is the collimator (geometrical) resolution and R_i^2 is the intrinsic spatial resolution of the camera head. Typically R_i^2 is on the order of 4 mm. Hence, for most collimators, particularly for the value of Rg at 10 cm from the collimator face, Rg is the limiting factor in the overall spatial resolution of the camera. In the above equation it is important to use the value of Rg at 10 cm or greater from the face of the collimator (Table 10-4), since this is where the organ or tumor of interest will normally be located.

Other, more specialized multichannel collimators have been designed for the gamma camera, such as the diverging-hole and converging-hole collimators. These collimators introduce geometrical distortion into the image obtained. With the advent of cameras of larger fields and greater sensitivity, these collimator types are not now widely accepted. The single pinhole collimator, as illustrated in Figure 10-5, has still been retained for imaging the thyroid gland and other superficial organs that are close to the skin surface. For short imaging distances from the pinhole collimator, it exhibits good spatial resolution. However, at greater imaging distances this collimator is significantly limited due to geometrical distortions and degraded resolution.

Instrumentation Choice and Bid Specifications

The first decision area in the choice of a gamma camera is its physical size and the thickness of the crystal required. The one most commonly used is the large field of view (LFOV) (typically 40 cm), since it can be

Table 10-4. Examples of Collimator Design Types and Their Dimensions and Response Characteristics

Collimator	Energy (keV)	No. of Holes (K)	Hole Size (mm)	Hole Length (mm)	Relative Sensitivity	Collimator Resolution (Rg) FWHM (mm) Distance from Collimator	
						0 cm	10 cm
General purpose	140	16	2.4	35	1.0	3.0	10.0
High sensitivity	140	5	3.3	35	2.2	4.5	15.0
Ultrafine resolution	140	41	1.4	25	0.7	2.0	8.5
Medium energy	360	12	2.0	45	0.8	4.5	9.5
High energy	500	4	3.3	75	0.3	3.5	8.5

applied to most imaging procedures such as liver and spleen, bone, brain, and renal imaging and, to a limited extent, to cardiac imaging if an optical or computer software zoom is available. This system is also employed in SPECT and whole-body scanning, as its FOV is sufficient to include most organs of interest. However, a dual-pass motion is required for the whole-body scanning mode, as the camera's LFOV is not sufficient to include the full width of the patient. In order to image a wide range of photon energies, a general-purpose camera should have a ⅜-inch-thick crystal. Earlier camera designs were made with a ½-inch crystal because of the greater use of higher photon energies such as ^{131}I (365 keV). However, the ⅜-inch crystal is now the optimum choice, since it permits better spatial resolution, as explained earlier. If the camera is to be restricted to low-photon-energy imaging, the ¼-inch crystal may be used, which will allow further improvement in spatial resolution.

More specialized cameras are designed with either a small field of view (SFOV) (typically 25 cm in diameter) or an extra LFOV such as the jumbo camera (50 cm in diameter). Recently, large rectangular crystal systems (typically 50 × 36 cm) have been made. The SFOV systems are designed with the ¼-inch crystal, since their shielding requirements and weight considerations are less. Most mobile cameras are built using this configuration. These cameras are particularly suitable for cardiac imaging because of the ease of positioning patients undergoing rest and exercise studies. The ¼-inch-thick crystal is the optimum choice for imaging ^{201}Th (80 keV) because of its low photon energy. A particularly convenient feature is the built-in data processor systems in the mobile camera, as one technologist can essentially perform mobile cardiac work.

The extra LFOV cameras are generally aimed at more conventional imaging modes—for example, bone and gallium imaging—but with the capability of reduced imaging times. They are also particularly suitable for whole-body scanning, since their FOV is sufficient for a single pass along the patient, reducing imaging time by half. These cameras generally use a ⅜-inch crystal, so if cardiac imaging or small-organ imaging is not being considered, they could serve the role of a general-purpose camera. However, the response characteristics for the larger cameras tend to be somewhat degraded compared to the LFOV or SFOV systems.

The second decision area involves the electronics package contained in the camera console. These factors are listed in the representative bid specifications for the gamma camera outlined in Table 10-5, and the more important ones will be discussed briefly. First, the console should be equipped with a minimum of three energy windows for gallium or dual-isotope imaging. An autopeak facility with preselected (auto-

Table 10-5. Representative Specifications of
Camera, Console Electronics, and Display System

Microprocessor or real-time computer correction system (uniformity, linearity, and energy)

Autopeak facility for six preselected energy windows

Simultaneous triple-energy windows, both manually and automatically selectable

Energy spectrum display (MCA 128 channels) and autopeak tracking facility

Information density, optical zoom, anatomical marker, and image orientation

High-resolution displays: multiformatter, nonpersistence scope, persistence scope, video convertor, and single- or triple-lens Polaroid camera

Single- or dual-pass whole-body electronics

Interface to computer

matic) energy windows for ^{99m}Tc, ^{201}Tl, ^{67}Ga, ^{123}I, and so on. A useful feature is peak tracking to correct for any drift in the photopeak settings during the study. A microprocessor correction facility or equivalent should be included for energy, linearity, and uniformity correction. Autoexpose features using information density over a specified area of the image should be available.

Simultaneous dual exposures with different settings are useful to avoid repeat scans, particularly in whole-body scanning or dynamic studies. The console should be equipped with a cardiac gate for gated imaging, and particular emphasis as to the nature of R-wave tracking should be stressed. Many cheaper models do not track the R wave, since they operate solely on a pulse height threshold concept. They often are problematic in many cardiac investigations where the R-wave pulse height varies.

The third decision area is the human engineering aspects of the detector system, such as the ease in detector positioning anterior, posterior, or lateral to the patient; the relative position of the control box and motor speed; and the ease of collimator exchanges, which cannot be overemphasized. In addition, the positioning of the image display system relative to the camera head is important for patient positioning and monitoring of the study.

Finally, one must balance the above factors with the NEMA performance specifications and the terms of the service contracts on the system. Cost evaluations should be based on capital costs and three- to five-year service contracts. One last point is that cameras from various vendors should not be compared using quality control tests such as the bar phantom, but with strict adherence to NEMA specifications. Furthermore, acceptance testing using NEMA criteria after installation is encouraged, as the service contract performance characteristics are legally binding for the contract's time period.

Single Photon Emission Computed Tomography (SPECT)

Interest in single photon emission computed tomography (SPECT) has been renewed since the early work of Kuhl [2] because of the successful application of transmission x-ray CT. However, the design problems in SPECT instrumentation should be clearly differentiated from those of CT. For example, the source of radiation is distributed in three dimensions in the patient, and the emitted photon flux is very small because of the limitations imposed by patient dose considerations and the effects of photon attenuation, which are significant for low-energy gamma emitters. In contrast to CT, in the SPECT detection mode the effects of photon attenuation must be corrected for in order to improve image contrast. This correction is also required for the calculation of functional parameters based on the observed counts in the tomographic plane of interest. In addition, considerable attention to the collimation of the detected gamma rays is necessary in order that the principles of the back-projection method of reconstruction may be applied. The detected gamma rays are assumed, to a first approximation, to be collimated parallel and perpendicular to the face of the detector. Finally, it is more important in the SPECT mode to minimize the detection of small-angle scattered photons, to ensure their directional nature is not lost. This section will review the various methods of SPECT, the image artifacts observed, and the design constraints on the instrumentation involved. The software and computer hardware requirements will be discussed later.

SPECT System Types

SPECT systems can be subdivided into two imaging types: longitudinal sectional imaging and traverse sectional imaging, as illustrated in Table 10-6. The former imaging mode refers to those systems whose primary reconstructed planes are parallel to the longitudinal axis of the patient, for example, that obtained with the moving detector *phocon* scanner that transverses along the subject. These systems are often referred to as *limited angular sampling devices.* This is because the angular range over which views of the organ of interest are obtained is limited. Examples would be the small angular range of the collimator holes in the rotating slant-hole collimator, the seven-pinhole collimator, or the focused collimators of the phocon scanner. The images obtained therefore suffer from geometrical distortion and tomographic blurring between each reconstructed plane and hence are generally not suitable for obtaining quantitative functional information. In some instances, however, such as the phocon scanner, this instrument is useful for longitudinal whole-body tomography where quantitative aspects are not as important. The instrument that has received the most attention is the *rotating gamma camera.* It is capable of complete angular sampling by rotating 360 degrees around the patient. Because of the

Table 10-6. System Types Involved in SPECT

Longitudinal Sectional Imaging (limited angular sampling)	Traverse Sectional Imaging (complete angular sampling)
Rotating slant-hole collimator	Translate/rotate scanner
Seven-pinhole collimator	Detector ring
Phocon scanner with focused collimator	Rotating gamma camera (SPECT)

camera's large sensitive area, several traverse reconstructed planes can be obtained simultaneously. The basic principles of gamma camera tomography, the possibilities of quantitative measurement, and the design constraints on the system will be reviewed briefly.

Basic System Design

Figure 10-6 illustrates the basic concept of gamma camera tomography, or SPECT. The camera head is supported by a gantry that rotates the detector around the patient, through either 180 degrees or 360 degrees. For each discrete angular step of 2 degrees or more, the camera acquires an image of the patient in the form of an image matrix, typically 64 × 64, and the image matrix is stored in a computer. The filtered back projection method is similar to x-ray CT and will not be discussed in detail here. However, the choice of the reconstruction filter is important. Although it is primarily applied to remove the common star artifact, it also influences the relative count distribution in the reconstructed slice, the extent of reconstruction noise introduced, and the absolute count obtained. It is anticipated that specific filter types will be tailored for each type of clinical investigation and photon energy because of the factors just discussed.

The filter algorithm used should also include a correction for photon attenuation for each image projection obtained of the patient. It is difficult to obtain an exact mathematical solution to the problem of photon attenuation because the attenuation coefficient varies for different tissue types and air spaces, such as in the thorax. Photon attenuation correction is under intensive research at present, and it is the most significant limiting factor in the objective of obtaining quantitative or absolute measurements in vivo (μCi/cm^3). At present, reconstruction software supplied by most imaging companies employs a constant value for the attenuation coefficient for the photon energy of the radiotracer being imaged.

Imaging Parameters

Typical data acquisition parameters are shown in Table 10-7. The linear sampling in SPECT refers to the size of the image matrix used for

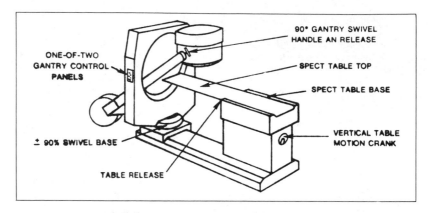

Fig. 10-6. Rotating camera SPECT system. This system can be used for both planar and whole body imaging.

each camera view and is related to the overall spatial resolution of the camera/collimator system. Since the radius of the rotation of the camera in Figure 10-6 is large (typically 15–25 cm), the spatial resolution of the system at this distance is on the order of 1 cm or greater. Hence, an image matrix of 64^2 for a LFOV camera is typically employed, or a matrix of 128^2 if the spatial resolution is improved with specialized collimation. The requirements of angular sampling are not as critical as those of linear sampling. It is typically on the order of 2 degrees or greater (60–120 views per 360-degree rotation). The choice of angular sampling is dependent on the total count per view required and the limitations imposed on the imaging time, which is typically about 20–30 minutes.

Image Artifacts

The specific types of image artifacts observed in SPECT are shown in Table 10-8. First, the reconstruction algorithm, as mentioned earlier, amplifies any noise in the original image projection obtained. Because of the random nature of the radionuclide decay and the low photon flux, significant statistical variations in each image projection are observed. It is therefore necessary to acquire a larger number of counts per view—more than the number required for planar imaging—in order to reduce this reconstruction noise. In general, greater than 2 × 10^6 counts are necessary for the total SPECT acquisition to obtain satisfactory images.

Second, the camera detector suffers from regional sensitivity variations or nonuniformity of response. Since the effects of these sensitivity variations are amplified through the reconstruction algorithm, they must be corrected for. This is achieved either through the use of a real-time microprocessor correction system in the camera head or by ac-

Table 10-7. Data Acquisition Parameters Involved in SPECT

Rotational angle:	180 or 360 degrees
Image matrix size (linear sampling):	64^2, 128^2
No. of image frames (angular sampling):	30, 60, 120
Time per image:	2 sec

Table 10-8. Sources of Image Artifacts Observed in SPECT

Statistical noise:	Noise amplification during reconstruction
Gamma camera:	Regional sensitivity variations
	Sensitivity variations with rotation
	Incorrect center of rotation
	Collimator spatial resolution variation
Software:	Influence of filter choice on resolution/contrast
	Choice of attenuation correction method

quiring a uniform flood field into the computer and computing an appropriate correction matrix for each camera view. Recent research work has suggested that very accurate flood corrections are required, with an accumulated count in excess of 30×10^6.

Third, variations in sensitivity of the camera with rotation have been observed. This sensitivity has mainly been attributed to the very small changes in the earth's magnetic field or to locally generated fields during detector movement. It should be realized that PM tubes are high-gain devices and sensitive to external magnetic field changes. Modern SPECT cameras now shield the PM tubes magnetically and contain specialized electronic circuitry to avoid changes in the X,Y signal amplitudes with changes in the position of the detector head.

The fourth artifact relates to variation in the collimator spatial resolution with distance from its face. This factor is reduced if opposed views are acquired as in a 360-degree rotation or when using dual opposed detectors on the SPECT gantry. The partial solution to this problem is to redesign the gantry to allow the use of elliptical orbits or body-contouring methods. The camera therefore remains closer to the patient during rotation. Alternatively, the use of long-aperture collimation could be employed, where the spatial resolution variation with distance is reduced. Both approaches are currently under investigation.

The next artifact listed in Table 10-7 is the use of the incorrect center of rotation of the camera. This refers to the computed center of rotation, based on the series of image matrices acquired by the computer. Its position can change due to variations in the mechanical movement of either the gantry, the detector arm, or the head. It can also vary due to the instability of the electronics in the computer line amplifier and the analog-to-digital circuitry. The analog-to-digital converters (ADCs) define the size (pixels/mm) of the image matrix acquired. Small variations of less than 1 pixel (64^2 matrix) can degrade the contrast obtained in reconstructed images. This artifact can be reduced through the use of a gear-driven gantry, detector arm, and head, all of which should be digitally encoded. Careful calibration of the line amplifiers and analog-to-digital converters is required by computer service personnel.

The remaining artifacts are essentially software related, such as the choice of reconstruction filter and the choice of attenuation correction. Computer software aspects will be discussed later. The data processing parameters (tabulated in Table 10-13) reflect all of the factors above.

Instrument Choice and Bid Specifications

At present, the design of SPECT systems is still under development, both in instrumentation and computer hardware and software additions. Some tentative general guidelines for purchasing a SPECT system will be briefly discussed, and these are tabulated in Table 10-9. The constraints on the camera design in SPECT are greater than those in planar imaging, and more emphasis should be placed on the camera's intrinsic response characteristics, that is, intrinsic energy resolution, linearity, and uniformity. For example, scatter reduction is more important because the organ of interest is viewed through a 180-degree or 360-degree rotation. The extent of scattering will therefore be greater in some projections, such as in liver tomography. Good intrinsic energy resolution is necessary. In addition, the ability to image with the energy window positioned on the high side of the photopeak for improved scatter rejection is useful. This is referred to as *off-peak imaging.* Im-

Table 10-9. Tentative Specifications for the
Purchase of a Rotating Gamma Camera SPECT System

Intrinsic characteristics:	Energy resolution 11–12%
	Uniformity ± 4%
	Linearity ± 1%
Rotational stability:	Magnetic shielding
	Z ratio circuits
Temporal stability:	Real-time turning circuits
	Reloadable microprocessor
	Stable electronics
Gear driven:	Uniform motion, incremental, or continuous mode
Digitally encoded:	Camera, arm, and gantry motion for noncircular orbits
Camera support:	Sufficient for different collimator weights or dual system
Center of rotation:	Independent of camera position (noncircular orbits) or collimator type
Scan times:	Adjustable, 2 min or greater

provement in the camera's linearity or uniformity of response will minimize artifacts in the reconstructed images. Finally, the temporal stability of the camera should not be overlooked. The additional constraints on camera performance must hold for long time periods if frequent service visits are to be avoided. This explains the requirements for microprocessors or real-time PM tubing tuning circuits as previously discussed.

Specifications for the gantry assembly, listed in the same table, should include digitally encoded gear-driven mechanisms that will allow for noncircular orbits. A gantry of robust design is required that will support collimators of different weights, without changes in the center of rotation. With the availability of more short-lived radionuclides, the gantry movement should be fast enough to achieve imaging times of 2 minutes or less.

In summary, most nuclear medicine departments will purchase a SPECT system for both planar imaging and tomography. The latter imaging mode has not been fully developed, nor is it likely to fully replace planar imaging. Appropriate emphasis should therefore be placed on the camera's characteristics for the planar mode. In addition, the SPECT hardware should not place restrictions on planar imaging. Some system designs permit both planar and whole-body imaging and make the system more cost effective. At present, most SPECT systems (excluding computer hardware) can be purchased for 25–40% more than a conventional camera. As a result, many centers are purchasing SPECT systems or at least gamma cameras that can be easily upgraded for a small additional cost. The progress made in SPECT instrumentation will remain one of the most interesting areas of development in nuclear medicine for some time.

Data Processing Equipment

In digital radiography or x-ray CT, the computer is primarily used to form a digital image of the patient. Since x-ray systems are high-resolution devices, particular attention is paid to the use of large image matrices (512–1024 pixels) and their fast transfer rate through computer memory to disc. In nuclear medicine the problems related to interfacing the gamma camera with the computer are much simpler. The discrete X,Y positional pulses for each scintillation event are easily digitized by ADCs. The data rates are generally less because of the lower photon flux and the smaller image-matrix sizes involved. However, the software requirements for nuclear medicine are more extensive, particularly in the area of postprocessing. This is because of the necessity to calculate a range of parameters related to organ function. The objectives of this section are therefore to outline only the computer hardware and software requirements specific to nuclear medicine.

Basic Hardware Requirements

A schematic diagram of the hardware requirements for a nuclear medicine computer, interfaced with a single gamma camera, is drawn in Figure 10-7. The camera's analog X, Y, and Z pulses are transferred immediately to a line amplifier to ensure impedance matching and clean transfer of signals to the mainframe computer. The latter is normally located at some remote position. The X and Y pulses are passed to the ADCs, and their signal pulse height is digitized to represent an I,J location in the image matrix as illustrated. The ADCs are normally 8 bit, and hence matrix sizes of 64^2, 128^2, and 256^2 are possible. The larger matrix sizes are used for the LFOV cameras. The Z pulse, which is a logic-on pulse, is presented to the ADCs so that only events accepted by the PHA are digitized. The matrix coordinates I,J for each accepted event are presented to the *direct memory access* (DMA). An image buffer (4K for a 64^2 matrix) is reserved in the *central processor unit* (CPU) memory. The DMA function is simply to add one count to the memory buffer location corresponding to the I,J matrix location. Hence, an image in the CPU memory can be accumulated on either a preset count or preset time basis. To facilitate transfer of image to disc, two image buffers in memory are generally used. While one is accumulating, the other is "dumping to disc."

A general summary of the important computer hardware components is listed in Table 10-10. First, in order to store large image-matrix sizes (256^2), a 128K work memory buffer is required. Large CPU or remote memory is therefore necessary. Note that word mode storage is often required, as the accumulated count per matrix or pixel point may exceed byte mode storage (that is, it may be greater than 255 counts per pixel). For fast dynamic work, such as in first-pass cardiac investigations, framing rates of about 40 frames per second are necessary.

Fig. 10-7. The basic hardware components: Line amplifier (L. Amp) analog to digital converters (ADCs), direct memory access (DMA), central processing unit (CPU), memory buffers, and hard copy disc.

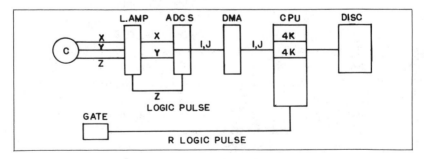

Table 10-10. Representative Example of Important
Hardware Components for a Three-Terminal Computer System

CPU memory, minimum 256 Kbyte

Three remote acquisition and processing terminals (simultaneous operations) with minimum of 256-Kbyte memory, menu, and image displays

Array processor or additional 2-Mbyte remote memory for enhanced data processing speed

High-speed disc drives: 30–300 Mbyte for data storage, 10–30 Mbyte for software operation and upgrades

Cartridge magnetic tape for long-term data storage

Color video display

High-resolution image formatter

Three camera interfaces with one either/or switch

This is possible only if image buffers are used as just described, and modern hard discs are employed as opposed to floppy discs. Table 10-11 shows a comparison of maximum frame rates for different types of disc drives. The main limiting factor is disc transfer speed. Computer memory transfer is very rapid. For prolonged dynamic studies, such as in cardiac or renal investigations, up to 100 image frames may be stored per study. In SPECT investigations, more than 300 image frames are required to be stored on disc for raw and reconstructed data. The minimum disc size recommended for a single camera/computer system is a 10-megabyte (Mbyte) disc. If there are two or more cameras interfaced with the computer, the minimum disc size recommended is 30–80 Mbyte.

More-specialized hardware requirements are as follows. In first-pass cardiac studies, it is often desirable to acquire in list mode as opposed to frame mode. In list mode, each accepted scintillation event is stored individually with a real-time marker on computer disc. The data can then be reformatted in frame mode with different cardiac cycle intervals. Appropriate hardware for list mode is required that will allow for rates on the order of 200K per second. For SPECT or cardiac investigations, where processing of data is extensive, the inclusion of an array processor will significantly reduce processing time. An example of the improvement obtained in SPECT investigations is illustrated in Table 10-12.

Acquisition Software

The acquisition programs are fairly straightforward. However, these programs should contain some special features. For example, multi-

Table 10-11. *Representative Maximum Acquisition Framing Rates for Different Disc Drive Systems*[a]

Frame Size (Word Mode)	Framing Rates (Frames/sec.)		
		Hard Disc	
	Floppy Disc	Single Platter	Multiplatter
64 × 64	4	25	40
128 × 128	1	6	10

[a]Framing rates may be increased if several image buffers are used (see Figure 10-7).

Table 10-12. *Additional Computer Hardware That Reduces SPECT Processing Time*

Hardware Additions	Computation Speed[a] (sec)	Total Study[b] (min)
None	40–60	40
Floating Point	20–40	25
Array Processor	1–2	5

[a]64^2 matrix
[b]Liver tomogram, three views

phase dynamic studies are useful for flow measurements, where the acquisition framing rates may vary automatically from a fast phase to a slow phase. For cardiac investigations, software is required to acquire gated blood pool studies, by gating the data transfer to the computer using the R wave from a cardiac electrocardiogram (ECG) gate system. In essence, the computer first samples several cardiac cycle intervals and determines the average cycle time. This time interval (T) is then typically divided into 32 equal segments, with an image frame (64 × 64) acquired for each corresponding time interval ($\Delta T = T/32$). Each image frame is stored in the computer memory. For each subsequent cardiac cycle, the data for each frame is then summed. This process is generally repeated for 300–400 cardiac cycles, to accumulate sufficient counts for each frame, typically 300K. The total study is then stored on disc for later analysis of wall motion or ejection fraction.

Specific problems arise, however, if the patient has arrhythmia. If the arrhythmia is significant, the study will be invalidated. One way to correct for this problem is to have sufficient memory (remote or CPU memory) to temporally store the frames of the previous cycle and specifically reject this data if the cardiac cycle interval is not within certain defined time limits of the average cardiac cycle. Many computer systems do not have this capability. Some systems perform a less acceptable correction, such as measuring the time of the previously acquired data cycle. If this time is incorrect the subsequent data cycle being stored is then rejected or a number of later cycles, some or all of which may be valid data. Another approach to this problem in gated blood pool imaging is to use list or serial mode acquisition. Although this method permits a more flexible approach to processing the data, it uses large areas of the disc storage.

Processing Software

The computer programs used for analysis of the data can be again divided into two main areas: general processing software and the more-specialized clinical software package. The former refers to the programs used to prepare the data for calculation of functional parameters, such as image manipulation, smoothing or subtraction, flagging of regions of interest (ROIs) for curve generation, and finally curve additions and subtractions. A specific aspect that should be emphasized is the flexibility of these programs. One should be able to execute them in a semiautomatic manner, with little operator input, using predefined sequences. These predefined sequences are simply a set of command files that may be specifically tailored for a given clinical investigation. They greatly enhance the efficiency in operating the computer. The methods used to generate ROIs are important, as often irregular-shaped areas of small size are required, for example in flagging the area of the renal cortex. The methods available should include rectangular area selection, using cursors, and irregular area selection using a light pen or preferably a joystick that can be more finely controlled. These controls should be tested for their ease of operation and speed.

The file structure under which these programs operate should be carefully designed in order to uniquely identify the patient study. The stored data should include acquisition parameters, all data generated from the processed study, and finally the parameters used in processing. It is necessary to stress that the above parameters should ideally be displayed and recorded on film for reading by the physician. Any one of them could significantly influence the final functional parameters calculated.

Analysis of the decision factors in the choice of the more-specialized clinical software package is more difficult. It is necessary to examine in some detail the theory used in the generation of the software and the relevant literature where investigators have used this same software/computer system. This is necessary as there are many subtle differences in clinical software between computer manufacturers. Examples of these differences can be found in the filtering methods used in preparation of cardiac data, such as in ejection fraction, regional ejection fraction determinations, or phase analysis of heart motion. The methods used for calculating effective renal plasma flow (that is, the particular mathematical algorithms used) may differ.

In many instances, computer manufacturers do not supply all the clinical software required. For example, they resort to the use of "user-generated" programs developed at various clinical research institutions. The computer manufacturer does not normally accept responsibility for the correct operation of these programs, at least over the short term. The full clinical software package and the possibility of future upgrading of clinical software over the useful lifetime of the computer should be carefully examined by the potential user.

Tomography Software

The software specifications for SPECT are more complex than for CT. The reconstruction algorithms include additional corrections, for example, photon attenuation, nonuniformity of response for the gamma camera, and the center of rotation for the gantry. This software is still in the development stage by various imaging companies, and the details of its operation are beyond the scope of this chapter. However, some general guidelines as to what factors should be included in the tomographic software package are listed in Table 10-13. The most important of these is the inclusion of both constant and variable attenuation correction methods. These methods are required for quantitation of the radiotracer distribution in the reconstructed plane. A method is also required to estimate the cross-sectional profile of the patient. Several complex methods have been proposed, but the simple geometrical estimation may remain the one of choice. Much developmental work on filtering methods is required, and it is expected that a wide range of reconstruction filter types will be necessary. It is important that the user be able to modify the filter parameters. The imaging

Table 10-13. Representative Software Requirements for SPECT

Attenuation correction algorithms: Fixed
                                                             Variable
Patient profile measurement: Geometrical estimation
                                             Transmission source
                                             Compton scattering
Gantry motion: Noncircular, elliptical, or body contouring
Reconstruction filters: Create or user modified
Dual-isotope or dual-head data acquisition
Gated tomography (dual gate)
Real-time acquisition/reconstruction
Real-time uniformity correction during gantry motion

and processing time is of some concern in SPECT. Software will soon be commercially available that allows simultaneous acquisition and processing of current data. This software development, together with the use of an array processor, should permit SPECT studies to be completed shortly after scanning is complete.

Instrumentation Choice and Bid Specifications

The nuclear medicine computer is generally a mainframe minicomputer built by one of the larger computer companies (such as Digital Equipment Corporation or Data General) with a specialized gamma camera interface, display console, and software supplied by the imaging company. During the last decade there have been significant improvements in the design of minicomputers and their reliability. The cost of various components such as CPU memory, and more recently hard-copy disc drives, has decreased. Computers supplied by the imaging companies are now very competitive in hardware design if the user elects to purchase such options as an expanded memory, array processor, and large-disc storage (80–300 Mbyte).

The minimum specifications for nuclear medicine computers have already been outlined in Tables 10-10 through 10-13. It is anticipated that apart from increased system speed and storage capability, the primary research and development effort over the next few years will be the further expansion of specialized clinical software and tomographic software. The potential purchaser should therefore determine the company's commitment to the development of software for specific diagnostic tests, such as effective renal plasma flow, blood flow, and cardiac investigations; quantitative thallium rest and stress testing; or phase analysis (first and second order).

Recent factors that have increased computer efficiency have been the use of multiple terminals that can operate independently. Software has recently been made available to allow each terminal to simultaneously acquire and process clinical studies. Batch processing of clinical studies is now possible. These factors are very important, as one of the primary problems in a busy nuclear medicine department is system access time for data processing.

The service of the computer system is a very important factor, particularly if the system is equipped with multiple terminals. Any delay in service of the computer's mainframe can seriously affect the daily work objectives of the entire nuclear medicine department. The service from most companies is far from ideal and varies significantly with location. Any bid specification should clearly state the computer service requirements and the service response time requirements for when the system is fully down and when a particular component is malfunctioning.

References

1. Muehllehrer, G., Wahe, R., and Sano, R. Standards for performance measurements in scintillation cameras. *J. Nucl. Med.* 22:72, 1981.
2. Kuhl, D. E. *Principles of Radionuclide Emission Imaging.* New York: Pergamon, 1983.

Suggested Reading

Bernier, D. R., Langan, J. K., and Wells, L. D. (eds.). *Nuclear Medicine Technology and Techniques.* St. Louis: Mosby, 1981.

Chandra, R. *Introductory Physics of Nuclear Medicine*, 2nd ed. Philadelphia: Lea & Febiger, 1982.

Parker, E. P., Smith, H. S., Taylor, D. M. *Basic Science of Nuclear Medicine.* New York: Churchill Livingstone, 1978.

11

Nuclear Magnetic Resonance Imaging Instrumentation

Laurence P. Clarke

In conventional radiology, three methods are primarily used for medical imaging, all involving the use of high-energy electromagnetic radiation: x-rays (80–120 kiloelectrovolts [keV]), gamma rays (80–511 keV), or ultrasound waves (1–15 megahertz [MHz]). These imaging methods allow anatomical information to be obtained with various degrees of spatial resolution. They may permit, to a much more limited extent, the differentiation of normal and pathological tissues. This differentiation may be achieved, for example, if the denatured tissues exhibit small changes in photon attenuation or acoustical impedance. In nuclear medicine imaging, a more direct measure of the presence of pathological tissues may be determined. However, this technique requires a radioactively labeled compound to be found that is selectively accumulated in or excluded from the tumor site. Only a limited number of such labeled compounds have been identified. Therefore these conventional imaging methods are far from optimum for identifying tumors, as they often do not provide sufficient image contrast between normal and abnormal tissue sites. More importantly, they often fail to provide a direct measure of the metabolic activity of the tumor or its response to therapy.

Nuclear magnetic resonance (NMR) is a radically different approach to medical imaging. It uses static magnetic fields and very low-energy electromagnetic waves in the *radiofrequency* (RF) range. More importantly, however, it is a resonance phenomenon associated with the energy state of protons or other nuclei in the patient's tissues and the frequency of the RF electromagnetic waves. By selective excitation of these nuclei at a precise frequency, a very low RF energy pulse can cause discrete RF emissions from these nuclei that reflect their immediate electrical environment. The electrical environment of the nuclei is related to the bonding of the nuclei with surrounding molecules. The electrical environment in turn is indirectly related to the chemical state of the tissues within which the nuclei are present. Hence, NMR can not only provide anatomical images but may also obtain indirect chemical information of the tissues being imaged.

NMR has many advantages over alternative radiological imaging techniques. It does not use ionizing radiation, since RF waves are of very low energy (few milliwatts). To date it has not shown any biological risk at the RF field strengths used for imaging. Since the first introduction of commercial systems in early 1981, NMR has been demonstrated to be superior to x-ray computed tomography (CT) head scanning. This new imaging technology is still in the developmental stage. For this reason, only an elementary discussion of the basic principles of this imaging method will be covered. A comparison of the different systems used for proton imaging and NMR spectroscopy in vivo will also be discussed briefly. For further information on the topics discussed in this chapter, the reader is referred to the Suggested Reading at the end of the chapter.

Basic Physical Concepts

Resonance Phenomena

NMR is related to the magnetic properties of the nucleus. Nuclei that contain an odd number of protons, neutrons, or both possess two important and interrelated physical properties. These are *spin* and *magnetic moment.* The magnetic moment is related to the spinning charge of the nucleus; that is, any moving charge (or current) has an associated magnetic field. The nucleus therefore behaves as a small bar magnet, as illustrated in Figure 11-1, with a north and a south pole, spinning on its own axis (a magnetic dipole).

There are more than 100 stable nuclei that possess both magnetic moment and spin. More importantly, all of the chemical elements have at least one naturally occurring magnetic isotope. Examples of those that are of medical interest are ^1H, ^{13}C, ^{19}F, ^{23}Na, ^{31}P, and ^{127}I. Particular emphasis has been placed on hydrogen because of its large abundance in biological tissue (approximately 70% of human tissue contains water) and because of the larger signal strength obtained from protons. It is therefore possible to obtain anatomical images by *proton imaging*, using relatively low magnetic field strengths (500–1500 gauss). For in-vivo NMR spectroscopy, to be discussed later, nuclei such as ^{31}P and ^{23}Na are important. For simplicity, the basic physical principles of the interaction of the static magnetic fields and RF electromagnetic fields will be initially discussed for protons only.

The basic concepts of the interaction between magnetic and RF fields with a magnetic nucleus (proton) is perhaps best explained using classical physics theory. Consider a small sample of tissue within a patient as illustrated in Figure 11-1B. In the absence of a magnetic field, the magnetic dipoles are oriented at random within the sample and do not result in a net magnetization vector. If a uniform magnetic field B (Figure 11-1C) is applied to the sample, the magnetic dipoles with the sample will reorient themselves in such a manner as to create a net magnetization vector M, parallel to the applied field. The vector M is often referred to as the "macroscopic magnetization vector" associated with a small "sample" of sensitive nuclei. Using the three-dimensional

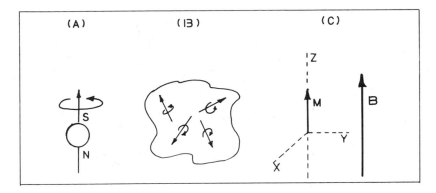

Fig. 11-1. Interaction of uniform magnetic field B with a proton: (A) Proton as magnetic dipole spinning on its own axis; (B) Random orientation of dipoles within tissue sample without an applied field; (C) Reorientation of dipoles resulting in a net "macroscopic magnetization vector" *M* in the direction of the applied field B (Z direction) as indicated.

coordinate system (X, Y, Z) as illustrated, the vector M is designated to be along the Z axis, perpendicular to the X,Y plane.

Consider a weak RF field (B_1), generated by an RF transmitter coil positioned along the X axis as illustrated in Figure 11-2A. If the frequency of the oscillating RF magnetic field is appropriately chosen, some of the proton nuclei within the sample will experience a torque that will displace them away from the applied field in the Z direction. In the more macroscopic view of the sample, the RF field will displace the net magnetization vector away from the Z axis. Because of the properties of each spinning nucleus, any torque applied to them will result in precessional motion around the main applied field B (i.e., analogous to the spinning gyroscope when it is displaced from the vertical direction of the earth's gravitational field). Again, in the macroscopic view of the sample, the vector M will precess around the applied field B (Z axis) (Figure 11-2A). If the frequency of the RF field is appropriately chosen, the vector M will experience a continuous torque while precessing, causing it to be further displaced from the Z axis as long as the RF pulse is applied, or in other words, resonance occurs. The condition for resonance is directly related to the applied field B, in accordance with Larmor's theory.

$$\omega = \gamma B$$

where W is the angular frequency of the precessional motion and γ is the gyromagnetic ratio, defined as the ratio of the magnetic moment of the nucleus to its angular momentum or spin.

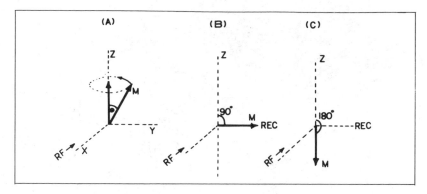

Fig. 11-2. Interaction of the small RF field applied at the resonance frequency and macroscopic magnetization perpendicular to the field B (X direction): (A) The vector M is displaced from the Z axis by the angle θ and precesses around the Z axis; (B) Displacement of M by 90° resulting in detected oscillating EMF by the receiver coil at the resonance frequency since magnetic flux lines cross the receiver coil; (C) Displacement of M by 180° inverting the net macroscopic magnetization vector. The transmitter/receiver coils are drawn separately for illustrative purposes. Their configuration is different for imaging systems where phase quadrant detection is normally employed.

If the RF field is applied long enough, the vector M in Figure 11-2B can be displaced by 90 degrees onto the X,Y plane. Consider an NMR receiver coil placed in the same plane along the Y axis. The magnetic lines of force, due to the net magnetization vector, cross the receiver coil and induce an *electromagnetic force* (EMF). Because of the oscillating nature of the vector M, the detected EMF and hence the induced oscillating or measured alternating current (AC), will be at the same resonant frequency. Again, if the RF field is applied for a longer period, the vector M can be further displaced by 180 degrees along the −Z axis; that is, the vector is inverted.

When the RF field is turned off, the net magnetization vector M returns to the equilibrium position along the Z axis, parallel to the applied field B. For example, the magnitude of the oscillating signal observed by the receiver coil for the 90-degree pulse will therefore decay as the vector M returns to the equilibrium position. The relaxation phenomenon will be briefly referred to later.

In medical imaging the patient is positioned inside a large magnet that generates a uniform magnetic field B, as illustrated in Figure 11-3. To obtain spatial information on proton spin density in tissue, magnetic field gradients are generally applied in orthogonal directions (X, Y, Z) such that each location in the patient will have a uniquely defined field strength. Each location will therefore resonate at a given fre-

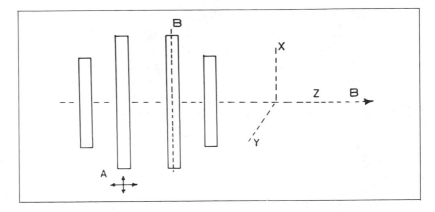

Fig. 11-3. A four-coil resistive magnet. The coils are centered along the same Z axis in an approximate complex Helmholtz arrangement to yield a uniform magnetic field *B*. Passive shimming (*A*) involves fine controlled geometrical positioning of the coils with respect to each other to improve field homogeneity. Active shimming (*B*) involves the use of small current-carrying coils in addition to the main coils illustrated. Gradient coils not shown are positioned orthogonally to each other along the X, Y, and Z axis to yield linear field gradients along the same axes.

quency (within a narrow frequency bandwidth). It is therefore possible to form an image of a patient if the received signal is Fourier transformed into discrete frequencies associated with each position.

Image Interpretation

The NMR signal detected in proton imaging is a function of several parameters which include the following:

NMR signal $\propto \eta, T_1, T_2, v$

where η is the proton or spin density, T_1 is the longitudinal or spin-lattice relaxation time, T_2 is the spin-spin relaxation time, and v is the velocity of the medium (fluid) being imaged, where appropriate. The exact functional relationship will depend on the imaging method used and the RF pulse sequence, to be discussed later. The T_1 parameter is related to the time required for the component of the net magnetization vector M in the Z direction (Figure 11-2) to return to its original value after the RF pulse has been turned off. In more physical terms, it is related to the reorientation of the spinning nuclei with respect to surrounding lattice environment, hence the term *spin-lattice relaxation time*. The *spin-spin relaxation time*, T_2, is related to the free induction decay (FID). It is physically related to the interaction between the spinning nuclei. The return of the net magnetization vector M in Figure 11-2 to the equilibrium position parallel to the applied static field is characterized by both time constants T_1 and T_2. The detected NMR signal is therefore related to both of these time constants.

For fluids and soft tissues, T_1 and T_2 are approximately equal. For solids such as bone, $T_2 << T_1$. For certain imaging conditions and pulse sequences, the NMR signal detected approximates to the following relationship:

NMR signal $\propto T_2/T_1$

Image contrast may be enhanced if this ratio varies for each tissue type. For example, it is possible to differentiate white from gray matter in the brain despite the fact that the proton density is similar in each area.

The objective of NMR imaging is to maximize the image contrast between normal and pathological tissues. This goal can be achieved by developing unique RF pulses that will maximize one or more components of the NMR signal detected. Hence the terms *proton image*, *T_1 weighted image*, and *T_2 weighted image*. The pulse sequences in the NMR imaging software generally includes *repeated free induction decay* (RFID), *inversion recovery* (IR), and *spin echo* (SE). The dependence of the NMR signal strength on η_1, T_1, and T_2 for each pulse sequence is illustrated in Table 11-1.

Spectroscopy

NMR spectroscopy has been a useful analytical laboratory technique in physics and biochemistry for several decades. However, recent interest has been expressed in in-vivo spectroscopy of important elements such as ^{31}P, ^{23}Na, ^{19}F, and ^{13}C. Particular interest has been expressed in phosphorous since it is a primary component of metabolic molecules such as adenosine triphosphate (ATP), phosphocreatine, inorganic phos-

Table 11-1. Dependence on NMR Signal Strength on η, T_1, and T_2

	Image Parameter		
	η	T_1	T_2
Repeated free induction decay	proportional to or dependent on	reduced if T_1 very long	—
Inversion recovery	proportional to or dependent on	decreases as T_1 increases	—
Spin echo	proportional to or dependent on	reduced if T_1 very long	Increases as T_2 increases

phates, and sugar phosphates, which are related to transfer of energy in cells. NMR spectroscopy refers to the ability to obtain resonance spectra, with individual resonance peaks being identified that are characteristic of the above compounds. The immediate electrical environment or bonding of ^{31}P is unique for each chemical compound. Each compound will therefore give rise to resonance peaks at a characteristic frequency. Since the resonance peaks are very close in frequency, very good magnetic field uniformity and strong magnetic fields are required to resolve them. For a laboratory NMR spectroscopy system, these physical criteria can be readily met for the small measuring volumes involved. However, for in-vivo spectroscopy, in particular whole-body measurements, much larger magnets and more stringent design criteria are required. In-vivo NMR spectroscopy has tremendous potential for measuring both chemical and metabolic activity of tumor sites and for identifying denatured muscle in regions such as the myocardium.

NMR Imaging Systems
Basic Hardware Components

The basic components of an NMR imaging system and their primary functions are listed in Table 11-2. These components include the magnet system, the imaging system, the NMR spectrometer, and the computer. The purpose of the magnet system is to generate a magnetic field B as shown in Figure 11-3 that will remain uniform throughout the volume of interest. A uniformity of 10–40 parts per million (ppm) is typically required for proton imaging. This field should remain stable with time. The imaging system includes gradient coils that generate orthogonal directed magnetic fields along the X, Y, and Z axes of the system. These coils uniquely define each point in space within the magnet with a given field strength. Each point in the patient can be uniquely defined by a corresponding frequency as described earlier. These gradient coils should be designed to yield linear field gradients. Specialized power supplies are required to permit the fields of these coils to be changed at high frequency, that is, in order to rapidly define each point in space.

The RF spectrometer is operated under computer control software. It is used, for example, to generate the pulse sequences listed in Table 11-1 and to analyze the signal from the Z receiver coils. The computer is interfaced with all of the NMR hardware components. Control software is used, for example, to drive the gradient coils, to control the operation of the spectrometer, and to monitor the stability of the power supplies to the magnet and gradient coils. The exact functions of the basic hardware components are quite complex and are beyond the scope of this chapter. The magnet system is a major component of the

Table 11-2. Basic Components of an NMR
Imaging System and Their Primary Functions

Components	Purpose
Magnet system	
Static field generation coils	Uniform magnet field
DC power supply	
Cooling system	
Shimming coils	
Imaging system	
Gradient coils X, Y, Z	Hardware for formation of image
Coil power amplifiers	
RF transmitter coils	
RF receiver coils	
NMR spectrometer	
Stable RF transmitter and receiver	RF signal generation and analysis of received signal
Computer	
Interface to NMR components	Generation of pulse sequences controlling all imaging components
Control software for signal	
Generation and analysis	

imaging system, however, and this component will be discussed further.

Magnet Types

There are three magnet systems currently used in NMR imaging: resistive, superconducting, and permanent magnets. Their construction, cost, and installation problems have significant differences, and each type will be reviewed separately. The characteristics of each system are summarized briefly in Table 11-3.

Resistive Magnets. Resistive magnets generally consist of four to six current-carrying coils or electromagnets. The geometrical configuration (complex Helmholtz arrangement) is designed to yield a uniform magnet field along their common Z axis as schematically shown in Figure 11-3. The uniformity or homogeneity of the magnetic field can be improved by active or passive shimming. *Active shimming* refers to the use of additional coils (of only a few turns) (Figure 11-3) to generate additional small magnetic fields to correct for inhomogeneity in the main field. *Passive shimming* refers to mechanical adjustment of the relative positions of the main coils to improve field homogeneity, also schematically shown in Figure 11-3. A well-designed resistive magnet can obtain a uniformity of 20–40 ppm over a spherical volume of 30-cm diameter or greater.

Table 11-3. Comparison of Magnet System Types

	Resistive	Superconducting	Permanent
Field strength	0.04–0.2 T	0.35–2.0 T	0.04–0.3 T
Power dissipation (magnet)	6–60 kW	—	—
Cooling	closed-cycle water (±0.2°C) heat exchanger	liquid helium liquid nitrogen	—
Homogeneity (25-cm diameter)	20–40 ppm	10–20 ppm	50–100 ppm
Stability (short-term)	10^{-5}–10^{-6}/hr	$< 10^{-6}$/hr	$< 10^{-6}$/hr
Warm-up time	1–2 hr	none	none
Weight	10,000 tons	15,000 tons	100,000 tons

The primary advantage of the resistive magnet is its low cost compared with other systems. Its design has been well established. Hence, the lifetime of the magnetic system should be greater than eight to ten years with low service-cost requirements. One specific advantage that has not been fully developed is the ability to "ramp" up or increase the field strength of the magnet. It is possible that for certain NMR pulse sequences, an optimum field strength would maximize the signal output. The ability to vary the magnet field strength may therefore be important.

The resistive magnet has some disadvantages, however. The maximum field strength that can be obtained is on the order of 0.15 telsa because of the limitations in cooling the magnet. The magnetic coils are wound in such a manner as to include a water cooling system within the coil structure. Despite novel designs of the cooling system, the significant heat generated by the currents involved is difficult to dissipate rapidly. A second disadvantage of this magnet type is the instability of both the power supply and the temperature of the main coils. Both factors will cause a change in the homogeneity of the magnetic field. Novel feedback mechanisms under computer control are required to improve both short- and long-term stability. A good resistive magnet design will have a short-term stability on the order of 1 in 10^{-6} per hour. One final disadvantage is the warm-up period before the magnet reaches an equilibrium operational condition. The latter is a problem if the power of the system is reduced or turned off overnight to minimize the running costs. A typical warm-up period is 1–1½ hours.

Superconducting Magnets. The interest in developing whole-body superconducting magnets is twofold: first, to determine if proton imaging at the higher magnetic field strengths is possible and if it results in improved spatial resolution, signal to noise, and hence image contrast; second, to determine the possibility of obtaining spectral information about other nuclei such as phosphorous in vivo or the possibility of multinuclide imaging. Multinuclide imaging relates to forming an image of each spectral peak such as in phosphorous spectra or other nuclei such as ^{23}Na. As a result, whole-body superconducting magnets have been built with magnetic field strengths of 3500–20,000 gauss (0.35–2.0 tesla).

The coil configuration of a superconducting magnet is similar to the resistive type illustrated in Figure 11-3. However, the relative positions of the main coils and current-carrying shimming coils are constrained to be within the Dewar system containing liquid nitrogen and helium. The windings of the main coils are made from a specialized alloy (typically niobium-titanium). When the alloy is reduced to a very low temperature (supercooled to $-269°$ C, near absolute zero temperature), it conducts electricity with very little resistance. As soon as the conductor is brought to the required low temperature, it does not require any further electrical power to drive the current. The current decreases at a very slow rate, typically 1 in 10^6 per day. Hence, the generated magnetic field remains very stable. The homogeneity of the superconducting magnet is generally better than that of the resistive magnet and is typically 10–20 ppm for a spherical volume of 30 cm.

The primary advantage of this magnet type is the higher field strengths obtained, the excellent stability of the magnet, and the homogeneity that can be obtained. For small-bore prototype systems at 30,000 gauss (3.0 tesla), homogeneity on the order of 10^{-1} ppm has been achieved. The latter specification is required for obtaining higher-resolution phosphorous spectra or other nuclear spectra in vivo. Whole-body 2.0-tesla magnets have recently been designed with homogeneity approaching 1 ppm and may be suitable for obtaining high-resolution phosphorous or sodium spectra in the body. Under careful laboratory conditions, it is possible in theory to ramp the magnetic field strength. This may allow both proton imaging and, for example, phosphorous spectra to be obtained at their optimum field strengths.

The major disadvantages of the superconducting magnet systems are the large quantities of liquid helium and nitrogen required for cooling. The annual cooling costs range from $40,000 to $65,000. The estimated costs are dependent on the size of the magnet system and the design of the Dewar system. An initial filling of the magnet system to reduce its temperature to $-269°$ C costs approximately $10,000. A second disadvantage is the possibility of a magnet "quench," where it rapidly warms to room temperature. This can arise from failure to refill the cryogens, a leak in the vacuum or liquid cooling container, or a violent mechanical shock. Quenching will result in loss of the liquid helium and possible damage to the magnet system, although good magnet designs allow for the problem of quenching to avoid mechanical damage. Nevertheless, it takes considerable time to bring the magnet down to superconducting temperatures. The latter assumes that a full supply of liquid helium and nitrogen is available.

Ramping of the magnetic field may prove to be unrealistic for two reasons. The process could take several hours to complete, as the homogeneity at the modified field strength may have to be optimized. Frequent ramping procedures will require additional cooling, and the cost of the liquid coolants required may prove to be prohibitive.

One important potential development in the operation of whole-body superconducting magnets is the use of closed-cycle helium refrigerators. These systems have been successfully used for small magnets, and they are expected to be successfully modified for the large whole-body magnets. The helium refrigerators will allow the system to be operated with minimal liquid helium requirements. Hence, the running costs of superconducting systems could be significantly reduced.

Permanent Magnets. Permanent magnets have been designed with magnetic field strengths of 400–30,000 gauss (0.04–0.3 tesla). They are built from large quantities of magnetized material such as iron or, more recently, ceramic material. Permanent magnets have many advantages. They do not require electrical power for their operation, and hence running costs of the total imaging system are less than for resistive or superconducting types. The magnet remains stable provided the temperature of its environment is kept constant. Field homogeneity of less than 75 ppm has been reported for a spherical volume greater than 30 cm in diameter. The fringe field of the permanent magnet is very small, as it remains very close to the conducting material. The costs of installation are expected to be comparable to a resistive system.

The primary disadvantage of this magnet type is its weight. The system designed with a 0.3-tesla field by one manufacturer weighs approximately 100 tons (200,000 lbs) and must be supported by several feet of concrete. This weight factor limits the possibility of constructing permanent magnets of greater field strength. However, preliminary research has suggested that alternative lightweight materials such as barium ceramic may be used to design higher-field-strength magnets. Also novel changes recently proposed in the shape of the magnet may reduce its weight. A second disadvantage involves the constraints on the temperature of the magnet. The temperature should remain with 0.1° F to ensure stability and uniformity of the magnetic field. Additional limitations relate to the operation of switching field gradients adjacent to the magnetic conducting material, caused, for example, by hysteresis problems. Specialized shielding of gradient coils is required. The design of lightweight permanent magnets exhibiting higher field strengths is currently under intensive research because of its obvious advantages. However, if the design of helium refrigerator systems is successful for superconducting magnets, the latter may be the system of choice because of the much higher field strengths currently obtained.

Site Specifications

At present, there is an inconsistency in the recommended NMR site specifications of each manufacturer. This is particularly true for problems associated with external RF interference, magnetic field shielding of the magnet room, and the required magnet shimming for each particular site environment. Site specifications will vary with the type of magnet system to be installed, the magnetic field strength, and the homogeneity required. The latter specification is dependent on whether a proton imaging system or an in-vivo spectroscopy system is to be installed. The limited experience to date has demonstrated that the image quality obtained for various site locations varies, despite the fact that in many instances the systems installed are manufactured by the same company. Since at present only four to five systems of each type and manufacturer have been placed at clinical sites worldwide, it is too early to be very specific on recommendations.

A summary of the problems encountered in installing NMR imaging systems and examples of proposed solutions are listed in Table 11-4. The magnet fringe field is dependent on magnet size, field strength, and magnet type. Sensitive electronic equipment, such as computer hardware and display systems, should be kept at large distances from the center of the magnet. Fortunately for solenoid-type electromagnets, the field strength decreases in approximate proportion to the radius cubed. For permanent magnets, the magnetic flux is essentially contained between the magnet pole pieces, and the space requirements above are reduced.

The uniformity of the magnetic field is influenced by the presence of magnetic materials in the building, floors, light fixtures, and plumbing. The best but most costly approach is to remove these materials from the retrofit installation or to exclude them in the design of a new building. The alternative approach is to shim the magnet to minimize the effect of fixed magnetic structures in the building. The science of magnet shimming has recently been improved and involves obtaining a simulated map of the magnetic structures, using a computer, and using active and passive shimming, as discussed earlier, to achieve homogeneity. Active shimming using current-carrying coils is now routinely performed under computer software control with some success. Shimming requirements are more critical for resistive magnets because the influence of the presence of magnetic materials in the magnet's environment is greater at its lower field strength (0.15 tesla). Permanent magnets also require shimming and generally employ current-

*Table 11-4. Example of Problems and Proposed
Solutions for Installation of NMR Imaging Systems.*[a]

Problems	Proposed Solutions
Magnet fringe field	Additional space allocation Magnetic field shielding of room
Magnet field uniformity	Removal of magnetic materials Magnet shimming (active and passive)
RF and electrical noise interference	RF receiver coil shielding RF room shielding Specialized electrical grounding system
Magnet cooling (resistive)	Room air conditioning Closed-cycle cooling system Heat exchanger to exterior
Magnet cooling (superconducting)	Room air conditioning Liquid helium and nitrogen Liquid helium refrigerator Cryogen venting system to exterior

Stability of magnet and gradient coil
  power supply
	Power conditioner
	Computer-controlled power feedback
	  mechanism

Floor loading
  Resistive (10,000 lbs)
  Superconducting (20,000 lbs)
  Permanent (200,000 lbs)
	Ground floor with additional concrete
	  support

[a]For full explanation, see Suggested Reading.

carrying conductors. Stable power supplies are required for the latter. Furthermore, small temperature variations of less than 0.1° F may degrade the homogeneity of permanent magnets. A temperature-controlled room or magnet chamber is necessary.

RF shielding is required from external radiofrequency sources to maintain image quality and a good signal-noise ratio. The RF source interference will be a problem if it falls inside the frequency range of the imaging system (typically 5 MHz for the resistive type). Although most RF radio transmissions are at a high frequency, interference may still be possible from other sources such as fluorescent lighting, citizen's band and police radio, and various illegal radiosignals. Two forms of RF shielding have been proposed: limited shielding of the NMR receiver coils within the magnet assembly and room RF shielding. RF shielding is a very empirical science, and it is difficult to build a perfect room shielding system. Some companies use steel for shielding; others use copper mesh or thin copper plate. Irrespective of the method used, problems arise because of necessary gaps in the shielding for power cables, venting, and air-conditioning systems.

Magnet cooling devices are system dependent. Resistive magnets are water cooled by a closed-cycle system, the heat from which is expelled by a chiller system to the building's exterior. The design of these systems is important for magnetic homogeneity and for minimizing field drift. Chiller systems should maintain water temperatures to within 0.2° C. They should be dedicated to the magnet system and, if possible, should be under computer software control for optimum operation. Additional room air conditioning is required, particularly for resistive electromagnets, which dissipate more heat to the room than superconducting systems. The superconducting magnet requires cooling by liquid helium and nitrogen. The efficiency and cost of operating these systems is inherently related to the Dewar system design. The latter determines the boil-off rate of the liquid coolants.

The stability of the direct current (DC) power supply for resistive magnets and the power supplies for gradient coil systems is important. Some manufacturers are recommending the installation of power conditioners for the magnet, the imaging hardware, and the computer. Power conditioners may be particularly required for climates where lightning storms are frequent and in areas where electrical power varies.

Instrumentation Choice and Bid Specifications

NMR imaging and in-vivo spectroscopy instruments are currently investigational type devices, a small number of which are being evaluated under federal guidelines for approval in the United States. Both the Food and Drug Administration and local health insurance agencies are required to give approval before reimbursement of professional fees is possible. One aspect that may be addressed by these agencies is the cost effectiveness of NMR. It is possible that a qualified approval for certain study types only will be recommended. This may place restrictions on instrumentation use or instrument type. Those institutions that are not funded by external grant sources must closely monitor the development of market-approval guidelines, as this factor is probably the major one in choice of NMR instrumentation. A related factor is the current low volume of patients that can be scanned with these imaging systems as compared to x-ray CT scanners, but this situation may improve as imaging protocols are more fully developed.

The diagnostic potential of NMR is directly related to instrument cost. For example, results to date indicate that magnets with higher field strengths (0.5 tesla superconducting systems) have better spatial resolution, image contrast, and signal-noise ratios for proton imaging than lower-field resistive systems (0.15 tesla). The optimum field strength has yet to be determined but it is possible that the optimum field strength will vary with the imaging pulse sequence used and clinical protocol (see Table 11-1). To be more specific, image contrast is dependent on several parameters in addition to field strength, such as T_1 and T_2. Their magnitude are also field strength dependent.

The diagnostic potential of the 1.5 telsa magnet system may be superior because of the dual possibility of multinuclide imaging and NMR spectroscopy. It is clear that spectrographic measurements will be more definitive than proton imaging. However, the cost of this system, site requirements, and service contracts are significantly greater than lower field systems. In addition the time to complete both imaging and spectroscopy measurements will be greater, with loss of patient throughput. It is still not clear whether proton imaging is feasible at 1.5 tesla. Good spatial resolution has been observed for proton imaging, but improvement in image contrast has yet to be fully demonstrated. Whole-body spectroscopy measurements using the 1.5 tesla magnet have employed surface coils that have a restrictive measurement (depth) range related to coil diameter. The use of orthogonal field gradients for whole-body multinuclide imaging and spectroscopy measurements is an important development to monitor before system purchase.

Suggested Reading

Foster, M. A. *Magnetic Resonance in Medicine and Biology.* New York: Pergamon Press, 1984.

Gadian, D. G. *Nuclear Magnetic Resonance and Its Applications to Living Systems.* Oxford: Clarendon Press, 1982.

Mansfield, P., and Morris, P. G. *NMR Imaging in Biomedicine.* New York: Academic Press, 1982.

Partain, L. *Nuclear Magnetic Resonance and Correlative Imaging Modalities.* New York: The Society of Nuclear Medicine Inc., 1984.

12

Diagnostic Ultrasound

Michael M. Raskin

Although its characteristics have been known for many years, it was not until recently that the first practical application of sound was made for imaging the body. Much present sound technology is due to military research during World War II in developing sonar. Successful medical applications of ultrasound began shortly after the war but have really burgeoned in the last decade. Ultrasound does not utilize ionizing radiation and has no known adverse side effects when used for diagnostic imaging. It has rapidly become embraced by both patient and physician, especially in the fields of radiology, obstetrics, gynecology, cardiology, and peripheral vascular disease.

Propagation of Sound

Particle vs Wave Theory

The only similarity between an x-ray beam and a sound beam is that both are waves transmitting energy. However, unlike x-rays, ultrasound cannot travel through a vacuum and requires a medium for its transmission. The speed of transmission of the sound depends upon the characteristics of the medium.

For purposes of understanding sound propagation, it is helpful to consider a medium as consisting of an arrangement of particles or molecules in which each particle is connected to its neighbor by elastic bands. When a particle is set into vibration, it also vibrates the neighboring particles via these elastic band connections. Even though vibrational energy is transmitted through the medium, there is no net movement of the medium because the total displacement of each particle is limited to very short distances about its resting position. While individual particles only move a few microns, the net effect of their motion is transmitted by their neighbors through a much longer distance. This sets up a chain reaction in which the speed of sound transmission is determined by the rate at which the force is transmitted from one particle to another (Figure 12-1).

Sound may also be considered analogous to light waves and similar to the propagation from other electromagnetic sources. It is often easier to explain propagation of ultrasound using mechanical vibration of particles in a medium. However, it matters not whether we assume sound is a "wave" or a "particle" for the derived equations are identical whichever method is used.

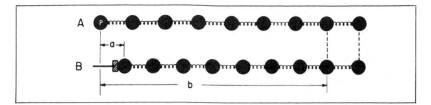

Fig. 12-1. (A) Particles *(P)* are in their resting state connected by small "springs." (B) When the first particle is displaced distance *a*, the effect of the motion is transmitted over distance *b*. (Modified from T. S. Curry III et al., *Christensen's Introduction to the Physics of Diagnostic Radiology* [3rd ed.]. Philadelphia: Lea & Febiger, 1984. With permission.)

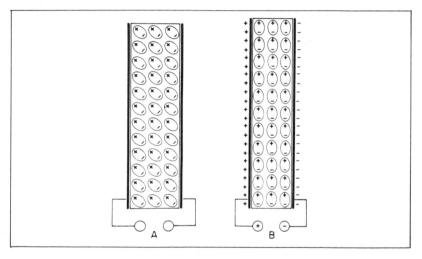

Fig. 12-2. (A) No charge is applied. (B) When an electrical charge is applied, the positive and negative ends of the electrical dipoles realign and change the dimensions of the crystal. (From T. S. Curry III et al., *Christensen's Introduction to the Physics of Diagnostic Radiology* [3rd ed.]. Philadelphia: Lea & Febiger, 1984. With permission.)

Piezoelectricity

When a mechanical stress is applied in certain materials, a voltage appears across the material in direct proportion to the applied stress. This was discovered by the Curie brothers in 1880 and called the *piezoelectric (pressure-electric) effect.* Piezoelectric materials contain "electrical dipoles" which will realign and cause the dimension of the crystal to change when an electric field is applied (Figure 12-2). If the voltage is applied in a sudden burst or pulse, the material will vibrate, producing ultrasound. This is actually called the *"reverse" piezoelectric effect.* Certain natural occuring crystals have piezoelectric properties, but

in ultrasound we primarily are concerned with certain synthetic ceramic materials known as *polarized polycrystalline ferroelectrics.*

Transducer Construction

The ultrasound transducer or "probe" contains the vibrating crystal. The transducer is both a transmitter and receiver but it does not send and receive at the same time. A short burst of electrical energy causes the crystal to vibrate and produce an ultrasound beam. The beam enters tissues and is partially reflected back to the transducer causing the crystal to vibrate and produce an electrical signal. The crystal must be "damped" to stop it from vibrating after the initial electrical burst in order that it may be "listening" for returning echoes. This is called the *pulse-echo mode* and it is most commonly used in imaging. The resonant frequency or natural frequency of a piezoelectric crystal is determined by the thickness of the crystal and is that frequency producing internal wavelengths twice the thickness of the crystal.

Focusing of Sound

The optimal size for the transducer depends upon its application. In some cases, it is desirable to have a focused beam while at other times it is desirable that the sound beam cover a particular distance. The directivity of an ultrasound beam may be altered by focusing. Although there are several methods of focusing, the most commonly used is an acoustic lens. A focusing lens may be advantageous for certain applications in diagnostic ultrasound as a concentrated small diameter beam can be produced over a predictable region.

A new type of focusing overcomes the limitations of fixed-focus transducers by utilizing an electronically focused annular array. This annular array is formed by creating concentric rings of transducer materials that are electrically isolated from one another (Figure 12-3). Each array element, or ring, has its own electronic amplifiers. The focusing control system combines signals from each array element, resulting in an electronically variable lens (Figure 12-4). While fixed-focus transducers confine the region of maximum sensitivity to a small zone about the focal point, dynamically focused annular arrays extend the region of maximum sensitivity throughout the entire image (Figure 12-5).

Characteristics of Ultrasound

Longitudinal Waves

Longitudinal waves consist of alternate depressions of rarifactions that have a particle motion that is parallel to the propagation of the sound wave. The repetition of this back-and-forth motion is called a *cycle*, and

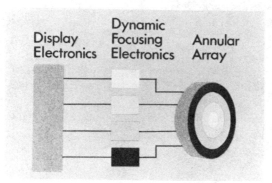

Fig. 12-3. Block diagram of an annular array system. (From Technicare Ultrasound Corp. *Imaging Properties of Dynamically Focused Annular Arrays.* Englewood, CO, 1982. With permission.)

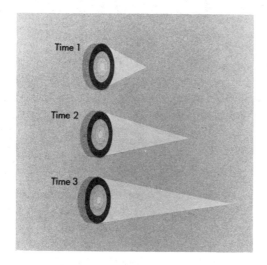

Fig. 12-4. With dynamic focusing, the array is always focused at the point where the echoes originate. At *Time 1*, the array is focused in the near field while it is focused at maximum depth at *Time 3*. The refocusing occurs automatically and continuously, resulting in a large number of focal zones. (From Technicare Ultrasound Corp. *Imaging Properties of Dynamically Focused Annular Arrays.* Englewood, CO, 1982. With permission.)

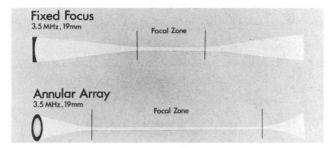

Fig. 12-5. Comparative beam profiles. (From Technicare Ultrasound Corp. *Imaging Properties of Dynamically Focused Annular Arrays.* Englewood, CO, 1982. With permission.)

each cycle produces a new wave. The length of the wave is the distance between two bands of compression or rarifaction and is appropriately called *wavelength* (λ). The velocity of propagation is dependent upon the elastic properties and the density of the medium. The longitudinal velocity for water at 25°C is 1495 meters per second. The velocity in body tissue approaches that of water but will vary according to the structure or consistency of that tissue (Table 12-1). A velocity of 1540 meters per second is used as an average for body tissue.

Transverse Waves

Although all materials can support longitudinal waves, only solids can support transverse waves. Bone may support transverse wave propagation that is substantially slower than that of longitudinal waves. The particle motion is perpendicular to the direction of propagation of the longitudinal wave. Transverse waves are seldom of importance in medical ultrasound except to explain the increased absorption of sound by solid material.

Frequency and Wavelength

Our ears usually can detect sound if the frequency of vibration is between sixteen and sixteen thousand cycles per second (Hz). Sound with a frequency above twenty thousand Hz is called *ultrasound.* However, this distinction is quite arbitrary, and medical diagnostic ultrasound is most often in the range of 2,000,000 Hz (2 MHz) to 10,000,000 Hz (10 MHz). The velocity of sound depends upon the elasticity and density of the medium through which it travels. Elasticity is the measure of "stiffness" of the elastic connections between the particles of material. Since the density of tissue remains fairly constant, the velocity of sound is virtually independent of frequency.

Table 12-1. Sound Velocities in Various Materials

Material	Velocity (m/sec)
Air	331
Distilled Water	1430
Fat	1450
"Average" Soft Tissue	1540
Kidney	1561
Muscle	1585
Skull Bone	4080

Frequency is the rate at which the particles in the medium vibrate. It is the number of pressure peaks that pass a given point in one second. The *wavelength* is the distance from one pressure peak to the next. Thus wavelength and frequency are inversely related as follows:

$$v = f\lambda$$

where v is the velocity of sound in the conducting media, f is the frequency, and λ is the wavelength.

Pulse Rate

The *pulse rate* represents the number of times the crystal is excited each second. The commonly used pulse rate is 1,000 pulses per second so the transducer is a receiver almost 1,000 times more than it is a transmitter. The pulse rate determines the total number of echoes returning to the transducer in a unit of time. If the receiving time is too short, echoes from more distant structures may still have not returned when the next pulse starts.

Sound Beam

The type of transducer element does not determine the property of the propagated sound beam. Sound beam characteristics are determined by the size, shape, and operating frequency of the transducer element. For a given size of transducer, the higher the frequency, the more directional the sound beam. Lower frequencies are used for their penetrating capability, while higher frequencies for their better resolving capacity. Depending on the construction of the transducer element, the sound beam remains parallel for a certain distance. The parallel component is called the *near zone* or *Fresnel zone*. The diverging portion of the beam is called the *far zone* or *Fraunhoffer zone* (Figure 12-6). At the point of transition (x'), the sound beam has maximum phase coherence. The beam divergence is determined as follows:

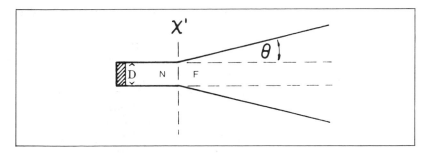

Fig. 12-6. The transition point (*x'*) is where the sound beam begins to diverge and distinguishes the near (*N*) zone from the far (*F*) zone. There is no divergence in the near zone and the width of the beam is each to the diameter (*D*) of the transducer.

$$\sin \theta = 1.22 \, \frac{\lambda}{D}$$

where θ equals the dispersion area of the far zone, and D is the diameter of the transducer.

Since we want to keep the angle of dispersion as small as possible, the Fresnel zone is longest with a large diameter transducer and high frequency sound. Although high frequency beams have superior depth resolution, tissue absorption increases with increasing frequency. Larger diameter transducers improve phase coherence, but they deteriorate lateral resolution. Many of these problems can be circumvented using the new annular array transducers.

Doppler Effect

The apparent frequency of a constant frequency source is dependent both upon the motion of the source and that of the receiver. The measurement of frequency is performed by counting the number of pressure peaks that pass an observer in one second. However, if the observer is moving toward the sound source while making the measurement, he will encounter a greater number of pressure peaks than if he were standing still. Similarly, if he had been moving away from the source, the measured frequency would be lower than the true frequency. This apparent change in frequency was first described by Christian Doppler in 1843 and is called the *Doppler effect*; it can be mathematically derived by

$$f = f_o \left(\frac{v - v_o}{v - v_s} \right)$$

where f is the observed frequency, f_o is the frequency of the sound source, v is the sound velocity, v_o is the velocity of the observer, and v_s is the velocity of the sound source.

This effect can readily be observed when a train whistle or siren is coming towards us and the frequency sounds higher pitched than when it is going away. The Doppler effect can be utilized in the detection of moving structures within the body, using a transducer that employs two crystals: one transmits a continuous ultrasound signal at a fixed frequency; the other continuously receives the returning echoes. The greater the difference in frequency between the transmitted sound and the returning sound, the greater the Doppler shift. The magnitude of the frequency of change is directly proportional to the velocity of the moving surface and is given by the equation:

$$\Delta f = \left(\frac{2f_o}{c}\right) v \cos \theta$$

where Δf is the change in frequency, θ is the angle between the sound beam and the direction that the reflecting surface is moving, v is the velocity of blood, and c is the velocity of sound.

The smaller the angle between the sound beam and the flow direction, the greater the vector of motion towards the transducer and the greater the Doppler shift. The most important application of this technique is in the detection of blood motion. Returning echoes originate primarily from the red blood cells rather than the motion of the vessel wall. This technique is extremely sensitive to determine blood flow even in very small vessels. It should be noted, however, that Doppler instruments actually measure blood velocity and not blood flow. However, actual blood flow can be closely estimated by determining the cross-sectional area of the blood vessel using pulse echo techniques:

$$F = A\bar{v}$$

where F is blood flow, A is cross-sectional area of the blood vessel, and $\bar{v}$ is the average blood velocity.

Interaction with Tissue

Attenuation

As sound propagates through a medium, the intensity of the sound progressively decreases. Factors that contribute to attenuation of the sound beam include divergence of the beam, deflection of the sound, and absorption of the energy.

Absorption of the sound is directly proportional to the distance traveled through the medium, the density of the medium, and the square of the frequency. Although higher frequency may have greater resolution, there is also greater absorption of the sound and higher power levels may be needed.

Power Levels

High power levels can cause tissue damage. However, the average power levels used in diagnostic ultrasound are several order of magnitude below these levels. Sound intensity is measured in *decibels*. The *bel*, in honor of Alexander Graham Bell, is ten times the size of a decibel and is the logarithm of the ratio of two intensities and is expressed as:

$$dB = 10 \log \frac{I}{I_o}$$

where dB stands for decibels (a relative unit), I is the intensity of sound, and I_o is the intensity of the reference.

Acoustic Impedance

The characteristic acoustic impedance (z) is the product of the density and the sound velocity through the medium. The impedance of an interface between two tissues determines the amount of energy that is reflected from or transmitted through various structures of the body. The acoustic impedance is independent of the frequency. The proportion of sound reflected when a beam strikes an interface depends upon the difference in characteristic acoustic impedance of the two adjacent media and the angle of incidence of the sound beam. When the sound beam is perpendicular to the interface, the ratio of the amplitude of an echo to the amplitude of the incident pulse, called the *amplitude reflection coefficient* (R_c), can easily be determined:

$$R_c = \left(\frac{z_2 - z_1}{z_2 + z_1} \right)^2$$

Reflection, Refraction, Transmission

Unlike x-rays, which use transmitted radiation to produce an image, ultrasound images are produced by virtue of the reflected sound wave. The transmitted sound wave contributes nothing to the image formation. The proportion of the beam reflected depends upon the differences in the characteristic acoustic impedance on both sides of the interface and the angle of incidence of the ultrasound beam. The angle of reflection is equal to the angle of incidence (Figure 12-7). The transducer must be carefully positioned so that the sound beam is perpendicular to an interface.

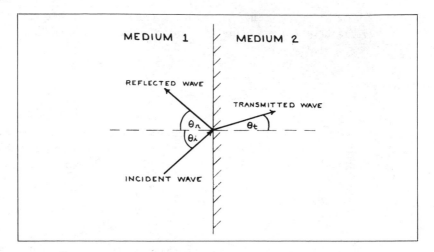

Fig. 12-7. The angle of the reflected wave *(θ_r)* equals the angle of the incident wave *(θ_i).* The angle of the transmitted wave *(θ_t)* can be determined using Snell's law. (From B. B. Goldberg, et al. *Diagnostic Uses of Ultrasound.* New York: Grune & Stratton, 1975. With permission.)

The energy that is transmitted through the interface will be influenced by the second medium. When energy strikes at an angle other than perpendicular, the resultant angle of propagation will be dependent on the ratio of the velocities of the media. If the velocities are different, the sound beam in the second medium will refract in accordance with Snell's law:

$$\frac{\sin \theta_1}{v_1} = \frac{\sin \theta_2}{v_2}$$

Resolution

Resolution is the ability of the ultrasound beam to separate two objects as distinct entities:

1. *Axial resolution,* often called *depth resolution,* is the ability to distinguish two objects on a line parallel to the sound beam. The distance between the two objects must be greater than one half of the pulse length, or the pulse length must be less than twice the distance between the two objects, in order for objects to be resolved as separate entities (Figure 12-8). The wavelength determines the theoretical limit for axial resolution.
2. *Lateral* or *horizontal resolution* is the ability to distinguish objects in a line perpendicular to the axis of the sound beam. In order to recognize the objects as discrete entities the sound beam must be narrower than the space separating the objects (Figure 12-9). Lateral

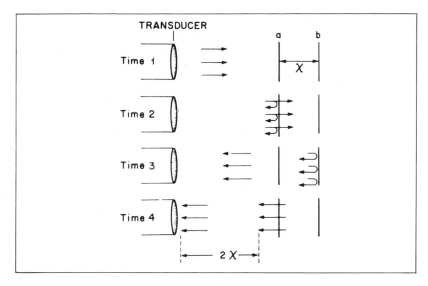

Fig. 12-8. Time 1: Ultrasound pulse is resolving surface *a* and *b* separated by distance *x*. *Time 2:* Part of the beam is being reflected from surface *a*. *Time 3:* Part of the beam is being reflected from surface *b* and both sets of reflected pulses are separated from each other. *Time 4:* The separation of the pulses is twice the distance of the two surfaces. (From T. S. Curry III et al., *Christensen's Introduction to the Physics of Diagnostic Radiology* [3rd ed.]. Philadelphia: Lea & Febiger, 1984. With permission.)

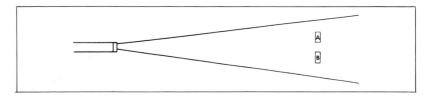

Fig. 12-9. The two objects, *A* and *B*, cannot be recognized as being separate because the beam straddles both objects and the space between.

resolution is inversely proportional to beam width. The beam width is dependent primarily on the diameter of the transducer, but also depends upon the frequency of the sound, and the distance from the source.

Clinical Applications
Abdomen and Pelvis

A large number of abdominal and pelvic abnormalities can be diagnosed by ultrasound. The primary limitations of this modality are overlying bowel gas and patient obesity.

1. *Biliary tree.* In many instances, ultrasound has replaced the oral cholocystogram as the primary method of evaluating the gallbladder. Although the accuracy of both modalities are similar, ultrasound is preferred since there is no contrast reagent administered nor is there any delay in performing the procedure. Furthermore, the entire biliary tree can be evaluated, and this can be extremely important for the patient with acute biliary symptoms. This procedure is often performed in conjunction with a radionuclide procedure for determining patency of the biliary system. Ultrasound provides the most accurate technique for determining the presence of gallstones.

2. *Guided biopsy.* Ultrasound can be used to great advantage to biopsy intrabdominal organs as well as to localize areas for external drainage. Although this can often be done under computed tomography guidance, ultrasound is often quicker to perform, more cost effective, and can be done at the patient's bedside.

3. *Pancreas.* Ultrasound can usually visualize the entire pancreas and determine the size, shape, and texture. Furthermore, a dilated pancreatic duct can usually be detected. The major difficulty is in visualizing the body and tail of the pancreas when there is overlying gas. This can be especially difficult in obese patients and computed tomography will usually provide more information in these individuals. It is emphasized, however, that both modalities usually detect pancreatic carcinoma only by altered morphology of the gland. Both modalities have been somewhat disappointing in detecting a small pancreatic carcinoma.

4. *Kidneys.* Ultrasound has always been efficacious in determining whether a renal mass was cystic or not. Furthermore, ultrasound can confirm the presence of solid masses as well as perinephric fluid collections. It can be performed in patients that are anuric and patients who are allergic to iodine and cannot have an intravenous urogram. Ultrasound can distinguish between obstructive urography and renal parenchymal disease in patients with anuria.

5. The *aorta.* Ultrasound can detect the presence of an abdominal aortic aneurysm as well as determine the diameter and extent of the aneurysm. Thrombis within the aneurysm can easily be identified. Ultrasound is the procedure of choice for following the patient with an abdominal aortic aneurysm who is not a surgical candidate by serial measurements of the diameter and extent of the aneurysm. Furthermore, since ultrasound can determine the presence of a mural thrombis, it can detect progressive obliteration of the true lumen on serial ultrasound studies even though the external diameter has not changed.

6. *Obstetrics and gynecology.* When a pelvic mass is clinically present, not only can ultrasound determine whether the mass is solid or cys-

tic, but it can often determine whether the mass originates from the uterus or the ovaries. Ultrasound has been used in obstetrics for some time and is the procedure of choice for evaluating the fetus as well as determining the location of the placenta. The growth and development of the fetus can be conveniently monitored by ultrasound by assessing biparietal diameter, femur length, head circumference, and abdominal circumference. Ultrasound is extremely accurate in determining fetal viability and gestational age. Needle guidance for amniocentesis is frequently performed using ultrasound for determination of placental localization.

Cardiology

Two-dimensional echocardiography has the ability to examine the heart in motion in a variety of planes. Wall motion abnormalities and aneurysms can be detected as well as determination of the mitral valve area. Echocardiography is accurate in determining the presence of mitral stenosis, although the degree of stenosis cannot be accurately predicted. It is, however, particularly useful in the evaluation of patients with mitral regurgitation and can distinguish mitral valve prolapse from ruptured chordae. It is also useful in detecting other valvular abnormalities of the various heart valves including infection. Two-dimensional echocardiography is the procedure of choice in determining the presence and degree of pericardial effusions. When used in conjunction with M-mode echocardiography, quantitative measure of chamber size and wall thickness can be accurately determined. Hypertrophy of the heart wall or dilatation of the cardiac chambers can easily be detected.

Two-dimensional echocardiography has become particularly useful in recognizing complicated congenital heart lesions in infants. Using contrast injections, intracardiac shunts at the atrial level can be detected. Duplex scanning combining Doppler ultrasound with two-dimensional echocardiography is able to provide quantitative measurements of blood flow and is fairly accurate in determining the degree of mitral regurgitation or aortic insufficiency.

Peripheral Vascular Disease

1. *Cerebrovascular disease.* Directional Doppler ultrasound can determine the direction of blood flow. This is especially important to determine the direction of flow in the ophthalmic artery as this vessel serves as a potential collateral bridge between the internal carotid and external carotid arteries. Retrograde ophthalmic artery flow is a highly accurate indicator of a hemodynamically significant lesion in the internal carotid artery system. More quantitative data is often supplied by *ophthalmopneumoplethysmography*, a technique used to directly measure the ophthalmic artery pressure by using small eyecups placed on the sclera of the eye.

Doppler ultrasound can also be used to produce a "flow image" of the carotid artery bifurcation. "Conventional" pulse-echo ultrasound can directly visualize the carotid artery bifurcation and often detect narrowing or the presence of plaques. The ultrasound examination is often supplemented by measurements of spectral analysis to determine variations in velocity of the blood through an area. Turbulent flow can easily be detected and the spectrum of ultrasound frequencies can be useful in providing a qualitative assessment of the degree of narrowing.

2. *Lower extremities.* Doppler ultrasound is accurate in determining blood pressure in the lower extremities and can determine the presence of a pulse even when one is not clinically palpable. Pressures can be measured in various areas of the lower extremities and pulses of each artery can be qualitated and compared. Additional valuable information can be obtained after the patient has exercised on a treadmill. Measurements performed immediately after exercise give additional information in regard to the level, severity, and number of occlusions.

Ophthalmology

Various types of ultrasound are used to determine not only abnormalities of the eye itself, but abnormalities in the retrobulbar area. Special digital readout instrumentation using frequencies of at least 20 MHz are used to measure the thickness of the cornea as well as the depth of the anterior chamber or thickness of the lens.

Real time ultrasound is the only method of accurately mapping vitreal pathology of the eye since optical visualization is usually not possible. Ultrasound can determine the extent of hemorrhage as well as retinal detachments and tumors.

Breast

Ultrasound of the breast should not be considered as a primary screening procedure by itself. Ultrasound may be of value in determining whether a mass is cystic or solid. It can also be helpful in aspiration of a nonpalpable breast mass. The procedure itself can be technically variable when contact scanning is performed. More accurate and reproducible visualization of the breast may be obtained with breasts suspended in a water bath with the patient in the prone position. However, this position makes the procedure more difficult to perform.

Selection of Equipment

High-resolution dynamic scanning can only be performed using real-time systems which have the ability to rapidly scan in multiple planes. The articulating scanning arm, gray scale B-mode contact scanner that is still around in some institutions, cannot be justified as the only ultrasound imaging equipment in a modern department.

The newer instruments use microprocessors to provide digitized amplitude, frequency, and phase information. This capability allows off-line echo analysis, such as tissue attentuation and other quantitative ultrasound measurements. Further increase in dynamic range and resolution may be provided through annular phased array transducers that will allow images to remain in focus throughout the field of view and provide increased sensitivity to low level echoes.

A hospital considering the purchase of a major piece of ultrasound equipment should consider a real-time B-mode scanner that has the capability of duplex scanning with Doppler. A machine so configured could be used for cardiology and peripheral vascular studies as well as abdominal and pelvic imaging. With the addition of a high frequency probe of 7.5 MHz or 10.0 MHz, small parts such as the thyroid and testes would be easy to examine.

Ultrasound equipment used for the abdomen, pelvis, and heart cannot be used for the eye. Ultrasound of the eye requires some very specialized equipment that is of no use for the rest of the body.

A specialized ultrasound machine using a water-bath cannot be justified in a small department, as ultrasound of the breast should not be used as a primary screening modality. However, a high frequency probe will be quite accurate in determining the nature of a palpable mass or a mass seen by mammography.

Summary

The general purpose real-time B-mode scanner provides the best compromise when a single machine is to be obtained. Many of the currently available real-time scanners allow the addition of two dimensional echocardiography as well as duplex Doppler. These upgradable instruments should be considered if one contemplates adding these modalities at a later date. The rapid obsolescence of many specialized ultrasound machines behooves the prudent purchaser to obtain an instrument with maximum flexibility and upgradability.

Suggested Reading

Curry, T. S., III, et al. *Christensen's Introduction to the Physics of Diagnostic Radiology* (3rd ed.). Philadelphia: Lea & Febiger, 1984.

Goldberg, B. B., et al. *Diagnostic Uses of Ultrasound.* New York: Grune & Stratton, 1975.

Raskin, M. M. *Comparative Abdominal and Pelvic Anatomy by Ultrasound and Computed Tomography.* West Palm Beach: CRC Press, 1979.

Sarti, D. A., and Sample, W. F. *Diagnostic Ultrasound - Text and Cases.* Boston: G. K. Hall, 1980.

Wells, P. N. T. *Physical Principles of Ultrasound Diagnosis.* London: Academic, 1969.

13

Bid and Performance Specifications and Warranty and Service Contracts

Jack S. Krohmer

Whether one is interested in acquiring a simple replacement for a single radiographic room or outfitting an entire new radiology department, it is recommended that a set of bid specifications be prepared. As is true with any purchase of any commodity, the proper time for establishing standards relative to the performance, warranties, and future service is *prior* to the actual purchase. It is at this time that the interest of the vendor in the purchaser's problems is at a maximum. Two major types of bid specifications are used: The first is a set of specifications designed in such a fashion as to assure the purchase of specific equipment; the second is a set of specifications designed to assure the purchase of equipment that will perform the tasks required at the most effective cost. This chapter will describe the latter of these two types of bid specifications.

The specification guidelines included in this chapter are designed to be relatively universal in that they can be used for the purchase of a single piece of equipment or, with expansion, an entire department of equipment. It is true that the documents may be more compact and simpler for purchase of a single piece of equipment; however, the included material can be used as a template for all future bid specifications for any amount of equipment.

The Bid Specification Packet

The bid packet consists of three portions: (1) a sample schedule of radiological equipment to be installed in specific rooms; (2) bid submission forms; and (3) detailed equipment specifications describing radiological equipment included on the bid request.

Item 1, Schedule of Radiological Equipment, will include the list of equipment required for each room or installation and will refer to Item 3 for actual specifications. Item 2, Bid Submission Forms (only one is included), will establish the form in which bids are to be submitted and the types of bids that are requested. Item 3, Detailed Equipment Specifications, will include general requirements, performance requirements, installation and relocation requirements, guarantees and warranties desired, and future service requirements.

It is obvious that the bid specification packet is a legal document, and thus great care must be exercised in preparing it. There is no question that a thorough understanding of the material presented in the earlier portions of this book must be acquired before the assembly of this packet is undertaken. It is recommended that the initial step in the preparation of the bid packet be the establishment of the specific radiological techniques and procedures that are required to be carried out for each installation. After this has been done, and after meetings with the various equipment vendors have been held to determine what sorts of equipment are available, the actual preparation of the bid packet can take place.

Keep in mind that a major reason for the preparation of a bid specification packet is to assure the purchaser that he will receive equipment that will carry out the tasks required and will protect him against future problems. It is also true that the document, if designed properly, will protect the vendor against future problems associated with the purchase. An important consideration in this regard is the assurance to the vendor that a proper and adequate power supply will be available for each piece of equipment required. Also, the vendor must be protected against possible future breakdowns in equipment due to improper use.

The bid specification packet (see Appendix, 294–315) is simply a sample that will undoubtedly require modification to meet the needs of the institution preparing a packet. However, many of the included items should and will be included in all bid specification packets. It is the author's hope that the material in this chapter will help to assure 1) the purchase of equipment that will allow performance of the required tasks, 2) cost-effectiveness, 3) maximum guarantees and warranties, 4) trouble-free installation and relocation, and 5) effective and reasonable on-going service of the equipment. It should be kept in mind that the most cost-effective purchase is not necessarily associated with the lowest bid submitted. An abrogation of the requirement to accept the lowest submitted bid is included in the very first paragraph of item 3, Detailed Equipment Specifications.

The request for the total purchase price is broken down into the cost of equipment and the other costs associated with the purchase; for example, installation charges, shipping and insurance charges, warranty charges, and training charges. It should be noted that the latter items can be written off as an expense for the year of purchase, whereas the cost of the equipment itself must be amortized over a number of years. This type of bid request allocates the various costs of the purchase to the most advantageous time period. This again leads to a more cost-effective acquisition of equipment.

Guidelines for Purchasing Equipment

Delivery Date

Most radiographic equipment is delivered from 90 to 180 days after award of the contract. Unforeseen delays resulting from strikes or inability to obtain raw material may occur, and there should be provisions in the contract to protect both parties in case of such events. The contract should also contain some provision as to what recourses the purchaser may have if the equipment is not delivered on time. In addition, the contract should specify to whom the equipment belongs until the time of its acceptance by the purchaser. In most cases it is wise to accept the equipment from a freight carrier on behalf of the vendor but to let the vendor accept the responsibility if the equipment has been damaged in transit or if parts are missing. This way the equipment belongs entirely to the vendor until it has been installed, inspected, and accepted.

Methods of Payment

Normally a vendor will require a certain percentage of payment with the purchase order (either in cash or in a lease agreement), another percentage when the equipment is delivered, and the remainder after acceptance of the equipment. The contract should contain some provisions for reimbursement in case the equipment is not acceptable under the terms of the contract. Installment payments provide some leverage in the case of equipment malfunction or other problems.

Installation of Equipment

One of the common complaints from purchasers is that once installation has begun, the work is periodic rather than continuous. This problem can be alleviated if a timetable is specified in the contract and penalties agreed on by both parties. Whenever there is a penalty clause in a contract, the purchaser should be aware that the vendor will make every effort to meet the contract specifications and to avoid the penalty. On the other hand, a vendor will take into consideration the possible loss of revenue and will adjust his price accordingly.

The first responsibility for the installation of radiographic equipment is on the part of the purchaser. The purchaser is normally responsible for preparing the site to meet specifications provided by the vendor, and the final site preparation must be acceptable to the vendor. Lack of adequate site preparation is a common cause for delays in installation. The purchaser is also responsible for providing the proper voltage supply to the area, which includes the provision of a feeder transformer with adequate kilovoltage-amperage (kVA) capacity, length and size of wiring, and "breakers" ("disconnects" in the electrical engineering jargon). To avoid any delay in installation, the purchaser should have the site inspected by the vendor, with the vendor specifying in writing any nonacceptable deficits or indicating that the site preparation is acceptable. Adequate site preparation will eliminate a number of future problems.

After installation has begun, it should proceed 8 hours a day without interruption until the installation has been completed and the equipment has been inspected and accepted by the purchaser. The contract should clarify what holidays will be observed during the working schedule, since the purchaser's and the vendor's holidays may not coincide. One of the major problems throughout the United States is that service personnel are also used as installation crews. This often means that the installation must be interrupted so that the service personnel can make emergency calls elsewhere. There are occasions when a trainee is left at the installation site to continue the installation under the terms of the contract while the regular competent installers are making a service call.

A reasonable way to approach the problem of installing equipment is to state in the contract that the equipment will be installed and inspected by a specified date—period. How many hours are required and what personnel are used for the installation is left up to the vendor, since it is his responsibility to have the equipment installed by the specified date. A penalty clause can be inserted into the contract for reimbursement if installation is not completed on time.

Performance Requirements and Acceptance of Equipment

The basic performance requirement for any piece of equipment is that it must work as intended and according to purchase specifications. In addition, its performance is important. Any radiographic equipment should be inspected for performance prior to acceptance by the purchaser. This inspection can be done either by the vendor or by a representative of the purchaser. Documentation of the performance specifications can be easily done by the vendor, since it usually means supplying the purchaser with data obtained by the vendor when he was doing his own inspection. Whether or not to accept the vendor's data is entirely up to the purchaser's confidence in the capability and integrity of the installation crew and the vendor's reputation. In any case, there should be documentation of all requirements noted in the purchase specifications. If there is any question, a qualified radiation physicist should be retained to document the performance specifications.

If deficiencies are found in the acceptance inspection, the vendor should be given a reasonable time to correct the problems—normally 30 days from the date of the inspection. Fortunately, most problems can be corrected at the time of the inspection. Since the vendor knows beforehand that an inspection is required, he usually makes sure there are no problems prior to the inspection.

Warranty Considerations

A synonym for *warranty* is *guarantee*. The reader is probably aware of the national turmoil over warranties and what they really mean. A warranty can be written in such nonspecific terms that any provision contained in the warranty is legally nonenforceable. In the radiological profession, a warranty usually means that the manufacturer or his representative will repair a piece of equipment for a specified time at no cost to the purchaser. One basic point that should be clarified is the date at which the warranty actually begins—after installation, after inspection, or after acceptance of the equipment by the purchaser. This point should be made clear in the purchase contract.

There are two points to ponder when considering the provisions of warranty services. On the one hand, the manufacturer wants to have a satisfied customer, and there are many occasions when a manufacturer will repair equipment that is clearly not covered under the written terms of the warranty. On the other hand, the manufacturer may follow the specific terms of the warranty when the customer is overbearing and unreasonable. The best warranty is one in which two reasonable parties make a common agreement on a reasonable procedure. One of the biggest problems with warranties occurs when equipment is purchased that does not meet the requirements of the purchaser but that meets the performance specifications of the manufacturer. Almost all such problems can be eliminated by purchasing equipment in terms of specifications so that the purchaser knows what he wants and expects, and the manufacturer knows exactly what is expected of his equipment.

In any case, a warranty should state that the equipment will operate as specified for a given period of time and will be essentially troublefree. However, it is unreasonable to expect any piece of equipment to be totally troublefree for any given length of time. A record should be maintained on each failure, and specifically, recurring failures should be recorded. A warranty should never be ended while recurring major component failure exists. In the case of recurring failures, the purchaser and vendor should discuss an extension of the warranty period to assure the purchaser and vendor that the equipment is indeed essentially trouble free. The obvious problem lies in the definition of "trouble free."

The warranty should also specify when service is to be rendered and how quickly a request for service is to be answered. Again, these requests should be reasonable. A distinction should be made between annoying malfunctions that do not actually interfere with use of the equipment and those failures that make a machine totally inoperable.

Service Contracts

All radiographic equipment should have preventive maintenance performed on a periodic schedule. As with an automobile, the better a piece of equipment is maintained, the less trouble one can anticipate. Well-maintained equipment will also last longer. In addition to routine maintenance, all applicable factory modifications should be included in the preventive maintenance program.

Different Systems of Service

Preventive maintenance and repair service can be performed by a repairman employed by the radiology department or the hospital, by a repair service on a per call basis, by a specific manufacturer on a per call billing, or under service contracts. There are advantages and disadvantages to each system. The problem of having a departmental or hospital repairman is that the knowledge of these servicemen rapidly becomes out-of-date, and they do not receive notification of modifications as vendors do, unless specific arrangements are made with the manufacturer. Some manufacturers will arrange training courses on specific pieces of equipment for hospital repairmen. It is the feeling of most of those involved in hospital maintenance that they cannot obtain repair parts as quickly as a vendor can obtain them. The reason is felt to be that a manufacturer normally will supply his own personnel before making available parts to someone who is in competition with his service organization. Another disadvantage is that very competent repairmen are extremely difficult to recruit. The good ones are highly paid by the manufacturer, and most hospital pay systems are not flexible enough to meet salary demands. The advantages of having a hospital repairman are that one normally does not have to wait for service, and that overall the cost of maintaining equipment is probably lower than through service contracts with specific vendors.

Service contracts with maintenance firms who repair many different types of radiographic equipment have the same drawbacks as having a hospital repairman. Most radiology departments having a maintenance program will have a service contract with a specific manufacturer for equipment made by the manufacturer. As a general rule, one should be cautioned against having maintenance contracts with one manufacturer for equipment made by another firm. In fact, most company policies will not allow their servicemen to maintain equipment made by another company, unless their organization sold that particular piece of equipment. The same holds true for purchasing radiographic equipment—one should not mix equipment made by different manufacturers in the same room unless it is to be maintained by a single firm. This principle will eliminate "passing the buck" when equipment failure occurs and the cause of the failure cannot be positively identified.

There are numerous manufacturers of radiographic equipment, and each company has its own policies in regard to service contracts. This makes it impractical, if not impossible, to spell out in specific terms what a maintenance contract should include. However, there are pitfalls and generalities to be discussed that may be beneficial to a potential customer. No one should enter into a maintenance contract without investigating the firm's performance in the local area.

Cost

The cost of a maintenance contract varies according to the services provided under the contract. As a general rule, a full service contract, excluding glassware, will cost from 5 to 10% of the purchase price of the equipment per year. A lower cost may be negotiated if a large volume of equipment is to be maintained under the contract. The purchaser can usually negotiate one of the options noted below.

Labor Contracts. Labor contracts provide for routine and emergency maintenance service for a set number of dollars per month or per year. The advantage of having a labor contract is that customers with maintenance contracts usually have higher priority for service than those who do not have such contracts. In addition, the vendor can schedule maintenance much more effectively if he can make an estimate of exactly what the service needs of his territory will be.

Labor Plus Parts Contract. For larger institutions, labor plus parts contracts are generally cheaper in the long run than pure labor contracts. Few people realize that it costs a large institution a minimum of $25 to process a purchase order. Under a typical straight labor contract the vendor bills the customer for parts used, with the result that the customer pays for the parts plus the cost of processing the bill or purchase order. These hidden costs can skyrocket if one is not aware of them; there is also an inevitable delay involved in gaining approval for a purchase order.

The parts problem can be handled in three ways. One way is to include in the contract all parts except radiographic tubes, image intensifier tubes, television pickup tubes, and major components. This method essentially eliminates the frequent processing of purchase orders for small, relatively inexpensive parts. The second method is to have an agreement with the service organization to provide parts up to a specified dollar limit per month, quarter, or year. The vendor would then accumulate the costs of parts provided up to the specified limit and then bill the customer for the excess over the limit on a quarterly or semiannual basis. This second method is somewhat cheaper than the first, since the customer takes some of the risk along with the vendor. The third method is to have the contract pay for all parts except for those specifically excluded. Expendable supplies are not usually included in any contract.

Equipment Maintenance Records

One of the secrets of maintaining a good repair service is to develop and maintain service records on every piece of equipment in a radiology department. As a bare minimum, the service record should identify the equipment by listing the serial number, the name of the vendor, the date installed, and the date accepted. Every service call made for a specific piece of equipment should be recorded in the service log. The service log should indicate the date and time when service was requested, the date and time service was performed, how long it took for the repairman to respond and to repair the equipment, and what was wrong with the equipment. This record should be maintained regardless of whether a hospital-based repairman or a representative of the vendor works on the equipment. A copy of every service report should be attached to the equipment log.

The service or maintenance log serves a variety of purposes. First of all, it allows the acquisition of hard data concerning the problems involved with an individual piece of equipment and readily allows recognition of recurring problems. Second, it serves as a record of "down" time and the response time from the serviceman. Third, service records are invaluable in negotiations of warranty contract extensions. Although seldom used for this purpose, service logs can also be used to identify the capability (or lack of capability) of specific servicemen who repair the equipment.

Minimum Service Requirements

In addition to routine maintenance items such as lubrication, cleaning, counterweight cable inspections, and routine adjustments, the following items should be checked at least semiannually, and preferably monthly:

1. Large and small focal-spot size per NEMA specifications
2. X-ray tube coast times and rotation speeds
3. Milliamperage, kilovoltage, and time settings (check and adjust)
4. Line voltage supply and line resistance
5. Tube overload circuits

6. Backup timers
7. Collimator alignments and positioning indicators
8. Phototiming tracking.

A service contract should specify the hours during which service will be provided; normally this includes the administrative work week. Specific holidays should be identified, since holidays may vary from organization to organization. Provision of service outside of administrative working hours involves extra cost and is normally subject to availability of personnel.

Most service contracts place a limit on the liability of the vendor; such contracts normally exclude those things over which the company has no control. On a monetary basis, the liability of the vendor is usually limited to the actual cost of a year's contract. For example, if a machine is damaged by a serviceman because of negligence and the cost of repairing that equipment is excessive, the vendor may cancel the contract, which will raise the question of breach of contract. Under a number of contract provisions, the most money for which the vendor will be legally liable is the annual maintenance charge for the particular item or items of equipment in question. The persons who develop and sign service contracts should be aware of these restrictive provisions on liability.

Penalty clauses of any type should be approached with extreme caution. The presence of a penalty clause in any type of contract automatically indicates to the vendor a potential loss of revenue, and prices are adjusted accordingly. If penalties are included, they should be reasonable. Most vendors will not enter into a contract under which they cannot reasonably expect to be able to meet the provisions. A wise approach is to ask for a contract bid with and without a penalty clause and to compare the prices. The most common penalty provisions include an extension of the contract at no cost to the purchaser, or the provision to call other repairmen whose bill will be paid by the vendor under contract.

All in all, service contracts can be very beneficial to the purchaser, provided the contract is entered into with some reasonable assurance of what the purchaser's needs really are and what the vendor is reasonably able to provide.

Suggested Reading

Hendee, W. R., and Rossi, R. P. Performance specifications for diagnostic x-ray equipment. *Radiology* 120:409, 1976.

U.S. Department of Commerce, National Bureau of Standards. *Methods of Evaluating Radiological Equipment and Materials.* Handbook 89. Recommendations of the International Commission on Radiological Units and Measurements. Washington, D.C., 1963.

Appendix: Bid Specification Packet
Schedule of Radiological Equipment to be Installed in the Radiology Department of —————— *Hospital*

Solicitation is hereby made for bids on the following schedule of Radiological Equipment to be procured for and installed in the Department of Radiology at —————— Hospital —————— Address —————— .

The following schedule of equipment lists various components which are described in the accompanying "SUPPLEMENTARY DETAILED SPECIFICATIONS DESCRIBING X-RAY EQUIPMENT TO BE INSTALLED IN THE RADIOLOGY DEPARTMENT OF —————— HOSPITAL."

For convenience, paragraph numbers from detailed specifications referring to requested items are listed in parentheses.

It should be noted that bids are requested for several groups of items or individual items. A separate "bid sheet" is included for each group of items or individual item. It is intended that eligible bidders submit a bid on each sheet. In addition, combination bids on any combination of requested equipment will be accepted on extra pages. It is understood that some bidders may not be able to bid on all groups of items or individual items. If bidder chooses not to bid on any sheet, he should enter "No bid" on the sheet.

Bids are to follow conditions and instructions in the detailed specifications.

DATE OF SUBMISSION: ——————

(NOTE: Only one "Bid form" is included. Actual bid requests would include a "Bid form" for each room or piece of equipment. Extra forms would be included for "Combination bids.")

Room Number:
Room Name: Radiographic/Fluoroscopic (Digital)
Room Function: Radiography and fluoroscopy with digital fluoroscopy and fluorography capabilities.

Schedule of Equipment
1. 1 each generator and control system I for overtable and undertable tubes with atuomatic exposure termination capabilities (3.1 to 3.17 and 3.2.1 to 3.2.1.4)
2. 1 each fluoroscopic automatic brightness control system (18.1 to 18.4)
3. 1 each radiographic-fluoroscopic table (9.1 to 9.12)
4. 1 each overhead tube crane for overtable tube (4.1 to 4.8)
5. 1 each image intensifier system (13.1 to 13.9)
6. 1 each automatic collimator for overtable tube (5.1 and 5.2)
7. 1 each automatic collimator for undertable tube (5.1 and 5.2)
8. 1 each television system (14.1 to 14.8)
9. 1 each overtable x-ray tube unit, type II, 0.6–1.2 mm, 15 degrees, 50/100 kW (7.1 to 7.5)

   10. 1 each undertable x-ray tube unit, type I, 0.6–1.2 mm, 12 degrees, 50/100 kW with oil circulator heat exchanger (7.1 to 7.6)
   11. 4 each high tension cables (6.1 to 6.5)
   12. 1 each vertical cassette holder (11.1 to 11.9)
   13. 1 each 105-mm or 100-mm spot-film camera with 6-exposure-per-second capability (18.4)
   14. 1 each digital fluoroscopy/fluorography system (17.1 to 17.5)

Room Number:
Room Name: Catheterization Lab
Room Function: Cardiac catheterization procedures including biplane cineradiography and single plane rapid filming techniques

Schedule of Equipment
   1. 1 each generator and controls system I for "C" arm tube *with* automatic exposure termination capabilities (3.1 to 3.17 and 3.2.1 to 3.2.1.3)
   2. 1 each generator and controls system I for overhead tube *without* automatic exposure termination capabilities (3.1 to 3.15, 3.17, 3.2.1 to 3.2.3)
   3. 2 each automatic brightness systems (16.1 to 16.5)
   4. 1 each floor mounted suspension system (4.1 to 4.9)
   5. 1 each special procedures table (10.1 to 10.2.3)
   6. 1 each overhead tube crane (4.1 to 4.8)
   7. 1 each overhead crane for overtable image intensifier (4.1 to 4.8)
   8. 2 each 9"/6" image intensifiers (13.1 to 13.9)
   9. 2 each 35 mm cine systems with 35 mm cameras (biplane system) (15.1 to 15.8)
  10. 1 each 100-mm sheet-film or 105-mm roll-film camera for mounting on "C" arm image intensifier (16.4)
  11. 2 each television systems (14.1 to 14.8)
  12. 1 each rapid film changer for 14" × 14" cut film (12.1 to 12.11, 12.2.2)
  13. 2 each x-ray tubes type II, 0.6–1.2 mm, 12 degrees, 40/100 kW (7.1 to 7.5)
  14. 1 each manual collimator for overhead tube (5.1 & 5.2)
  15. 4 each high tension cables (6.1 to 6.5)

Room Number:
Room Name: Fracture Surgery
Room Function: Orthopedic Surgery, fluoroscopy and verification radiography

Schedule of Equipment
1. 1 each mobile "C" arm unit (18.1 to 18.13)
2. 1 each video disc recorder (20.1 to 20.3)
3. 1 each Polaroid camera for recording video disc images

Room Number:
Room Name: Radiography
          Radiography
          Head Radiography
          Chest Radiography
Room Function: Four rooms to operate on one shared generator with satellite controls. (*NOTE: Rooms may be bid with separate generators and controls that meet the specifications for the shared generator and satellite controls.*)

Schedule of Equipment

GENERAL

1. 1 each generator and control system IV with automatic exposure termination capabilities (3.1 to 3.16, 3.2.4 to 3.2.4.6)

RADIOGRAPHY #1

2. 1 each horizontal radiographic table (8.1 to 8.1.5, 8.2.1 to 8.2.1.4)
3. 1 each overhead tube crane (4.1 to 4.8)
4. 1 each x-ray tube, type II, 0.6–1.2 mm, 12 degrees, 40/100 kW (7.1 to 7.5)
5. 2 each high-tension cables (6.1 to 6.5)
6. 1 each automatic collimator (5.1 and 5.2)
7. 1 each control, master or satellite

RADIOGRAPHY #2

8. 1 each radiographic-tomographic table with integrated tube stand (8.1 to 8.5, 8.2.2 to 8.2.2.4)
9. 1 each x-ray tube, type II, 0.6–1.2 mm, 12 degrees, 40/100 kW (7.1 to 7.5)
10. 2 each high-tension cables (6.1 to 6.5)
11. 1 each automatic collimator (5.1 and 5.2)
12. 1 each control, master or satellite

HEAD RADIOGRAPHY

13. 1 each Franklin head unit (existing, to be moved)
14. 1 each x-ray tube, type III, 0.6–1.2, 12 degrees, 40/100 kW (7.1 to 7.5)
15. 2 each high-tension cables (6.1 to 6.5)
16. 1 each control, master or satellite

CHEST RADIOGRAPHY

17. 1 each automatic chest changer with integrated tube mount and integrated automatic film processor and with automatic exposure termination detectors (19.1 to 19.11)
18. 1 each x-ray tube, type III, 0.6–1.2 mm, 12 degrees, 40/100 kW (7.1 to 7.5)
19. 2 each high-tension cables (6.1 to 6.5)
20. 1 each manual
21. 2 each intensifying screens as per choice of Director, Department of Radiology
22. 1 each control, master or satellite

Bid Submission Form

_____ HOSPITAL,
Address
City, State, and Zip Code
Room Number: _____ and Bid Sheet Name: Radiographic/Fluoroscopic (Digital)
Bid Specifications
Any Deviations?   NO _____   YES _____
   NOTE: If yes, separate sheets must be attached as per bid specifications.

		Component description for each item requested in
Quantity	Unit	Equipment Schedule

Total Guaranteed Net Bids

1. Purchase (Installed):
   a. Cost of equipment only $ _____
   b. Installation cost $ _____
   c. Shipping, freight & ins. cost $ _____
   d. Warranty service cost $ _____
   e. Postwarranty service cost $ _____
   f. Training cost (incl. manuals) $ _____
   g. Cost of expendables in total cost $ _____
   h. TOTAL INSTALLED COST $ _____

2. Trade-in value of replaced equipment $ _____

3. Annual cost, complete service contract: 1 year $ _____ , 3 years
   $ _____ , 5 years $ _____ , 6 years $ _____ , 7 years $ _____

4. Monthly cost, lease/service contract: 5 years $ _____ , 6 years $ _____ ,
   7 years $ _____

*Supplementary Detailed Specifications Describing X-ray
Equipment to be Installed in the Radiology Department of
_____ Hospital*

1. General

1.1   The _____ Hospital _____ Address
_____ reserves the right to accept or reject any or all
bids or any portion or combination of equipment herein
specified and bid on.

1.2   The intent of these specifications is to indicate to the prospective
bidders the minimum standards for equipment systems to
perform procedures assigned to the indicated room(s).
Equipment systems that are bid on must meet or exceed these
specifications.

1.3   All systems that are bid on must include all interconnecting
devices and accessories that are necessary to allow the system to
operate and perform the procedures assigned to the indicated
room(s).

1.4   Bids shall be submitted on the bid forms that are supplied. The
bid form shall include a complete listing of items and quantities
as indicated in the Schedule of Equipment to be installed in the
indicated room(s). The form shall include the various requested
prices for the entire installation or piece of equipment. Any
deviations shall be handled as indicated in 1.5 below. The bids
must be typed or printed in ink. Original autograph signatures
in ink are required. Facsimile or rubber stamp signatures will
not be accepted.

1.5   Any deviations from these specifications must be separately
stated in writing and included with the bid. Written statements
must include item numbers, item descriptions, and a description
of the deviation(s).

1.6   All prices quoted shall be F.O.B. destination delivered inside and
installed in the indicated room(s) at _____ Hospital,
_____ Address _____ and shall be valid for
at least 60 days.

1.7   It is the intent of _____ Hospital, _____
Address _____ to determine the following costs for
each requested bid item *before* award:

    1. Total installed cost for purchase broken down as follows:
        a. Cost of equipment
        b. Cost of installation, shipping & insurance, training,
            warranty, and expendables included in total cost
    2. Annual cost for complete service contract only for the
        following periods:
        a. 1-year period
        b. 3-year period
        c. 5-year period

3. Monthly cost for lease/service contract for each of the
   following periods:
   a. 5-year period
   b. 6-year period
   c. 7-year period

Complete service cost as used herein is intended to mean entire
cost per year for complete service for the equipment including
cost of all parts (including x-ray and image tubes), labor, travel
and all other costs associated with said service.

Lease/service contract cost as used herein is intended to mean
entire cost per month for lease of the installed equipment and
complete service for the equipment including cost of all parts
(including x-ray and image tubes), labor, travel and all other
costs associated with said service.

Service of equipment covered by service contract or lease/service
contract shall be for failure or defective operation not
attributable to misuse by operating personnel.

1.8 All bidders shall submit layout drawings and wiring diagrams of
the indicated room(s) within 24 days after such request.
Drawings shall be complete and shall detail any and all changes
necessary in the room(s), all wiring runs, method of connection
to wall boxes, etc. Appropriate prints of project are included
with this document.

1.9 No item or any component of an item shall be bid unless it has
regularly stocked replacement parts available for immediate
shipment.

1.10 Submission of any proposal on all or any part of these
specifications infers full understanding of their meaning and
intent. Deviations from specifications shall be included in
writing with original bid as per paragraph 1.5 above.

1.11 Successful bidder shall supply complete schematic and wiring
diagrams, complete parts lists, instructions, and technical
manuals for each component.

1.12 Terms for equipment purchase shall be:
10% at placement of order
70% upon receipt of delivered equipment
20% upon final acceptance of installed equipment.
Terms for lease/service contracts shall be as specified by the
lessor.

1.13 Final acceptance of installed equipment shall be by
_____ , Director, Department of Radiology,
_____ Hospital, and _____ , Radiological
Equipment Consultant. Acceptance tests will determine that
equipment meets or exceeds equipment manufacturer's
specifications and all appropriate government requirements for
that equipment. Acceptance tests will be carried out by
manufacturer's representative using manufacturer's appropriate
test equipment. These tests will be scheduled so that either the

Director of the Department of Radiology or the Radiological Equipment Consultant or their appointed designee is present during tests. Equipment will not be put into operation for normal use until these tests have been completed and the equipment has been accepted in writing.

2. Conformance Standards

2.1   All items shall conform to the latest version of Regulations for the Administration and Enforcement of the Radiation Control for Health and Safety Act of 1968, Subpart C—Performance Standards for Electronic Products where applicable.

2.2   All equipment as installed shall conform to the Occupational Safety and Health Administration regulations as required.

2.3   All applicable items should be listed or have listing pending by Underwriters Laboratories or any other nationally recognized testing laboratory.

2.4   All items and installations shall comply with all codes and regulations as required by the City of _____ , _____ County, the State of _____ and the National Electrical Code.

3. High-Tension Generators and Control Systems

3.1   General.

3.1.1   Generators bid under these specifications shall be solid state and microprocessor or computer controlled.

3.1.2   All systems shall be three-phase, 12-pulse systems.

3.1.3   All systems shall be compatible with a nominal 480/277-V AC, three-phase, 60-Hz wye-connected power supply. Auxiliary power is a 208/120-V AC, single phase, 60-Hz wye-connected power supply.

3.1.4   Only solid-state high-tension rectifiers shall be included.

3.1.5   Radiographic potential available from the generator shall range from 50 to 150 kV in push-button steps of not more than 2 kV or be continuously variable through this range. Fluoroscopic kilovoltage shall include the range from 50 to 120 kV. An additional low kilovoltage range for mammographic studies is desirable, but not required.

3.1.6   Fixed milliampere stations or milliampere-second selection in combination with fixed milliampere stations are both acceptable formats.

3.1.7   For fixed milliampere system, a minimum of ten (10) milliampere stations shall be provided for radiographic techniques (including spot-film radiography). Any combination of mA and focal-spot size up to the maximum of the generator and/or x-ray tube rating shall be available, with actual values determined at or before time of installation.

3.1.8   Falling load systems shall be such that maximum milliamperage is automatically selected based upon x-ray tube rating, kilovoltage, and time. Additionally, not less than five (5) fixed mA stations (assigned any combination of mA and focal-spot size up to maximum of generator and/or x-ray tube rating) shall be available with actual values determined at or before time of installation.

3.1.9   The radiographic timer shall provide a range of times from 1 millisecond to at least 6 seconds. The timer shall be fully solid state.

3.1.10   Solid-state primary contacting for radiographic exposures (including spot-film techniques) is required. Fluoroscopic contacting can be either electromechanical or electronic.

3.1.11   Digital indication of kilovoltage, milliamperage, milliampere-seconds, and time both before and after exposure shall be provided.

3.1.12   Exposure counters for each tube shall be provided. Fluoroscopic hour counter shall be provided.

3.1.13   A high-speed rotor control shall be provided. The rotor control shall accelerate the x-ray tube rotor from 0 to at least 9000 rpm in not more than 5 seconds depending upon x-ray tube specified. Forced deceleration of the x-ray tube rotor shall be included. The x-ray control shall automatically initiate high-speed rotation when technique requires such rotation. Exposure shall be blocked (or terminated) unless anode is rotating at the proper speed.

3.1.14   An anode heat calculator and integrator shall be included with indication included on control panel.

3.1.15   Exposure shall be blocked when an overload single exposure technique is selected. Exposure shall also be blocked or a signal emitted when anode heat is at its preset value.

3.1.16   Automatic exposure termination (AET) for radiographic exposures (including spot-film techniques) shall be provided. Automatic exposure termination detectors (three) should be of the ionization chamber type. The AET device shall accurately reproduce film density except when limited by the minimum time specified for the system or by improper choice of detector. Any combination of the three (3) AET detectors shall be possible. Selection of the AET detector or detectors shall be provided on the control panel. Safety timing shall automatically terminate exposure before tube overload is reached, but shall allow maximum time for kV-mA combination selected.

3.1.17   An automatic brightness control (ABC) for fluoroscopic procedures shall be provided. This ABC shall automatically control image brightness by varying kV, mA or a combination thereof when fluoroscopic beam transmission is changed. Sensor for the ABC system will be in the TV system.

3.2   Specific system.

3.2.1   High-tension generator and control system I.

3.2.1.1 One hundred (100) kilowatt maximum output. Ratings shall be at least 1000 mA @ 100 kV, 800 mA @ 125 kV, and 600 mA @ 150 kV.

3.2.1.2 Exposure repetition rate shall be up to 12 per second or more.

3.2.1.3 Automatic line voltage compensation shall be included.

3.2.1.4 Anatomic programming capabilities are desired. A minimum of 12 basic anatomic programs shall be included. Modification of the basic programs shall be available for variation in patient size and other parameters.

3.2.2   High-tension generator and control system II.

3.2.2   At least eighty (80) kilowatt maximum output. Ratings shall be at least 800 mA @ 100 kV, 600 mA @ 125 kV, and 500 mA @ 150 kV.

3.2.2.1 Exposure repetition rate shall be up to 8 per second or more.

3.2.2.2 Automatic line voltage compensation shall be included.

3.2.3   High-tension generator and control system III.

3.2.3.1 High-tension generator may have a 6-pulse rectifier circuit.

3.2.3.2 At least fifty (50) kilowatt maximum output. Ratings shall be at least 600 mA @ 70 kV, 500 mA @ 100 kV, and 400 mA 125 kV. (Potentials above 125 kV are not required.)

3.2.3.3 Exposure repetition rate shall be up to 8 per second or more.

3.2.3.4 Automatic line voltage compensation shall be included.

3.2.4   High-tension generator and control system IV.

3.2.4.1 System IV shall consist of one high-tension transformer and four (4) control systems, equivalent in specifications to those elements of system I.

3.2.4.2 One single high-tension transformer shall supply radiographic tubes in four (4) separate rooms.

3.2.4.3 A separate control shall be provided for each of the aforementioned rooms.

3.2.4.4 Each control shall allow selection of the required techniques for its assigned room. Selection at one control shall not cancel selections at other controls. A single master control and three slave controls are acceptable.

3.2.4.5 Depression of the rotor switch on any one control shall initiate a visible signal on the other three controls to indicate that the first control is in use. Exposure on the other three controls shall be blocked at this time.

3.2.4.6 Interlocks shall be provided to prevent exposure in any room from the control panel of either of the other three rooms.

4. Suspension Systems

4.1 Overhead and other suspension systems shall be capable of safely and properly supporting all systems herein specified including radiographic x-ray tube and associated collimator, image intensifier system, television monitors, and any other system that may be indicated.

4.2 All stationary and movable rails, dollies, conveyors, pedestals, and columns shall be of sufficient size, strength, length, and capacity to allow for safe and convenient positioning of components without undue interference from or with other components.

4.3 All counterbalance systems shall incorporate a safety mechanism or back-up system to prevent free fall in case of spring or cable failure.

4.4 Counterbalance system, rail or pedestal leveling, and cable drape shall be such that no drift will occur with all locks in the "Off" condition and the component in the normal operating mode or position.

4.5 Electric locks are required. Lock control switches shall be located conveniently with respect to the steering handle. Separate locks shall be provided for vertical, longitudinal, transverse, and rotational modes of movement where applicable. Locks shall "give" under manual loading before damage to equipment occurs.

4.6 Centering locks or detents shall be provided where appropriate for radiographic x-ray tube and image intensifier system to ensure alignment with image receptor and/or radiation source.

4.7 The suspension systems shall allow all components to be "parked" clear of normal traffic areas.

4.8 Installation shall be made in such a manner to ensure that all required examination techniques are possible for the system. Maximum distance of radiographic x-ray tube focal-spot from image receptor shall be at least 40 inches for table Bucky in horizontal or vertical position and at least 72 inches for vertical Bucky stand.

4.9 In catheterization rooms, the suspension system for the x-ray tube/collimator and the image intensifier system shall be of the floor-mounted "C" or "U" arm type. The suspension system, along with the provided patient couch, shall allow complete flexibility or imaging positions with cranial and caudal angulations possible, as well as angulations around the long axis of the patient. The central axis of the x-ray beam and the axis of the image intensifier tube shall be aligned for all angulations of the "C" or "U" arm. Patient couch shall provide motorized vertical adjustment and manual longitudinal and transverse adjustment of the table top in order to facilitate positioning of the x-ray beam in the area of interest.

5. Collimators

5.1   All collimators and beam-defining and beam-confining devices shall conform to U.S. Department of Health & Human Services standards.

5.2   Types of collimators specified are automatic and/or manual.

6. High-Tension Cables

6.1   High-tension cables shall be shockproof and shall conform to federal, state, and local specifications.

6.2   The insulation rating shall be at least 75 kV per cable.

6.3   The outer covering of the cables shall be a washable plastic.

6.4   Cathode cable for grid-controlled x-ray tube shall be of four-conductor type with separate conductor for grid bias. Grid-type cable is inferred when application requires it.

6.5   Cable lengths shall be custom plus six (6) feet.

7. X-ray Tube Units

7.1   X-ray tubes and tube units herein specified shall have rotor speed characteristics (single or dual speed) which match the generator characteristics for the particular system.

7.2   Anode heat storage shall be as follows:

7.2.1 Type I   500,000 heat units minimum.

7.2.2 Type II   400,000 heat units minimum.

7.2.3 Type III   300,000 heat units minimum.

7.2.4 Type IV   200,000 heat units minimum.

7.3   Focal-spot combinations and dimensions shall be separately stated in the schedule of equipment. Focal spot dimensions shall conform with NEMA tolerances.

7.4   Target angles shall be separately stated in the schedule of equipment. Target angle tolerance shall be plus or minus 1 degree.

7.5   Maximum kilowatt ratings for each focal-spot size shall be separately stated in schedule of equipment for each tube.

8. Radiographic Tables

8.1   General.

8.1.1   All radiographic tables shall be provided with high-speed Bucky with 12 : 1 ratio, 103 lines per inch, aluminum interspaced, and nominal 40-inch focus. Exposure shall be interlocked with Bucky exposure switch.

8.1.2   All radiographic tables shall have Bucky automatic exposure termination detectors unless otherwise stated in equipment schedules or specifications.

8.1.3   Bucky, when movable, shall be movable from head to foot end of table or table base with positive lock in any position.

8.1.4   Bucky shall be counterbalanced on all tilt-type tables.

8.1.5   Radiolucent compression band shall be provided with each table. Footrest and patient step shall be provided with tilting table.

8.2   Specific tables.

8.2.1   Standard horizontal radiographic table.

8.2.1.1   Standard horizontal radiographic table shall have minimum longitudinal travel of 28 inches from center in each direction and a minimum transverse travel of plus or minus 4 inches from center.

8.2.1.2   Table top motions shall be manually operated.

8.2.1.3   Positive locks with conveniently located switches for both modes of travel shall be provided.

8.2.1.4   Linear tomographic attachment shall be available when requested. Tomographic bar shall be of quick-disconnect type. Motor-driven tomographic unit shall be integrated with table Bucky and tube support.

8.2.2   Integrated table-tube stand unit.

8.2.2.1   Table top shall be free floating with minimum longitudinal motion of plus or minus 22 inches from center and minimum transverse motion of plus or minus 5 inches from center.

8.2.2.2   Tube stand shall be fully integrated with table. Possibility of performing lateral and off-table radiography is desirable.

8.2.2.3   Bucky travel and tube support travel shall permit 40-degree tomography and 45-degree longitudinally angulated radiography.

8.2.2.4   Motor-driven tomographic unit fully integrated with table and tube stand shall be available. Tomographic bar, if required, shall be of the quick-disconnect type. Tomographic function shall be controlled by tableside control and shall include selection of angle from 5 to 40 degrees and fulcrum height from 0 to 250 mm in increments of 2 or more millimeters.

8.2.3   Integrated table-tube stand unit with automated film-handling system for cassetteless horizontal Bucky radiography. (Picker Rapido is exactly specified.)

8.2.3.1   The following accessories are to be supplied:
1. Lateral cassette holder
2. Two (2) tow handles and accessory clamps
3. Two (2) patient hand grips
4. 12 : 1 ratio, 103-line aluminum interspaced Bucky grid
5. One (1) extra 14 × 17-inch loading magazine
6. One (1) extra 17 × 14-inch loading magazine
7. One (1) extra 11 × 14-inch loading magazine
8. One (1) extra 10 × 12-inch loading magazine
9. Compression band assembly
10. One (1) spare receiving magazine

8.2.3.2   Unit is to be coupled directly to a Kodak model M-7 film processor.

---

9. Radiographic-Fluoroscopic Table

9.1   Table shall tilt from 90 degrees Trendelenburg to 90 degrees vertical with automatic stops at 15 degrees Trendelenburg and at horizontal.

9.2   Four-way power-driven table top is required.

9.3   Table top shall be flat. If curved, a filler must be provided to make flat.

9.4   Table Bucky shall be high speed with a 12:1 ratio, 103 lines per inch, nominal 40-inch focus grid. Exposure shall be interlocked with Bucky exposure switch.

9.5   Table Bucky shall be provided with three (3) ionization chamber-type automatic exposure termination detectors.

9.6   Accessories shall include compression device, shoulder support, footrest, Bucky slot closer, radiation side shield, leaded screen drape, mounting step, hand grips, and spot-film power assist.

9.7   Spot-film device shall include front (or front and rear) loading for 9.5 × 9.5-inch (24 cm × 24 cm) cassettes with variable spot-film programs.

9.8   Leaded screen drape shall be affixed to the spot-film device in a manner to allow hanging at front end or foot end of device.

9.9   Fluoroscopic shutters shall automatically limit beam size to image receptor or program mode selected on spot-film programmer. Shutters shall allow manual adjustment to limit beam to any size smaller than image receptor or program mode.

9.10   Spot-film device shall have automatic exposure termination detector(s).

9.11   Spot-film grid system should be capable of no-grid, single-grid, or double-crossed-grid usage. Single grids should be 6 : 1 ratio, 80 or 100 lines per inch. Double-crossed grids should be equivalent to a 12 : 1 ratio grid. Spot-film grids should be power driven into and out of position.

9.12   Spot-film device shall include a radiolucent compression device that can be moved into and out of the field.

---

10. Special Procedures Table

10.1   Spectrum catheterization table is herein exactly specified.

10.1.1   Table is to be provided with a flat table top with pad.

10.1.2   Table is to be provided with a cradle top with pad and straps.

10.1.3   Two (2) transfer carts and storage units are to be provided for top transfer and storage.

10.1.4   Table is to be provided with motor-driven programmable device for peripheral angiography. A kVp stepping device is required.

10.1.5   Table is to be provided with motor-driven raising and lowering device for table top.

10.1.6   The following accessories are to be provided:
1. Angulating headboard for flat top
2. Four (4) autoclavable handles for rail attachment
3. Foot switch with three (3) switches
4. Compression device for attachment to accessory rails with three (3) cloth bands
5. Arm board for flat tops, right or left
6. Arm board for cradle tops, right or left
7. Autoclavable instrument tray and mount
8. I.V. pole and mount for table tops
9. Removable undertable diaphragm device and mount for magnification
10. Top within a top
11. Center lock device
12. Lead protective apron and mount

10.2   Table to be installed in Catheterization Lab shall function in all projections specified for floor-mounted "C" or "U" arm system (4.9).

10.2.1   Fluoroscopy from head to groin shall be possible. System shall be such that any obstruction of fluoroscopy by table frame can be corrected without sliding patient on table top.

10.2.2   Table shall be provided with power-driven elevation.

10.2.3   Base-mounted fluoroscopic tube or image intensifier is not required.

11. Vertical Cassette Holder

11.1   Vertical cassette holder will be used for routine chest and other erect radiography.

11.2   Cassette holder shall be provided with sensors to accommodate automatic collimation.

11.3   Cassette holder shall be provided with three (3) field automatic exposure termination detectors.

11.4   Cassette holder shall accommodate various cassette sizes up to 14 × 17 inches (vertical or horizontal).

11.5   Vertical travel shall be at least 34 inches.

11.6   Cassette holder assembly shall be counterbalanced.

11.7   A grid slot shall be provided just on the tube side of the cassette.

11.8   A 12 : 1 ratio, 103 line per inch, nominal 72-inch focus grid shall be provided.

11.9   An overhead bar or other device for patient self support during vertical exposures shall be provided.

12. Rapid Film Changers

12.1   General.

12.1.1   All film changers bid on shall be for the size film specified or larger, roll or cut film, and shall consist of at least a three-stage programmer with film rates up to 6 frames per second or greater.

12.1.2   Screens shall be as specified.

12.1.3   AP grid shall be 12 : 1 ratio, 103 lines per inch, aluminum interspaced, 40-inch nominal focus. Lateral grid shall be 8 : 1 ratio, crosshatched, aluminum interspaced, 40-inch nominal focus.

12.1.4   Grids shall be easily removable, without tools, for magnification techniques.

12.1.5   Two (2) film supply magazines shall be supplied with each changer.

12.1.6   Two (2) film receiving magazines shall be supplied with each changer.

12.1.7   Individual film sequence and patient identification shall be provided.

12.1.8   When biplane operation is specified, one program selector shall interlock the film transports for either simultaneous or alternate exposures.

12.1.9   When magnification techniques are specified, stands shall be provided such that 2X magnification at 36-inch film-to-focal-spot distance is possible in both AP and lateral projections when used in combination with table specified.

12.1.10   When single plane operation is specified, stand shall be provided to allow operation in AP or lateral mode.

12.1.11   Stands shall be mobile within the room.

12.2   Specific systems.

12.2.1   The latest-model Franklin changer meeting the above requirements is herein exactly specified.

12.2.2   The latest-model Elema-Schonander AOT changer meeting the above requirements is herein exactly specified.

12.2.3   The latest-model Elema-Schonander Puck system meeting the above requirements (exclusive of frame rate) is herein exactly specified.

---

13. Image Intensifier System

13.1   Image intensifier shall be bimode with 9/6 or 10/7-inch input phosphor field diameters or trimode with 9/6/4.5- or 10/7/4.5-inch input phosphor field diameters.

13.2   Image intensifier tube input phosphor shall be cesium iodide only.

13.3   The intensifier system shall include power supply, controls, cables, spot-film tunnel adapter, suspension system, optics, and light distributor as required for television readout, spot-film camera, and/or cine system.

13.4   Minimum resolving capability shall be at least 2.8 line pairs per millimeter (specified at the input phosphor) at 0.1 modulation transfer fraction at the output phosphor.

13.5   Quantum detection efficiency should be greater than 40%.

13.6   Center-to-edge phosphor brightness fall-off should be less than 50%.

13.7   Signal decay from 90% to 10% shall take place within 10 milliseconds and be smooth without intermediate peaks.

13.8   Pulse defocusing shall be less than 30%.

13.9   Minimum conversion factors for the various input phosphor modes are as follows:
9- or 10-inch   70 cd/m^2/mR/sec
6- or 7-inch   35 cd/m^2/mR/sec
4.5-inch   25 cd/m^2/mR/sec

14. Television System

14.1   Each television camera system shall be fully solid state (except for Plumbicon or lead-oxide vidicon) and shall include camera, camera control, sync generator, selected Plumbicon or lead-oxide vidicon, cables, optics, housings, and mountings.

14.2   Each camera and control unit shall be 525 lines per frame, 30 frames per second with 2 : 1 interlace to be used on 60-Hz power source.

14.3   Cameras shall be a very low noise, wide dynamic range, and fast decay cameras suitable for use in digital fluoroscopy.

14.4   Each camera system should have a bandwidth of 5 to 7 MHz, an aspect ratio of 4 : 3 or 1 : 1, a signal-to-noise ratio of at least 1000 : 1 with RMS noise (no light output) of less than 7 millivolts.

14.5   Each optical system shall be provided with an externally adjustable aperture control. The camera lens iris or interlens iris control (preferably motor driven) shall adjust from minimum f-stop to at least f/22. Aperture area shall be reproducible to less than 5%.

14.6   Cameras shall be provided with dual-level adjustable gain control.

14.7   A compatible television monitor shall be provided with each system. Picture tube size shall be 17-inch diagonal measurement.

14.8   A monitor cart, ceiling suspension, or wall shelf shall be provided with each system as the system dictates.

15. Cine Systems

15.1   Cine systems shall be 35 mm with shutter synchronized to pulse system.

15.2   Selectable frame rates shall be approximately 7.5, 15, 30, and 60 frames per second.

15.3   Loading magazine shall accommodate a minimum of 400 feet of film.

15.4   Two (2) loading magazines shall be supplied with each cine system.

15.5   A remaining film footage indicator shall be provided.

15.6   Camera shall be capable of accepting a physiological trace either through its own optics or via additional optics through pressure plate.

15.7   Film acceleration to operating velocity shall be 1 second or less. Deceleration to stop shall be 1 second or less.

15.8 When cine is selected, cine exposure shall be automatically initiated by the second position of a two-position foot switch. (A preferred alternate includes separate switches for fluoroscopy and cine or spot-film exposures. Hand switch for cine or spot filming is preferred.)

16. Automatic Brightness Control System

16.1 The automatic brightness control system shall provide constant video brightness or constant film density. Adjustments shall be made at installation for best compromise of density or brightness, quantum mottle, and radiation dose.

16.2 The preferred system shall be the variable-mA x-ray tube current type with automatic kV control or override.

16.3 The sensing area shall be the center portion of the output field of the image intensifier tube.

16.4 The 105-mm roll-film or 100-mm cut-film spot-film camera (if purchased) shall use the generator-controlled automatic brightness control system to control film density. The camera shall provide repetition rates of up to 6 frames per second.

16.5 The cine automatic brightness system shall have a maximum milliamperage of at least 300. Cine density shall be controlled by pulse-width variation, mA variation, or kV variation. Cine shall be possible with both large and small focal-spots. Circuits for overload protection shall be provided.

17. Digital Fluoroscopy/Fluorography System

17.1 A digital system capable of providing the following procedures is required:
1. Mask mode subtraction fluorography (radiography)
2. Time interval difference subtraction fluoroscopy
3. Future dual-energy K-edge fluorography (radiography).

17.2 The system must be fully compatible with the radiographic/ fluoroscopic system with which it will be used.

17.3 The system shall provide 30 frames per second real-time acquisition and 256 level gray-scale display of information using a 256 × 256 matrix.

17.4 The system shall provide real-time analog-to-digital conversion of acquired information prior to disc storage and processing.

17.5 The system shall provide 2- and 4-frame averaging for mask mode subtraction studies.

17.6 The system shall provide image enhancement capabilities such as spatial filtering, temporal filtering, and Fourier transform filtering.

17.7 The system shall include a minimum of 474 Mbytes of Winchester digital disc storage. This should preferably be achieved with a 474-Mbyte digital disc with the possibility of the addition of a second of these discs. (NOTE: Analog storage is not desired.)

17.8 The system shall include two (2) CRT displays. One shall have light pen capabilities for outlining areas of interest for quantitative image analysis.

17.9 A multi-image camera system shall be included for image recording from CRT display.

17.10 A floppy disc or magnetic tape digital storage system shall be included for archival storage.

17.11 Capabilities for in-room operation of the system are desired.

17.12 The system shall include an array processor that allows correction for patient motion and other data processing.

17.13 Diagnostic software programs available for the system shall include cardiac gating and analysis, ejection fraction calculation, blood flow velocity, volumetric calculations, and linear distance measurement. User-developed programs shall be possible.

17.14 The output of the system should provide logarithmic, exponential, or user-defined gray-scale mapping.

17.15 The capability of using the digital system in conjunction with a second x-ray room is desirable.

---

18. Mobile "C" Arm Unit

18.1 Mobile "C" arm unit shall have an image intensifier tube of 6 or 7 inches input phosphor diameter.

18.2 Television system shall be as specified in paragraphs 14 with mobile monitor cart.

18.3 Automatic brightness control and/or video automatic gain control shall be provided to maintain constant brightness video image.

18.4 Mechanical structure shall provide for easy maneuverability.

18.5 Fiberoptics coupling between output phosphor and television camera is desirable.

18.6 Motor-driven vertical adjustment of approximately 18 inches shal be provided.

18.7 Movements shall be plus or minus 180 degrees about a horizontal axis and 90 degrees through "C" arm guide. Locks shall be provided for any intermediate position.

18.8 X-ray tube focal spot shall be 1.0 millimeter.

18.9 Potential shall be variable up to at least 100 kVp @ 15 mA.

18.10   Radiographic time from 1/30 to 5 seconds shall be provided.

18.11   A radiographic cassette holder shall be provided for attachment to image intensifier housing.

18.12   Unit shall be provided with a video disc recorder (20.1–20.3)

18.13   Power requirements shall be a nominal 120-V AC, 60-Hz @ 20-amp supply.

19. Automatic Chest Changer

19.1   This unit is to be fully automatic in moving 14 × 17-inch films from supply magazine to exposure position and then to integrated automatic film processor.

19.2   Supply magazine shall accommodate at least 100 sheets of noninterleaved 14 × 17-inch film.

19.3   Changer shall be provided with two (2) supply magazines and one (1) receiving magazine.

19.4   Device or method for positive film and patient identification shall be provided.

19.5   Intimate film-screen contact shall be provided by either vacuum or pressure device.

19.6   Chest changer shall be capable of complete integration with specified generator-tube–automatic collimator system.

19.7   Provision shall be made for accommodating automatic exposure termination detectors and automatic collimator sensors.

19.8   A removable 12 : 1-ratio, 103-line-per-inch aluminum interspaced, 72-inch focus grid and grid holder shall be provided. A "park" position for this grid shall be provided.

19.9   Intensifying screens to be provided will be identified in schedule of equipment.

19.10   Changer is to be provided with a fully compatible integrated automatic film processor.

19.11   An overhead bar or other device for patient self support during vertical exposures shall be provided.

20. Video Disc Recorder

20.1   Video disc recorder shall provide for storage and recall of up to 100 images.

20.2   Video disc recorder shall provide both fluoroscopy and single-image "electronic radiography" modes. Choice between these two modes shall be provided by a key-operated switch, so that unit can be locked in "electronic radiography" mode.

20.3   Disc recorder shall be controlled by fluoroscopy switch.

21. Installation

21.1   Successful bidder shall install all items bid on and delivered.

21.2   All conduit and individual copper conductors will be supplied by the bidder. Special cables shall be provided by the successful bidder.

21.3   Installation shall be scheduled through _____ , Hospital Administrator.

21.4   Installation and wiring shall be performed in a neat and orderly manner in accordance with all codes and regulations for electrical work.

21.5   All metallic parts shall be grounded through an equipotential grounding system. The system shall be such that potential differences of no more than 5 mV shall exist between any two exposed metallic surfaces of the installation.

21.6   Installation shall be accomplished without obstruction or undue inconvenience to performance of other hospital activities.

22. Guarantees and Warranties

22.1   All solid-state components and circuit boards shall be guaranteed by cost-free replacement for 12 calendar months from date of acceptance, except for high-tension transformer rectifiers, which shall carry a 5-year cost-free replacement guarantee.

22.2   All electromechanical devices and manually movable mechanical devices shall be guaranteed by cost-free replacement for 12 calendar months from date of acceptance.

22.3   Intensifier tubes shall be guaranteed by cost-free replacement against defects and/or blemishes for 6 calendar months and thereafter prorated over the following 18 months for warranty replacement costs.

22.4   Television pickup tubes shall be guaranteed by cost-free replacement against defects, blemishes, image retentions, loss of resolution, and/or smearing for 6 calendar months and thereafter prorated over the next 6 months for warranty replacement costs.

22.5   X-ray tubes shall be guaranteed by prorated replacement over a period of 1 year.

22.6   High tension cables shall be guaranteed by cost-free replacement for 12 calendar months.

22.7   All replacements of equipment in guarantee or warranty shall be for failure or defective operation not attributable to misuse by operating personnel.

22.8   For all items listed above, if the manufacturer's stated guarantees or warranties exceed those specified above, then the manufacturers guarantees or warranties will be accepted rather than those stated above.

23. Service and Support

23.1 Capabilities of bidders for performing prompt and effective service at a reasonable cost upon their own installations will be taken into account at the time of awarding contracts.

23.2 Guaranteed minimum time of response upon call for service is response on same day or early on the day following call for service.

23.3 A complete line of frequently required replacement parts shall be available for 24-hour delivery.

23.4 Bidders are requested to provide their responses to the following questions in written form and to submit them with bids:
1. How long has your organization been in the x-ray sales and service business under its present name? Under other names?
2. How many trained service personnel are in your employ?
3. List 1) names, 2) years of x-ray service experience, and 3) years with your organization for each service person in your organization.
*4. What is your hourly charge for service calls?
*5. What is your overtime charge for service calls?
*6. What is your basis for determining time spent and overtime spent in performing service?

23.5 Bidders are further requested to indicate the number of service personnel who will be present on-site for installation of equipment for which bidder has been awarded contract(s).

23.6 Bidders are requested to specify the time required for installation of the equipment to first patient use after all of the equipment has arrived on the hospital premises.

23.7 Bidders are requested to indicate the amount of time that will be provided for staff training and the methods of providing this training.

*NOTE: Items 4., 5., and 6. in paragraph 16.4 above will not be relevant if service contract or lease-service contract is awarded. However, items 4., 5., and 6. in 16.4 should be filled in.

Index

Abdomen, ultrasound evaluation,
  279–281
Acoustic impedance, in ultrasound,
  277
Air circulator, 74
Air dissipation time, 166
Aliasing, 185
Analog-to-digital converter, in digital
  fluorography, 206–208
Anatomically programmed
  generator, 41–42
Angiography
  cameras in, 159–160
  digital fluorography, 202–216. See
    also Fluorography, digital
  digital subtraction, 202–216. See
    also Fluorography, digital
  video disc recorder used in,
    123
  equipment considerations, 191–
    219
    for all purposes, 216–217
    basic components of standard
      laboratory, 217
    for biplane fluoroscopic/cine
      applications, 217–218
    buying add-on vs. integrated
      unit, 215–216
    compromise for all-purpose
      operation, 216–217
    for coronary arteriography, 200,
      202
    cost, 191
    decisions on purchasing, 216–
      218
    in digital fluorography, 202–216.
      See also Fluorography, digital
    grid vs. nongrid techniques,
      198–199, 201
    in magnification, 196–202
    speed relationships, 191–196
    video recording, 218, 219
  exposure/film rate requirements
    in, 192–194
  falling load generator in, 41
  field size coverage for rapid-
    sequence film changers in,
    160
  film changers in, 155–156. See also
    Film changers
  flow-rate injections in, 140

grid vs. nongrid techniques, 198–
  199, 201
heat exchanger for, 74
heat generated during, 60–61
image intensifier in, 159–160
injectors in, 139–153. See also
  Injectors, angiographic
kilowatt rating chart for, 197
magnification, 196–202
  basic principles, 196, 198
  equipment in, 199–202
  focal-spot size in, 197–198, 200
  radiation exposure in, 198–199
metal-center-section tubes in, 77–
  78
motion in, 191–194
patient position in, 160–161
peripheral, 158–160
pressure-controlled injection in,
  139–140
radiation requirements in, 194–
  195
  image intensifier related to,
    194–195
  in magnification, 198–199
recording method requirements
  related to anatomical area in,
  192–194
speed relationships in, 191–196
  generator capability, 195–196
  radiation requirements, 194–196
  stop motion, 191–194
stereoscopic tubes for, 78, 79
switching x-ray tube load in, 38
tube rating charts applied to, 60–
  61
types of injectors used in, 139–140
video recording in, 218, 219
Angiomat 3000 angiographic
  injector, 147–149
  optional features, 148–149
  standard features, 147–148
Anode, 53
  damage to, 63
  heel effect, 67, 68
Anticipation time, 163
Aortic aneurysm, ultrasound
  evaluation, 280
Aortography, patient position in, 160
Aperture field, selection in
  phototimers, 136

Archiving CT scans, 186–187
Area reduction factor, of image
    intensifier, 89
Arthrography, triple-focus
    radiographic tube for, 80
Artifacts
    in computed tomography, 184–186
        beam-hardening effects, 184
        motion effects, 185
        rings, 185
        streaks, 185–186
    in single photon emission
        computed tomography, 242–
        244
    star, 175
Attenuation
    coefficient of, and CT image
        display, 178–179
    of sound, 276
Automatic brightness control, 111–
    113. *See also* Brightness,
    automatic control
    bid specifications for, 311
Automatic chest changer, bid
    specifications for, 313
Automatic dose-rate control, 112
Automatic exposure controls, 127–
    136. *See also* Phototimers
Automatic gain control, of television
    camera, 109
Autotimers, 127–136. *See also*
    Phototimers

Back projection
    in CT image reconstruction, 174–
        175, 178
    filtered, 178
Bandwidth, of television monitor,
    108–109
Base-plug-fog density, 8
Battery-powered generator, 46–48
Beam-hardening artifacts, in CT, 184
Beam splitter, 96, 115
Bearings
    damage in x-ray tube, 63–64
    magnetically supported rotating
        anode tube to protect, 78–79
Bel, 277
Bid specifications, 285–315. *See also*
    Service contracts; Warranty
    contracts
    for angiography equipment, 216–
        218
    for automatic brightness control,
        386
    for automatic chest changer, 389
    for cassette holder, 381
    for cine system, 310–311
    for collimators, 304
    for computers in nuclear
        medicine, 251
    conformance standards in
        contract, 300
    delivery date, 287
    for digital fluorography, 295, 297,
        311–312
    for film changer, 308–309
    for fluoroscopic tables, 306
    for fluoroscopy, 295, 297, 311–312
    for fracture surgery, 295
    for gamma camera, 236, 238–239
    general provisions in contract,
        298–300
    for head and chest radiography,
        296
    for high tension cables, 304
    for high tension generators and
        control systems, 300–302
    for image intensifier, 309
    importance of, 285
    installation of equipment, 287–
        288, 314
    major types, two, 285
    methods of payment, 287
    for mobile C arm unit, 312–313
    for nuclear magnetic resonance,
        266–267
    performance requirements and
        acceptance of equipment,
        288–289
    purchase cost break down, 286
    purpose, 286
    for radiographic tables, 305–307
    radiological procedures for
        installation, 286
    service contracts, 290–293. *See also*
        Service contracts
    for single proton emission
        computed tomography, 244–
        245
    for suspension systems, 303
    for television system, 310
    for video disc recorder, 313
    warranty considerations, 289–290
    for x-ray tube units, 304
Biliary tree, ultrasound evaluation of,
    280

Biplane examinations, film changers in, 167
Bit, 208
Blood flow, Doppler effect in detection of, 276
Breast, ultrasound evaluation, 282
Brightness, automatic control, 111–113
  bid specifications for, 311
  combined systems, 113
  by controlling entrance radiation, 111–113
  electronic systems, 111
  in fluoroscopy, 113
  kilovoltage system, 112
  milliamperage system, 112
    with kilovoltage override, 113
  pulse width system, 113
  of television camera, 209
  of television monitor, 106
Byte, 208

Calcium tungstate intensifying screens, 1–5
Camera
  in angiography, 159–160
  cinecamera, 119–121
  gamma, 228–239. See also Gamma camera
  multiformat, in computed tomography, 187–190
    image-producing system, 188–189
    image recording, 189–190
    interface in, 188
  in nuclear medicine, 222, 228–239
  photospot, 117–119
  television, 98–104. See also Television camera
    comparison of tubes, 103–104
    image orthicon tube, 101–103
    vidicon pickup tube, 98–101
Capacitor discharge generators, 45–46
Capacitor leakage, 131–132
Cassette holder, bid specifications for, 308
Cassettes, 13–14
  in film changers, 156, 158–159
  and screen contact, 13–14
  in spot-film devices, 115
  standardization for use with phototimer, 131, 135

variation in x-ray absorption in front portion, 14
CAT scan. See Tomography, computed
Cathode assembly, 54–55
Cathode ray tube, 104
Central processing unit (CPU), 210
Cerebrovascular disease, ultrasound evaluation, 281
Cesium iodide, in image intensifier, 86
Chest, equipment for radiography, 296
Cine system, bid specifications for, 310–311
Cinecamera
  overframing in, 121
  radiation exposure with, 120
  in recording fluorographic image, 119–121
Cinefluorography, switching of x-ray tube load in, 37
Coast time, 63–64, 78–79
Collimators
  bid specifications for, 304
  in gamma camera, 234–237
Commutation, pulsed, 33–34
Comparator circuit, 111–112
Compton scattering, 224
Computed tomography. See Tomography, computed
Computers
  anatomically programmed generator, 41–42
  basic operation of microprocessor, 210–212
  in calculating CT image statistics in region, 180
  in controlling tube load, 40–41
  in Cordis angiographic injector, 146–147
  in digital fluorography, 202–216. See also Fluorography, digital
  in image processing, 207–209
  in storage, 209–213
  functional units of, 210–212
  image of CT storage in, 179–180
  in iteration method for CT image, 175–177
  and magnification in CT, 179–180
  memory in, 209–212
  microprocessor correction systems in gamma camera, 233–234, 242–243

Computers—*Continued*
in nuclear medicine, 245–251
acquisition software, 247–249
hardware requirements, 246–247
microprocessor correction systems in gamma camera, 233–234, 242–243
processing software, 249–250
tomography software, 250–251
programs for, 211
software influencing type of CT scanner, 187
storage by, 209–213
terminology with, 207–208
Cones, in retina, 83
Constant potential generator, 35–37
Continuous loop video tape recorder, 123
Contract, in bid specifications and purchasing, 294–315. *See also* Bid specifications; Service contracts; Warranty contracts
Contrast
definition, 83
image intensifier in enhancement of, 83, 90
single-phase vs. three-phase x-ray generator and, 51
of television monitor, 106
in television pickup tube, 104
Contrast medium, syringe heater for, 143–144
Contrast range, of image intensifier, 90
Contrast ratio, 90
Contrast resolution. *See* Resolution
Conversion factor, for image intensifier, 88
Cooling
characteristics of x-ray tube, 56–62
of magnet in nuclear magnetic resonance, 262–263, 265, 266
Cordis angiographic injector, 146–147
Core memory, 209–210
Coronary arteriography, equipment for, 200, 202, 217–218
Costs
of angiography equipment, 191
of cooling superconducting magnet in nuclear magnetic resonance, 262

of CT scanner, 186
of service contract, 291
CPU (central processing unit), 210
Crossover, in film, 7
Crystals. *See also* Scintillation crystal piezoelectric effect, 270–271
CT number, 179
CT scan. *See* Tomography, computed
Current flow, 26

Dark adaptation, for visualization at fluoroscopy, 81, 82
Dark current, 100
Data bus, 210
Data processing equipment. *See* Computers
Decibel, 277
Deflecting coils, 105
Delta configuration, 22–23, 26, 27–28
Design
of gamma camera, 232–237
in nuclear medicine instrumentation, 221–222
of scintillation crystal, 221–222
in single photon emission computed tomography, 240–242
of x-ray tube, 65–70
Detection efficiency, 227
Detectors
in computed tomography, 170–171
in nuclear medicine instrumentation, 222
in scintillation counter, response characteristics, 227–228
Digital fluorography, 202–216. *See also* Fluorography, digital
Digital subtraction, types, 213–214
Diode, 24
Discs, in computer storage, 211–212
Doppler effect
in detection of blood flow, 276, 281
in ultrasound, 275–276
in peripheral vascular disease, 282

Echocardiography, 281
Edison effect, 55
Electrical safety, of angiographic injectors, 140–142

Electrocardiogram
  angiographic injectors operating
    on basis of, 147, 149, 151
  film changer delay based on, 152
Electromagnetic radiation, in
    imaging techniques, 253
Electromechanical contactors, in
    switching x-ray tube load, 30
Electron cloud, 55
Electron demagnification factor, 88–
    89
Electron gun, 105
Electronic lenses, 86
Energy, intrinsic resolution
  in gamma camera, 230–231
  in scintillation counter, 227–228
Energy discriminator, in scintillation
    counter, 227
Energy spectrum, of scintillation
    counter, 225–227
Energy subtraction, 214
Equipment
  air circulator or blower, 74
  angiographic injectors, 139–153.
    See also Injectors,
    angiographic
  for angiography, 191–219. *See also*
    Angiography, equipment
    considerations
  automatic exposure controls, 127–
    136. *See also* Phototimers
  bid specifications for, 285–315. *See
    also* Bid specifications
  cassettes, 13–14
  in computed tomography, 169–
    173, 186–190. *See also*
    Tomography, computed
  for digital fluorography, 294–295
  electrical safety of, 140–142
  film changers, 155–168. *See also*
    Film changers
  film processors, 10–13
  for fluoroscopy, 294–295
  for fracture surgery, 295
  grids, 14–15
  heat exchanger, 74
  heat monitoring device, 74–75
  image intensifier, 81–126. *See also*
    Image intensifier
  installation provisions, 287–288,
    314
  intensifying screens, 1–7
  mobile radiography, 43–48

in nuclear magnetic resonance,
    259–267. *See also* Nuclear
    magnetic resonance
  magnets, 259–264
  site specifications, 264–266
in nuclear medicine. *See also*
    Nuclear medicine
    instrumentation
  data processing equipment,
    245–251
  gamma camera, 228–239
  photomultiplier tube, 225
  scintillation counter, 222–228
  single photon emission
    computed tomography, 240–
    245
radiographic film, 7–10
retrofitting of, 126
service contracts, 290–293. *See also*
    Service contracts
site preparation before installation
    of, 287–288
solid-state construction, 43
television cameras, 98–104. *See
    also* Television camera
television monitors, 104–109. *See
    also* Television monitor
testing under load, 19
total radiographic system, 1–16
in ultrasound, 282–283
video disc recorder, 123–124
video tape recorder, 121–123
viewboxes, 15
warranty considerations, 289–
    290. *See also* Warranty
    contracts
x-ray generators, 17–51. *See also* X-
    ray generator
x-ray tubes, 53–80. *See also* X-ray
    tube
Equipotential ground, 141
Exposure, film
  automatic controls, 127–136. *See
    also* Phototimers
  determination of time in film
    changer, 161–166
  film-rate related to time in
    angiography, 192–194
  interrogation time, 163–165
  maximum available, 166
  maximum permissible, 166
Exposure load control, 39
Extinction, forced, 34–35

Eye
  as image system, 82–83
  ultrasound evaluation, 282

Falling load generator, 39–41
Fan-beam geometry, in computed
    tomography, 171–172
Fiberoptic coupling, 97
Field coverage, 66–67
Film changers, 155–168
  bid specifications for, 308–309,
      313
  biplane applications, 167
  cassette units, 156, 158–159
  cassetteless units, 156–158
  characteristics, 155
  cycle for, 161–162
  determination of film exposure
      times in, 161–166
  field size coverage for, 160
  film stationary time in, 163–165
  film transport time in, 162–163
  film used with, 157, 159
  future applications, 167–168
  in magnification angiography,
      199–200
  maximum permissible exposure
      time in, 166
  patient position with, 160–161
  peripheral, 158–160
  phase relationships in, 164–165
  phototimer applications with, 166
  programmers for, 156
  speed of, 155
  total film cycle time in, 162
  types, 156–160
  x-ray tube used with, 159
Film exposure. See Exposure, film
Film grain, 70
Film processing, in recording camera
    for fluorographic image, 118
Film processors, 10–13
  cleaning of, 11–12
  inadequate replenishing rates, 11,
      12
  problems with, 10–11
  temperature drifts in, 12–13
Film, radiographic, 7–10
  average gradient, 8
  base-plug-fog density, 8
  comparing different products, 7
  in computed tomography image
      recording, 189–190

crossover effect in, 7
  in film changers, 157, 159
  handling in film changer, 157
  processing, 10–13. See also Film
      processors
  overdevelopment, 13
  underdevelopment, 11–12
  reciprocity failure, 9–10
  sensitometric exposures in
      determining characteristic
      curves, 7
  speed, 8–9
  standardization for use with
      phototimer, 131, 135
  structure, 7–8
  transport systems in film
      changers, 156–158, 162–163
Filter, in single photon emission
    computed tomography, 241
Filtered back projection, in CT image
    reconstruction, 178
Filtering effect, in CT, 178
Flip-flop, 208
Flood field uniformity, 231–232
Floppy discs, 211–212
Fluorescent, definition of, 2
Fluorography, digital, 202–216
  analog-to-digital converters in,
      206–208
  basic principles, 202–204
  bid specifications for, 295, 297,
      311–312
  computer storage in, 209–213
  image intensifier in, 205, 215–
      216
  image processor in, 207–209
  manipulating digital image in,
      214–216
  technique, 203–204
  television cameras in, 205–206
  types of subtraction in, 213–214
    energy, 214
    hybrid, 214
    temporal, 213
  x-ray tube and generator in, 204–
      205
Fluoroscopic table, bid specifications
    for, 306–307
Fluoroscopic tower, 114
Fluoroscopy. See also Image
    intensifier
  automatic brightness control in,
      113

bid specifications for, 295, 297, 311–312
heat exchanger for, 74
heat generated during, 58–59
recording image, 113–126
    cinecameras in, 119–121
    continuous loop video tape recorder, 123
    photospot camera in, 117–119
    resolution of system, 124–126
    retrofitting equipment for, 126
    spot-film devices, 114–119
    video disc recorder, 123–124
    video tape recorder, 121–123
screens used before image intensifiers in, 81
television viewing in, 94–95
tube rating charts applied to, 58–59
viewing system in, 82
x-ray tube selection for, 73
Focal distance, 14
Focal range, 14
Focal track, 53
Focal-spot size, 67–70
    in angiography, 195–196
    energy distribution, 72
    in magnification angiography, 71–72, 197–198, 200
    measurement of, 69–70
    in stereoscopic angiographic tube, 78
    tolerance of, 68–69
Focusing, of ultrasound beam, 271
Fog density, 8
Forced extinction, 34–35
Fourier transforms, in CT image reconstruction, 178
Fracture surgery equipment, 295
Fraunhoffer zone, 274
Fresnel zone, 274
Full width half maximum, 228, 230–231
Full width tenth maximum, 231

Gadolinium intensifying screens, 5
Gain, of image intensifier, 88
Gallbladder, ultrasound evaluation, 280
Gamma camera, 228–239
    basic operation, 228–230
    collimator design and types, 234–237

comparing cameras from different vendors, 239
design factors influencing intrinsic response characteristics, 232–233
electronics package in, 238–239
field of view considerations, 238
human engineering aspects, 239
instrumentation choice and bid specifications, 236, 238–239
interfacing with computer, 245. See also Computers, in nuclear medicine
intrinsic response characteristics, 229–232
    count rate performance, 232
    design factors influencing, 232–233
    differential vs. absolute linearity, 231
    differential vs. integral uniformity, 232
    energy resolution, 230–231
    flood field uniformity, 231–232
    full width tenth maximum, 231
    spatial resolution, 231
microprocessor correction systems in, 233–234
physical size, 236, 238
rotating, 240–241
schematic diagram, 228–230
spatial resolution in, collimator type and, 234–237
thickness of crystal in, 236, 238
in tomography, 241–245. See also Single photon emission computed tomography (SPECT)
useful vs central field of view in, 230
Gamma rays
detector response characteristics for, 227–228
interaction with scintillation crystals, 224–225
Gated blood pool imaging, arrhythmias influence on accuracy, 248–249
Gaussian distribution, 77
Generator. See X-ray generator
Geographic unsharpness, 71
Gibb's phenomena, 185

Gray scales, in computed
    tomography, 178–179
Grid(s), 14–15
    classification, 14
    focused, 14
    linear vs cross-hatched, 15
    parallel, 14
    problems with, 15
Grid bias timing control, 36, 37
Grid cutoff, 15
Grid-focus x-ray tube, 38
Grid ratio, 14
Ground
    equipotential, 141
    of power distribution system, 141–
        142
    of radiographic equipment, 141–
        142

Head, equipment for radiography of,
    296
Heart, ultrasound evaluation, 281
Heat exchanger, 74
Heat unit, 56–57
    single-phase, 56–57
    three-phase, 56–57
Heating
    characteristics for x-ray tube, 56–
        62
    damage to x-ray tube by, 63–65
    monitoring device, 74–75
    new developments to counteract
        problem in x-ray tube, 75
Helical recording system, 121–122
Helium refrigerator, in nuclear
    magnetic resonance, 263, 264
Hertz, definition of, 17
High-speed electronic precontacting,
    34
High-tension cables, bid
    specifications for, 301
Horn angle, 55
Hounsfield units, 179
Housing
    of image intensifier tube, 87
    of x-ray tube, 55
        damage to, 64
        heat dissipation by, 56
Hybrid subtraction, 214

Image distributor, 115
Image intensifier, 81–126
    in angiography, 159–160

area reduction factor for, 89
automatic brightness control in,
    111–113
automatic exposure controls used
    with, 127–136. *See also*
    Phototimers
automatic gain control in, 109–111
basic components of, 84, 85
bid specifications for, 309
cesium iodide in, 86
characteristics of, 87–91
choice of size, 125–126
construction of, 84–92
contrast enhancement by, 83, 90
contrast range of, 90
conversion factor for, 88
degrading over time, 124–125
in digital fluorography, 205, 215–
    216
dual-field, 91–92
function of, 87
gain of, 88
housing of, 87
increase of brightness level by, 83,
    88
input area, 85
input window of, 92, 93
lag in, 90
larger field of view systems, 92, 93
in magnification angiography,
    199–200
minification factor for, 88–89
mirror-optic viewing system for,
    93–94
mishandling, 125
multiple-field, 91–92
phosphor of, 84, 86–87
    output phosphor area, 85
photocathode in, 86
pincushion defect in, 90–91
power supply of, 87
purpose of, 83
radiation requirements in
    angiography related to, 194–
    195
resolving capability of, 89–90
    complete system, 124–126
spot-film device used with, 114–
    117
television camera tubes in, 98–104
    comparison, 103–104
    contrast, 104
    cost, 104

image orthicon, 101–103
  vidicon pickup, 98–101
television coupling systems for,
    95–97
television monitor operation
    principles, 104–109
three types, 83
vapor deposition in, 86
viewing radiographic output from,
    82, 92–97
  mirror-optic system, 93–94
  television monitor, 94–95
vignetting in, 90
weight of, 125
Image isocon tube, 102–103
Image noise
  in computed tomography, 182–183
  in single photon emission
    computed tomography, 242–
    243
Image orthicon television pickup
    tube, 101–103
  construction, 101–102
  function, 102
  vidicon compared with, 103–104
Infinity beam, 95–96
  optical system, 115
Injectors, angiographic, 139–153
  Angiomat 3000, 147–149
  commercially available, 146–153
  components of, 142–143
  Cordis, 146–147
  delay control for, 144–145
  difficulties with pressure of, 139
  disposable vs. nondisposable
    syringes in, 146
  electrical safety of, 140–142
  electromechanical power ram, 142
    removable, 145–146
  flow-rate, 140
  general operation modes of, 145
  historical developments of, 139
  injection volume control for, 144
  interfacing of, 144
  mechanical stop of, 145
  Medrad Mark IV, 149–153
  mobile stand, 143
  operator controls of, 142–143
  pressure-controlled, 139–140
  remote control of, 144, 145
  safety circuitry devices of, 143
  syringe heater for, 143–144
  trigger control for, 144

types, 139–140
Instrumentation. See Equipment
Intensification factor, 88
Intensifying screen, 1–7
  care of, 7
  efficiency of, 2–3
  in film changers, 158
  fluorescent, 2
  mottle in, 71
  phosphorescent, 2
  phosphors in, 1–2
  poor contact due to cassette
    abnormality, 13–14
  probabilities of incidents with x-
    ray photon interactions, 2–3
  rare-earth, 4–6
    advantages, 5
    disadvantages, 5–6
    quantum mottle in, 6
  reflective layer, 2
  resolving power of, 4
  speed of, 1, 4, 5
  standardization of, 6
    for use with phototimer, 131,
    135
  structure, 1–2
  terbium phosphor in, 4–5
  thickness related to speed and
    resolution, 4
Interlacing, 107–108
Interpolation, in CT magnification,
    180
Interrogation time, 18, 163–165
  conditions for, 38
  definition, 38
  millisecond, 33–34
  of x-ray generator, 50
Interslice scatter, in computed
    tomography, 183–184
Inverse parallel circuit, 32–33
Inversion recovery, 258
Ionization chamber, 127, 129–130
Isocon tube, 102–103
Isowatt principle, 40
Iterative techniques, for CT image,
    175–177

Kidney, ultrasound evaluation, 280
Kiloelectron volt, 18
Kilovoltage automatic brightness
    control, 112
Kilovoltage compensation, 129
Kilowatt ratings, 18–20, 59–60

Kinescope, 104
  of x-ray generator, 49

Lag
  in image intensifier, 90
  of television camera, 103–104
  of vidicon, 100, 103
Lanthanum oxybromide intensifying
    screens, 5
Large-scale integrated (LSI) chip,
    212
Lens
  objective, 95
  in television viewing of image
      intensifier, 95, 96
Light gate, 128
Line pair, 89
Line rate, 108
Load, control methods for exposure,
    39–41
Load ripple factor, 20–21
Lower extremity, ultrasound
    evaluation, 282
Lucite, in phototimers, 127–128
Lucite paddles, 128

Magnet, in nuclear magnetic
    resonance, 259–264
  comparison of different types, 261
  permanent, 263–264
  resistive, 260–261
  superconducting, 261–263
Magnet fringe field, 264, 265
Magnetic field
  active and passive shimming and,
      260
  in nuclear magnetic resonance,
      253–257, 259, 264–265
    fringe field, 264, 265
    uniformity, 264–265
Magnetic moment, 254
Magnetic resonance imaging. See
    Nuclear magnetic resonance
Magnetically supported rotating
    radiographic tube, 78–79
Magnification radiography, 70–72
  angiography, 196–202
    basic principles, 196, 198
    equipment in, 199–202
    focal-spot size, 197–198, 200
    radiation exposure in, 198–199
  computed tomography, 179–180
  definition, 70

focal spot effects in, 71–72
photographic enlargement
    distinguished from, 70, 71
selection of x-ray tube for, 74
Matrix
  in computed tomography, 173–174
  computer capability and, 208–209
Mechanical stop, of angiographic
    injector, 145
Medium frequency converter
    generator, 29–30
Medrad Mark IV angiographic
    injector, 149–153
  ECG trigger modules I, II, and III,
      151
  oscilloscope module, 152–153
  pulsed ECG film changer delay of,
      152
  steps prior to injection, 150
  system monitor circuit for, 150
Memory
  core, 209–210
  random-access, 212
  read-only, 212
Metal-center-section tubes, 77–78
Metric system, 17
Microprocessors. See also Computers
  basic operation of, 210–212
  correction systems in gamma
      camera, 233–234
  in x-ray generators, 40–42
Milliamperage automatic brightness
    control, 112
Millisecond interrogation, 33–34
Millisecond termination, 34–35
Minification factor, of image
    intensifier, 88–89
Minimum reaction time, 132–133
Mirror-optic viewing system, 92–94
  advantages and disadvantages, 92–
      93
Mitral valves, ultrasound evaluation,
    281
Mobile C arm unit, bid specifications
    for, 312–313
Mobile radiography equipment, 43–
    48
  battery-powered generators, 46–48
  capacitor discharge generators,
      45–46
  line resistance variations within
      hospital, 44–45
  power supply to, 44

Modular component generator, 43
Monitor, television. *See* Television
     monitor
Motion
    in angiography, 191–194
    artifacts in computed tomography,
        185
    stop, 191–194
Mottle
    components of, 70–71
    quantum, 70–71
    in radiographic film, 70–71
    screen structure, 71
Multiformat camera, in computed
        tomography, 187–190
Multipurpose multiroom/multiple
        console generator, 43

Nuclear magnetic resonance, 253–
        267
    advantages over alternative
        radiological imaging
        techniques, 253–254
    basic physical concepts in, 254–
        259
    cooling of magnet in, 262–263,
        265, 266
    diagnostic potential of, 266–267
    equipment in, 259–267
        basic hardware components,
            259–260
        instrumentation choice and bid
            specifications, 266–267
        magnets, 259–264
        site specifications, 264–266
    floor loading for, 265
    image interpretation in, 257–258
    magnetic field in, 259, 264–265
    magnets in, 259–264
        permanent, 263–264
        resistive, 260–261
        superconducting, 261–263
    power supply in, 265, 266
    resonance phenomenon in, 254–
        257
    RF shielding in, 265
    shimming in, 264–265
    spectroscopy in, 258–259
    warm-up period for, 261
Nuclear medicine instrumentation,
        221–251
    cameras in, 222, 228–239

data processing equipment, 245–
        251
    acquisition software, 247–249
    arrhythmias influence on gated
        blood pool imaging, 248–249
    hardware requirements, 246–
        247
    instrumentation choice and bid
        specification, 251
    processing software, 249–250
    tomography software, 250–251
gamma camera, 228–239
scintillation counter, 222–228. *See
        also* Scintillation counter
scintillation crystals in, design
        considerations, 221–222
single photon emission computed
        tomography, 240–245. *See
        also* Single photon emission
        computed tomography
        (SPECT)

Off-focus radiation, 76, 77
Off-peak imaging, in single photon
        emission computed
        tomography, 244
Ophthalmology, ultrasound
        applications in, 282
Optical coupling, 95–97
Orthicon television pickup tube, 101–
        103
Oscilloscope, with angiographic
        injector, 152–153
Overframing, 121

Pancreas, ultrasound evaluation, 280
Parspeed, 1
Partial absorption event, 226
Peak kilovoltage, 18
Pelvic mass, ultrasound evaluation,
        280–281
Penumbra, 71
    in computed tomography, 183–184
Peripheral vascular disease,
        ultrasound evaluation, 281–
        282
Phase-in time, 18, 163–165
Phase-locked, definition of, 119
Phosphor
    of image intensifier, 84, 86–87
    in intensifying screen, 1–2
    lag of, 90
    terbium, 4–5

Phosphorescent, definition of, 2
Phosphorus, in nuclear magnetic
    resonance spectroscopy, 258–
    259
Photocathode, in image intensifier,
    86
Photoconductive layer, 99
Photomultiplier phototimer, 127–129
Photomultiplier tube, in scintillation
    counter, 225
Photon attenuation, in single photon
    emission computed
    tomography, 241
Photonflux, 90
Photo-optical, definition of, 127
Photopeak fraction, 227
Photospot camera
    disadvantages of, 118
    film frame rates of, 117
    large vs. small format, 117–118
    radiation required for exposure in,
        118
    in recording fluorographic image,
        117–118
    spot-film device compared with,
        118–119
Phototimers, 127–136
    aperture field selection in, 136
    capacitor leakage with, 131–132
    density control buttons, 135
    entrance-type, 128
    exit-type detector, 129
    federal government regulations on,
        132
    with film changer, 166
    ionization chamber, 127, 129–130
    limitations, 131–132
    minimum response time of, 132–
        133
    performance of, 132–134
    photomultiplier, 127–129
        construction, 127–128
        problems with, 129
    repeatability of, 133
    response to changes
        in kilovoltage, 129
        in tissue thickness, 133–134
    selection of, 136
    size related to magnification
        distortion, 131
    solid-state, 130–131
    standardization of film, cassettes,
        and screens used with, 131,
        135

technical considerations, 135–136
    types, 127–131
Piezoelectricity, 270–271
Pincushion defect, in image
    intensifier, 90–91
Pinhole camera, focal-spot size
    measured by, 69–70
Pixels
    in computed tomography, 174
    computer capability and, 208–
        209
    shifting in digital fluorography,
        215
    standard deviation of fluctuations
        in CT, 180
Plumbicon, 101
    extended red, 103
Portable radiography equipment,
    43–48. See also Mobile
    radiography equipment
Power levels, in ultrasound
    evaluation, 277
Power ram, of angiographic injector,
    142
    of Angiomat 3000, 147, 148
    removable, 145–146
Power source
    electrical grounding for, 141–142
    and selection of x-ray tube, 73
    single-phase equipment
        heat unit of, 56–57
        tube rating characteristics for,
            61–62
    three-phase equipment
        generator, 20–21. See also X-ray
            generator
        single-phase compared with,
            50–51
        heat unit of, 56–57
        tube rating characteristics for,
            61–62
        voltage transformer, 22–23
    and tube rating, 61–62
Power supply
    fluctuations in, 28–29
    of image intensifier, 87
    for mobile radiography
        equipment, 44
        line resistance variations within
            hospital, 44–45
    in nuclear magnetic resonance,
        265, 266
Precontacting, high-speed electronic,
    34

Pregnancy, ultrasound evaluation, 281
Processing, of radiographic film, 10–13. *See also* Film processors
Proton imaging, in nuclear magnetic resonance, 254–257. *See also* Nuclear magnetic resonance
Pulse height analyzer, in scintillation counter, 227
Pulse rate, in ultrasound, 274
Pulse width automatic brightness control, 113
Pulsed commutation, 33–34
Pulsed ECG film changer delay, of Medrad Mark IV angiographic injector, 152

Quantum mottle, 70–71
in rare-earth screens, 6

Radiation, off-focus, 76, 77
Radiation exposure
in angiography, 194–195
with cinecamera, 120
in magnification angiography, 198–199
for photospot camera, 118
Radiofrequency shielding, in nuclear magnetic resonance, 265
Radiographic tables, bid specifications for, 305–307
Radiography
angiography. *See* Angiography
applications of x-ray generators in, 50
for angiography, 50
for fluoroscopy, 50
for routine studies, 50
for skull film, 50
automatic exposure controls in, 127–136. *See also* Phototimers
cassettes in, 13–14
computed tomography, 169–190. *See also* Tomography, computed
conversion equations in, 17
digital. *See* Fluorography, digital
efficiency of, 3
electrical safety of equipment in, 140–142
equipment, bid specifications for, 285–315. *See also* Bid specifications

film changers in, 155–168. *See also* Film changers
film in, 7–10. *See also* Film, radiographic
film processing in, 10–13
fluctuating power supply in, 28–29
grids in, 14–15
image intensifier in, 81–126. *See also* Image intensifier
intensifying screens in, 1–7
magnification, 70–72. *See also* Magnification radiography
portable equipment, 43–48. *See also* Mobile radiography equipment
recording fluorographic image, 113–126. *See also* Fluoroscopy, recording image
selection of generator in, 48–51
televisions used in, 98–111. *See also* Television camera; Television monitor
total system, 1–16
video tape recorder used in, 121–123
viewboxes in, 15
x-ray generators in, 17–51. *See also* X-ray generator
x-ray tubes in, 53–80. *See also* X-ray tube
Radiofrequency waves, as basis for nuclear magnetic resonance, 253–257
Radionuclides. *See also* Nuclear medicine instrumentation
Rare-earth intensifying screens, 4–6
Raster, 99, 106
line rate, 108
scanning, 110
square or rectangular, 110–111
system in computed tomography, 188–189
of television monitor, 106, 107
Reciprocity failure, 9–10
Recording
of computed tomography image, 189–190
in fluoroscopy, 113–126. *See also* Fluoroscopy, recording image
Records, of equipment maintenance, 292
Rectifiers, high voltage
selenium, 23, 24
silicon, 23–25

Rectifiers, high voltage—*Continued*
   in x-ray generator, 23–28
Registration, in digital fluorography,
   203
Relaxation time, in nuclear magnetic
   resonance, 257–258
Remasking, in digital fluorography,
   215
Remote control operation, of
   angiographic injector, 144,
   145
Repeatability, of phototimers, 133
Repeated free induction decay, 258
Replenishing rates, in film processor,
   11, 12
Resolution
   axial, in ultrasound, 278
   with cinecamera, 120
   in computed tomography, 182–184
   contrast
      in computed tomography, 169,
      182–183
      in nuclear magnetic resonance,
      258
   focal-spot energy distribution
      related to, 72
   focal-spot size related to, 67–68,
      71–72
   horizontal, 108–109
   of image intensifier, 89–90
      complete system, 124–126
      in multiple-field tube, 91–92
   lateral, in ultrasound, 278–279
   in nuclear magnetic resonance,
      258, 267
   spatial
      of collimator in gamma camera,
         234–237
      in computed tomography, 182,
         183
      of gamma camera, 231
      in single photon emission
         computed tomography, 243
      star pattern in measurement of,
         69–70
   of television monitor, 108–109
   in ultrasound, 278–279
   vertical, 108
   of video tape recorder, 122
Resolution aperture, in CT image,
   176
Resolving capability. *See* Resolution
Resolving power, of intensifying
   screen, 4

Resonance, nuclear magnetic, 254–
   257
Rest image, 100
Retrace, horizontal and vertical, 106
Retrofitting, 126
Ring artifacts, in computed
   tomography, 185
Ripple, 20–21
Rods, in retina, 82–83
Rotor body, 54

Safety
   of angiographic injectors, 140–142
   circuitry devices for angiographic
      injector, 143
Scattered radiation
   CT signal free from, 170
   interslice radiation in CT, 183–184
Scintillation camera, 228–239. *See
   also* Gamma camera
Scintillation counter, 222–228
   basic components of, 222–223
   detection efficiency of, 227, 228
   detector response characteristics,
      227–228
   energy spectrum in, 225–227
   full width half maximum in, 228,
      230–231
   interaction of gamma rays with
      crystal in, 224–225
   intrinsic energy resolution in, 227–
      228
   photomultiplier tube in, 225
   photopeak fraction in, 227
   pulse height analyzer, 227
Scintillation crystal, 223–225
   design considerations, 221–222
   energy spectrum of, 225–227
   interaction of gamma rays with,
      224–225
   sodium iodide, 223–224
   thickness in gamma camera, 236,
      238
   total vs. partial absorption events
      in, 226
Scintillation event, 223–225
SCR, 31. *See also* Silicon-controlled
   rectifier (SCR)
Screen grain, 2
Screen, intensifying. *See* Intensifying
   screen
Scrub frame, 206
Selenium high-voltage rectifier, 23,
   24

Semiconductor, in switching x-ray
    tube load, 31–32
Service contracts, 290–293, 315
    costs of, 291–292
    different systems of, 290–291
    hours of service provision in, 293
    labor in, 291–292
    labor plus parts in, 291–292
    liability of vendor, 293
    minimum service requirements,
        292–293
    penalty clauses, 293
    personnel doing repairs, 290
    preventive maintenance, 290
    records on equipment
        maintenance, 292
Servopositioning, 115
Shimming
    active vs passive, 260
    of magnetic fields, 264–265
Signal plate, 99
Silicon-controlled rectifier (SCR)
    in forced extinction, 34–35
    in inverse parallel circuit, 32–33
    mechanism of action, 31–32
    in phototimer, 128
    in pulsed commutation, 33–34
    in switching x-ray tube load, 31
Silicon dioxide, in television camera,
    103
Silicon high-voltage rectifier, 23–25
Single-phase equipment
    heat unit of, 56–57
    tube rating characteristics for, 61–
        62
Single photon emission computed
        tomography (SPECT), 240–
        245
    basic system design, 241, 242
    correction of photon attenuation
        in, 240, 242–243
    design problems differing from
        CT, 240
    gantry assembly in, 245
    image artifacts in, 242–244
    imaging parameters in, 241–243
    instrument choice and bid
        specifications in, 244–245
    limited angular sampling devices,
        240
    longitudinal section imaging, 240,
        241
    off-peak imaging in, 244
    photon attenuation in, 241

    planar imaging in, 245
    reconstruction filter in, 241
    software for, 250–251
    system types, 240–241
    traverse sectional imaging, 240–
        241
    variation in camera sensitivity
        with rotation, 243
Snell's law, 278
Sodium iodide scintillation crystal,
        223–224. *See also* Scintillation
        crystal
Software, in nuclear medicine data
        processing equipment. *See
        also* Computers
    in data acquisition, 247–249
    in processing, 249–250
    in tomography, 250–251
Solid-state equipment
    automatic exposure controls, 130–
        131
    generators, 43
    video disc recorders, 124
Sound. *See also* Ultrasound
    attenuation, 276–277
    beam characteristics in
        ultrasound, 274–275
    interaction with tissue, 276–279
    longitudinal waves, 271, 273
    propagation, 269–271
        particle vs. wave theory, 269
Source-image receptor distance, 78
SPECT, 240–245. *See also* Single
        photon emission computed
        tomography (SPECT)
Spectroscopy, in nuclear magnetic
        resonance, 258–259
Speed
    and angiography, 191–196
    of intensifying screens
        calcium tungstate, 1, 4
        of rare-earth metals, 5
        standardization, 6
    of radiographic film, 8–9
Spin echo, 258
Spin-lattice relaxation time, 257–258
Spin, of nucleus, 254
Spin-spin relaxation time, 257–258
Spot-film devices, 117–119
    cassettes used in, 115
    photospot camera compared with,
        118–119
    in recording fluorographic image,
        114–117

Star artifacts, 175
Star pattern, 69–70
Stator, 55
Stem radiation, 77
Stereoscopic angiographic tube, 78, 79
Stop motion, in angiography, 191–194
Streak artifacts, in computed tomography, 185–186
Subtraction angiography, digital, 202–216. *See also* Fluorography, digital
Subtraction, digital
  energy, 214
  hybrid, 214
  temporal, 213
  types, 213–214
Suspension systems, bid specifications for, 303
Switching
  constant potential generator in, 35–37
  electromechanical contactors in, 30
  grid-controlled x-ray tube, 37–38
  nonsynchronous, 35
  primary, 30–35
  secondary, 35–38
  silicon-controlled rectifier in, 31–35
  synchronous, 35
  thyratrons in, 30–31
  of x-ray tube load, 30–38
Synchronous-type exposure, 33
Syringe heater, of angiographic injector, 143–144
Syringes
  in angiographic injector, 146
  disposable vs. nondisposable, 146

Table, bid specifications for, 305–307
Tape recorder, video, 121–123
Target
  angle, 53, 66–67
  diameter, 66
  speed, 65–66
  of vidicon, 99
Television camera
  automatic brightness control, 111–113
  automatic gain control, 109
  bid specifications for, 310
  brightness level of, 209

buildup time in fluoroscopic application, 206
computer capability, pixel, and matrix considerations, 208–209
in digital fluorography, 205–206
infinity beam coupling between image intensifier and, 115
special circuitry for x-ray use, 109–111
square or rectangular raster in, 110–111
vignetting in, 96–97
Television field and frame, 107–108
Television monitor, 104–109. *See also* Image intensifier
basic operating principles, 104–109
bid specifications for, 310
brightness control of, 106
in computed tomography, 188–189
construction, 105–106
contrast control of, 106
coupling systems for viewing, 95–97
  fiberoptic, 97
  optical, 95–97
electron gun in, 105
eliminating flicker in, 106–107
horizontal resolution of, 108–109
line rate of, 108
raster of, 106, 107
retrace in, 106
vertical resolution of, 108
in viewing image intensifier output, 94–95
  advantages and disadvantages, 93, 95
Temporal subtraction, 213
Terbium, as phosphor in intensifying screens, 4–5
T-grain technology, 7–8
Thermionic emission, 55
Three-phase equipment
  generator, 20–21. *See also* X-ray generator
  single-phase compared with, 50–51
  heat unit of, 56–57
  tube rating characteristics for, 61–62
  voltage transformer, 22–23
Thyratrons, in switching x-ray tube load, 30–31

Thyristor. *See also* Silicon-controlled rectifier (SCR)
  mechanism of action, 31–32
  in switching x-ray tube load, 31
Timing, of x-ray generator, 49–50
Tomography, computed, 169–190
  analog images in, 187
  archiving systems in, 186–187
  array of points in image, 173
  artifacts in, 184–186
    beam-hardening effects, 184
    motion effects, 185
    rings, 185
    streaks, 185–186
  attenuation coefficients of tissue related to display image in, 178–179
  calculation of image statistics in region, 180
  contrast resolution in, 169, 182–183
  costs of scanner, 186
  data acquisition for reconstruction of image, 169–173
  digital images in, 187
  display parameters for image, 178–181
  elimination of translate motion in, 171–172
  first generation scanners, 169–170
  fourth generation scanners, 172–173
  gray scale in, 178–179
  image generation in, 173–178
  image storage in computer in, 179
  interslice scatter in, 183–184
  magnification in, 179–180
  matrices in, 173–174
  modern developments in, 190
  multiformat camera in, 187–190
    image recording, 189–90
    image-producing system, 188–189
    interface in, 188
  multiple detectors in, 170–171
  off-axis reconstruction in, 180–181
  penumbra in, 183–184
  quality of image in, 181–184
    contrast resolution, 182–183
    patient dose and, 183–184
    spatial resolution, 182, 183
  reconstruction methods for image in, 169, 174–177
    analytic techniques, 178

  back projection method, 174–175
  filtered back projection, 178
  Fourier transforms, 178
  iterative techniques, 175–177
  off-axis program, 180–181
  second diagnostic console used with, 186
  second generation scanners, 170–171
  selection of equipment, 186–187
  single photon emission, 240–245. *See also* Single photon emission computed tomography
  software influencing type of scanner in, 187
  standard deviation of fluctuations in mean pixel value in, 180
  television monitor used with, 188–189
  third generation scanners, 171–172
  units of display in, 179
  viewing images in, 187–190
  windowing in, 179
Tomography gamma camera, 241. *See also* Single photon emission computed tomography
Total absorption event, 226
Transducer, in ultrasound construction, 271
  and sound beam characteristics, 274–275
Transformer, x-ray, 20–30
  AC power source for, 21–22
  and film exposure time, 163–164
  high voltage rectifiers, 23–28
  single-phase, 22
  six-pulse vs 12-pulse output, 27–28
  three-phase, 22–23
  wire winding configurations in, 22–23
Trigger control
  for angiographic injector, 144
  of Angiomat 3000, 149
  of Medrad Mark IV, 151
Triple-focus radiographic tube, 80
Tube
  insert, 55
  image intensifier. *See* Image intensifier
  rating charts, 57–60
  x-ray. *See* X-ray tube

Tungsten, reducing deposition in
    metal-center-section tubes,
    77–78

Ultrasound, 269–283
  annular array in, 271
  characteristics of, 271–276
  clinical applications, 279–282
    abdomen and pelvis, 279–281
    aortic aneurysm, 280
    biliary tree, 280
    breast, 282
    heart, 281
    kidney, 280
    in obstetrics and gynecology,
        280–281
    in ophthalmology, 282
    pancreas, 280
    peripheral vascular disease,
        281–282
  Doppler effect in, 276–277
  equipment requirements in, 282–
      283
  focusing of sound in, 271
  interaction of sound waves with
      tissue in, 276–279
    acoustic impedance, 277
    attenuation, 276–277
    power levels, 277
    reflection, refraction, and
        transmission, 277–278
  microprocessors in, 283
  piezoelectric effect in, 270–271
  pulse rate in, 274
  resolution in, 278–279
  sound beam characteristics in,
      274–275
  sound propagation in, 269–271
  transducer construction in, 271
  waves in
    frequency and wavelength, 273–
        274
    longitudinal, 271, 273
    transverse, 273
Underdevelopment of film, 11–12
Unsharpness, geometric, focal spot
    related to, 71

Vapor deposition, in image
    intensifier, 86
Video recorders
  in angiography, 218, 219
  disc recorder, 123–124
    bid specifications, 313

tape recorder, 121–123
  continuous loop, 123
  resolution of, 122
Vidicon
  image orthicon compared with,
      103–104
  pickup tube, 98–101
    characteristics, 100
    construction, 98–99
    dark current of, 100
    function, 99–100
    lag of, 100
    Plumbicon, 101
    target of, 99
  return beam, 102–103
Viewboxes, 15
Vignetting
  in image intensifier, 90
    in multiple-field tube, 91–92
  in television image, 96–97
Voxel, in computed tomography, 174

Warranty contracts, 314, 351–352
  clarification of beginning of, 289–
      290
  statement of rendering service, 289
Watt, 18
  heat unit conversion to, 57
Waves, in ultrasound, 271–274
  frequency and wavelength, 273–
      274
  longitudinal, 271, 273
  transverse, 273
Windowing, in computed
    tomography, 179
Wye configuration, 22–23, 26, 27–28

X-ray generator, 17–51
  anatomically programmed, 41–42
  in angiography, 50, 195–196
  anticipation time when used with
      film changer, 163
  applications in radiology
      department, 48–51
  battery-powered, 46–48
  bid specifications for, 300–302
  capacitor discharge, 45–46
  classification of capabilities, 49
  computerized control of tube load,
      40–41
  constant potential, 35–37
  current flow in, 26–27
  in digital fluorography, 204–205

electromechanical contactors in
switching tube load, 30
exposure load control logic in, 39
falling load, 39–41
film exposure time of, 163–165
fluctuating power supply for, 28–
29
in fluoroscopy, 50
high-voltage rectifiers in, 23–28
interrogation time, 18, 38, 163–
165
kilowatt ratings, 18–20, 49
medium frequency converter, 29–
30
mobile, 43–48. *See also* Mobile
radiography equipment
modular component, 43
multipurpose multiroom/multiple
console, 43
proper condition in, 38
proper operation of exposure
buttons, 64–65
purchasing correct equipment, 17
in routine radiography, 50
selection considerations, 48–51
silicon-controlled rectifier in
switching tube load, 31
six-pulse vs. 12-pulse output, 27–
28
in skull examination, 50
solid-state construction, 43
switching tube load in, 30–38
primary, 30–35
secondary, 35–38
terminology, 18
testing under load, 19
three-phase
and film exposure time in film
changer, 164–165
vs. single-phase, 20–21, 50–51
thyratrons in, 30–31
timing capability, 49–50
transformer for, 20–30. *See also*
Transformer, x-ray
X-ray transformer. *See* Transformer,
x-ray
X-ray tube, 53–80
air circulator or blower for, 74
anode assembly, 53
damage from improper
operation, 63
anode heel effect, 67, 68
auxiliary equipment for, 74–75
air circulator or blower, 74

heat exchanger, 74
heat monitoring device, 74–75
bearing damage to, 63–64
biased, 72
bid specifications for, 304
cable receptacle, 55
cathode assembly, 54–55
construction of, 53–56
cost-effectiveness, 75
damage from improper operation,
62–65
design characteristics, 65–70
in digital fluorography, 204–205
excessive filament currents causing
damage, 64–65
in film changers, 159
focal-spot size, 67–70
glass envelope of, 55
heat exchanger for, 74
heat monitoring device for, 74–75
heating and cooling
characteristics, 56–62
in angiography, 60–61
in fluoroscopy, 58–59
power source influence on, 61–
62
single-phase and three-phase
equipment, 56–57
tube rating charts for, 57–60
housing of, 55
damage to, 64
heat dissipation by, 56
identification tag for, 55–56
magnetically supported rotating,
78–79
in magnification radiography, 70–
72, 197–200
metal-center-section, 77–78
new developments in, 75–80
rating charts for, 57–60
rotating anode, 53–56
rotor body, 54
selection of, 73–74
film size or field size and, 73–74
power level of generator and, 73
proper method of, 53
in magnification radiography,
74
stationary anode, 53
stator, 55
stereoscopic angiographic, 78, 79
target, 53
damage to, 63
design characteristics, 65–67

X-ray tube—*Continued*
  thermionic emission in, 55
  triple-focus, 80
X-rays. *See also* Radiography
  absorption, variation in front of
    cassette, 14
  determination of energy of, 18

Zero time, 163